Parallel Supercomputing in SIMD Architectures

Author

R. Michael Hord
Advanced Technology Laboratory
General Electric Company
Moorestown, New Jersey

CRC Press
Boca Raton   Ann Arbor   Boston

Acquiring Editor: Russ Hall
Production Director: Sandy Pearlman
Coordinating Editor: Suzanne Lassandro
Cover Design: Chris Pearl

Library of Congress Cataloging-in-Publication Data

Hord, R. Michael, 1940—
    Parallel supercomputing in SIMD architectures/author, R. Michael Hord.
       p.   cm.
    Includes bibliographical references.
    ISBN 0-8493-4271-6
    1. Parallel processing (Electronic computers). 2. Computer architecture. I. Title.
QA76.5.H675   1990
004'.35—dc20
                                                          89-71253
                                                               CIP

    This journal represents information obtained from authentic and highly regarded sources. Reprinted material is quoted with permission, and sources are indicated. A wide variety of references are listed. Every reasonable effort has been made to give reliable data and information, but the author and the publisher cannot assume responsibility for the validity of all materials or for the consequences of their use.

    All rights reserved. This book, or any parts thereof, may not be reproduced in any form without written consent from the publisher.

    Direct all inquiries to CRC Press, Inc., 2000 Corporate Blvd., N.W., Boca Raton, Florida 33431.

© 1990 by CRC Press, Inc.

International Standard Book Number 0-8493-4271-6

Library of Congress Card Number 89-71253
Printed in the United States

FOREWORD

We live in a golden age of computer architecture innovation. The advent of parallel processing has produced a flood of architecture paradigms, some more successful than others.

The most well known taxonomy for parallel computers was proposed by M. J. Flynn in 1966. It is based on the multiplicity of the instruction and data streams, which identifies four classes of computers: (1) Single instruction stream single data stream (SISD) computers correspond to regular Von Neumann non-parallel computers that can execute one instruction at a time on one data item at a time. (2) Single instruction multiple data (SIMD) computers are based on a central program controller that drives the program flow and a set of processing elements that all execute the instructions from the central controller on their individual data items. (3) Multiple instruction single data (MISD) computers are based on a pipeline principle. A given data item passes from processor to processor and is acted on differently at each stage of this assembly line. (4) Multiple instruction multiple data (MIMD) computers consist of a number of SISD computers configured to communicate among themselves in the course of a program.

Examples of recent successful MISD computers include the Cytocomputer and the PIPE. Instances of the MIMD category are too numerous to list but include NCUBE, iPSC, PASM, WARP, ZMOB, transputer based machines, Encore, Sequent, Alliant and Datacube.

In this book we explore the SIMD category. After a chapter of background that includes a brief discussion of the PEPE, STARAN and SOLOMON early machines, seven important SIMD machines are examined in depth: Illiac IV, MPP, DAP, GAPP, Connection Machine, GAM and CLIP4. Other significant SIMD machines not covered here include the Adaptive Array Processor, GRID, Non Von, Hughes Wafer Stack, GF11, PAPIA and CLIP7.

All of the machines in these categories have distinct characteristics, differing in connectivity, word size, memory management, and other major parameters. But even then the field of parallel processing is not fully mapped because there are whole classes of architectures that do not neatly fit the Flynn taxonomy. What is one to do with neural network computers, logic enhanced memories, data flow architectures and heterogeneous machines? And Very Long Instruction Word machines? And optical computers? In the face of this abundance, the premise of this book is simple. SIMD computers are a successful category of parallel processors because they provide an effective balance between complexity and simplicity. They are rich in algorithmic opportunity yet conceptually clean and easy to understand. The material presented attempts to offer a rebuttal to both camps of SIMD critics: those that claim the range of applicability is narrow because many problems are not sufficiently parallel to use so many processors efficiently and those that claim that having all of the parallel processors execute the same instruction is too restricted to embody the complexities of real world problems. I say instead that the range of applicability is enormous and hence broad enough to make good general computing and economic sense. And I say that the conceptual clarity of the architecture provides an ease of programming that keeps software costs within reason. The contents of the book are intended to allow the reader to judge these issues for himself.

Much of the material included here is based on and adapted from work previously published in government reports, journal articles, books and vendor literature. A listing of the major sources is provided following the text.

PREFACE

Parallel Supercomputing in SIMD Architectures is a survey book providing a thorough review of Single-Instruction-Multiple-Data machines, a type of parallel processing computer that has grown to importance in recent years. It was written to describe this technology in depth including the architectural concept, its history, a variety of hardware implementations, major programming languages, algorithmic methods, representative applications, and an assessment of benefits and drawbacks.

The book is intended for a wide range of readers. Computer professionals will find sufficient detail to incorporate much of this material into their own endeavors. Program managers and applications system designers may find the solution to their requirements for high computational performance at an affordable cost. Scientists and engineers will find sufficient processing speed to make interactive simulation a practical adjunct to theory and experiment. Students will find a case study of an emerging and maturing technology. The general reader is afforded the opportunity to appreciate the power of advanced computing and some of the ramifications of this growing capability.

Although there are numerous books on parallel processing, this is the first volume devoted entirely to the massively parallel machines of the SIMD class. The reader already familiar with low order parallel processing will discover a different philosophy of parallelism—the data parallel paradigm instead of the more familiar program parallel scheme.

The contents are organized into nine chapters, rich with illustrations and tables. The first two provide introduction and background covering fundamental concepts and a description of early SIMD computers. Chapters 3 through 8 each address specific machines from the first SIMD supercomputer (Illiac IV) through several contemporary designs to some example research computers. The final chapter provides commentary and lessons learned. Because the test of any technology is what it can do, diverse applications are incoporated throughout, leading step by step to increasingly ambitious examples.

For Susan

THE AUTHOR

R. Michael Hord is presently Manager of the Processing Applications Laboratory at the General Electric Advanced Technologies Laboratories, Moorestown, New Jersey. In this capacity he directs research and development activities employing the Connection Machine, the Butterfly, and the Warp advanced architecture computers. The current emphasis is on acoustic signal analysis, military data/information systems, and future architectures.

Until the end of 1989, Mr. Hord was Head of the Advanced Development Center at MRJ, Inc., Oakton, Virginia, where he directed diverse computer applications using parallel architectures including two Connection Machines. Application areas included image processing, signal processing, electromagnetic scattering, operations research and artificial intelligence. Mr. Hord joined MRJ in 1984 where he also directed the corporate research and development program.

For 5 years (1980 to 1984) Mr. Hord was the Director of Space Systems for General Research Corporation. Under contract to NASA and the Air Force, he and his staff assessed technology readiness for future space systems and performed applications analysis for innovative on-board processor architectures.

SIMD parallel processing was the focus of his efforts as the Manager of Applications Development for the Institute for Advanced Computation (IAC). IAC was the joint DARPA/NASA sponsored organization responsible for the development of the Illiac IV parallel supercomputer at Ames Reaearch Center. Projects included computational fluid dynamics, seismic simulation, digital cartography, linear programming, climate modeling and diverse image and signal processing applications.

Prior positions at Earth Satellite Corporation, Itek Corporation and Technology Incorporated were devoted to the development of computationally intensive applications such as optical system design and natural resource management.

Mr. Hord's five prior books and scores of papers address advanced parallel computing, digital image processing and space technology. He has long been active in the applied imagery pattern recognition community, has been an IEEE Distinguished Visitor, and is a frequent guest lecturer at several universities. His B.S. in physics was granted by Notre Dame University in 1962 and in 1966 he earned an M.S. in physics from the University of Maryland.

BRIEF CONTENTS

Chapter 1
Introduction 1

Chapter 2
Background 5
2.1   SIMD vs. MIMD 5
2.2   Fine Grain vs. Coarse Grain 6
2.3   Connectivity 7
2.4   Early Machines 7
      2.4.1   PEPE 7
      2.4.2   STARAN 10
      2.4.3   SOLOMON I 11

Chapter 3
Illiac IV — The First SIMD Supercomputer 17
3.1   History 17
3.2   The Illiac IV System 30
3.3   The CFD Language 39
3.4   Language Review 48
3.5   Performance 56
3.6   Seismic Analysis Application 56
3.7   Landsat Analysis 62
3.8   Synthetic Aperture Radar 65
3.9   Fast Fourier Transform 69
3.10 Linear Programming Image Enhancement 70
3.11 Comments on Some Case Studies 75
3.12 The Effects of the Illiac IV System on Computing
      Technology 79

Chapter 4
The Massively Parallel Processor (MPP) 85
4.1   The MPP Design 85
4.2   Parallel Forth Language 90
4.3   Parallel Pascal Design 105
4.4   Ising Spin Exchange Simulation 121
4.5   Stereo Analysis 125
4.6   Sort 132

Chapter 5
The Distributed Array of Processors (DAP) 143
5.1   The DAP Design 143
5.2   Parallel Data Transforms 156

5.3 Solution of a Large System of Equations 159
5.4 Image Understanding 164

Chapter 6
The Geometric Arithmetic Parallel Processor (GAPP) 181
6.1 The GAPP Design 181
6.2 GAPP Programming 186
6.3 GAPP System 199

Chapter 7
The Connection Machine (CM) 205
7.1 The CM System 206
    7.1.1 Overview 208
    7.1.2 The Connection Machine 209
    7.1.3 The DataVault 219
7.2 Programming 221
    7.2.1 The *Lisp Language 223
    7.2.2 The C* Language 227
    7.2.3 The Paris Language 233
7.3 Applications 236
    7.3.1 Physics 237
    7.3.2 Operations Research 250
    7.3.3 Image Processing 268
    7.3.4 Miscellaneous 273

Chapter 8
Research SIMD Computers 301
8.1 GAM 301
8.2 CLIP4 308

Chapter 9
Commentary 329

Appendix
Fault Tolerant SIMD 347

Acknowledgments 365

Index 367

TABLE OF CONTENTS

Chapter 1
Introduction 1

Chapter 2
Background 5
2.1. SIMD vs. MIMD 5
2.2. Fine Grain vs. Course Grain... 6
2.3. Connectivity 7
2.4. Early Machines 7
    2.4.1. PEPE (Parallel Element Processing Ensemble) 7
    2.4.2. STARAN... 10
    2.4.3. SOLOMON I.. 11

Chapter 3
Illiac IV — The First SIMD Supercomputer 17
3.1. History 17
    3.1.1. The Design Concept.. 18
    3.1.2. Implementation Difficulties................................... 21
3.2. I4 System.. 30
3.3. The CFD Language.. 39
3.4. Language Review.. 48
    3.4.1. PE Variables 50
    3.4.2. PE Variables Memory Allocation........................... 51
    3.4.3. Array Addressing... 51
    3.4.4. Routing... 52
3.5. Performance 56
3.6. Seismic Analysis Application: A Three-Dimensional Finite Difference Code for Seismic Analysis on the Illiac IV Parallel Processor.. 56
    3.6.1. Empirical Evidence... 56
    3.6.2. The TRES Computer Program.............................. 58
    3.6.3. Implementation on the Illiac IV 60
    3.6.4. Results Obtained with the I4TRES Program............ 62
3.7. Landsat 62
    3.7.1. Landsat Data Analysis.. 63
    3.7.2. Illiac IV Implementation 63
3.8. Digital Processing of Synthetic Aperture Radar Data on Illiac IV 65
    3.8.1. Synethic Aperture Radar Concepts 65
3.9. Fast Fourier Transform 69
3.10. Linear Programming Image Enhancement 70
    3.10.1. Statement of the Problem................................... 70

	3.10.2.	Linear Programming Approach 73

        3.10.2. Linear Programming Approach 73
        3.10.3. Practical Issues 73
3.11. Comments on Some Case Studies 75
        3.11.1. Sparse Matrix Multiply 75
        3.11.2. A Model for Diaster 76
        3.11.3. Monte Carlo Methods on the Illiac 78
        3.11.4. Conclusions 79
3.12. The Effects of the Illiac IV System on Computing
Technology 79
        3.12.1. Component and Manufacturing Technology 80
                3.12.1.1. Major Impetus to ECL Development 80
                3.12.1.2. Test Bed for Design Automation 80
                3.12.1.3. New Contribution to Logic Circuitry 80
                3.12.1.4. First Significant Use of Semiconductor
Memory 80
                3.12.1.5. Definitive Contribution to
Interconnection Technology 80
                3.12.1.6. A Major Milestone in Multilayer PC
Cards 81
        3.12.2. Machine Architecture 81
                3.12.2.1. Definitive Demonstration of Array
Approach to Computation 81
                3.12.2.2. Synchronous Control to Focus Research
on Efficiency of Computation 81
                3.12.2.3. First Large Scale Computer to be
Microprogrammed 81
                3.12.2.4. Synchronizaton of Independent
Disk Drives 81
                3.12.2.5. Exhaustive Simulation as a Realistic
Diagnostic Tool 81
                3.12.2.6. Test Bed for Future Machines 82
        3.12.3. System Architecture 82
        3.12.4. Applications 83
                3.12.4.1. New Horizons in Solvable Problems 83
                3.12.4.2. Spurring the Development of New
Algorithms 83
                3.12.4.3. Rethinking Problems for Parallel
Processors Pays Dividends on Other
Processors 83
References 84

Chapter 4
The MPP 85
4.1. The MPP Design 85

		4.1.1.	MPP Software	85

- 4.1.1. MPP Software 85
- 4.1.2. MPP Hardware 87
  - 4.1.2.1. Array Unit 87
  - 4.1.2.2. Array Control Unit 88
  - 4.1.2.3. Staging Memory 89
  - 4.1.2.4. Host Processor 90
- 4.2. Parallel Forth Language 90
  - 4.2.1. Introduction 90
  - 4.2.2. Vocabulary and Data Definition 91
  - 4.2.3. Parallel I/O 92
  - 4.2.4. Memory Operations 92
  - 4.2.5. Array Stock Operations 92
  - 4.2.6. Arithmetic, Logic, and Comparison Operations 93
  - 4.2.7. Control Operations 94
  - 4.2.8. PECU and Mask Stock Primitives 95
  - 4.2.9. MPP Parallel Forth Word Reference 95
    - 4.2.9.1. Context Changing Words 95
    - 4.2.9.2. Arithmetic Words 95
    - 4.2.9.3. Logical Words 98
    - 4.2.9.4. Comparison Words 98
    - 4.2.9.5. Stack Operation Words 99
    - 4.2.9.6. Memory Operation Words 101
    - 4.2.9.7. Control Words 102
    - 4.2.9.8. I/O Words 103
    - 4.2.9.9. Defining Words 104
    - 4.2.9.10. Compiler Words 104
    - 4.2.9.11. PECU Primitive Words 104
    - 4.2.9.12. Mask Stack Operations 105
- 4.3. Parallel Pascal Design 105
  - 4.3.1. Motivation 105
  - 4.3.2. Parallel Pascal Specification 106
- 4.4. Ising Spin Exchange Simulation 120
  - 4.4.1. Introduction 120
  - 4.4.2. Algorithms 122
  - 4.4.3. Vector Machines 124
  - 4.4.4. Preliminary Results 125
- 4.5. Stereo Analysis 125
  - 4.5.1. Introduction 125
  - 4.5.2. Background 126
  - 4.5.3. Difficulties in Stereo Matching 126
  - 4.5.4. Matching Technique 127
  - 4.5.5. Matching Algorithm on the MPP 128
  - 4.5.6. Preprocessing of the First Image 128
  - 4.5.7. Determination of Matches 129

	4.5.8.	Removal of "Bad Match" Areas in the Disparity Function 129
	4.5.9.	Smoothing the Resulting Disparity Function 131
	4.5.10.	Warping the Test Image 131
	4.5.11.	Interactive Operations on the MPP 131
	4.5.12.	Interactive Turnaround Time 132
4.6.	Sort	 132
	4.6.1.	Massively Parallel Communications 132
	4.6.2.	Sorting as a Communication Primitive 133
	4.6.3.	Sorting Algorithms Implemented on the MPP 133
	4.6.4.	Sort Aggregation 136
	4.6.5.	Sort Distribution.. 137
	4.6.6.	Merge Aggregation and Merge Distribution 139
	4.6.7.	The Unmerge Operation 140
References 141

Chapter 5
DAP 143
5.1. The DAP Design 143
    5.1.1. The DAP Software 144
5.2. Parallel Data Transforms 156
    5.2.1. Introduction 156
    5.2.2. Computation on Processor Arrays 156
    5.2.3. A Complementary Approach 157
    5.2.4. An Overview of the PDT Approach 158
    5.2.5. Implementation 158
    5.2.6. Summary 159
5.3. Solution of a Large System of Equation on DAP Using a Hybrid Gauss/Gauss-Jordan Method 159
    5.3.1. Introduction 159
    5.3.2. Principles of the Equation Solver 160
    5.3.3. Implementation Details 162
    5.3.4. Performance 164
5.4. An Image Understanding Performance Study on the Distributed Array Processor 164
    5.4.1. Introduction 164
    5.4.2. Overview 164
    5.4.3. Image Capture 165
    5.4.4. DOG Convolutions 166
    5.4.5. Direct Application of Convolutions 167
    5.4.6. Segmentation 169
    5.4.7. Segment Labeling 172
    5.4.8. Feature Generation 173
    5.4.9. Classification 176
    5.4.10. Conclusion 177

References..178

Chapter 6
GAPP..181
6.1. The GAPP Design...181
    6.1.1. Introduction 181
    6.1.2. Background 181
    6.1.3. Cell Description 182
        6.1.3.1. General................................182
        6.1.3.2. Registers 183
        6.1.3.3. Full Adder/Substractor 183
        6.1.3.4. RAM..................................183
        6.1.3.5. Control/Clock 183
    6.1.4. GAPP Chip Descriptions 183
        6.1.4.1. Control................................183
        6.1.4.2. Shift Register Groups 184
        6.1.4.3. Chip Performance/Mechanics 184
    6.1.5. GAPP Array..185
        6.1.5.1. Assembly 185
        6.1.5.2. Input/Output...........................185
        6.1.5.3. Sizes 185
6.2. GAPP Programming...186
    6.2.1. Convolution and Correlation............................191
    6.2.2. Straightforward Convolutions...........................195
6.3. GAPP System 199
    6.3.1. Introduction 199
    6.3.2. Architecture 200

Chapter 7
The Connection Machine 205
7.1. The CM System 206
    7.1.1. Overview..208
    7.1.2. The Connection Machine 209
        7.1.2.1. Performance Specifications................217
    7.1.3. DataVault 219
7.2. Programming 221
    7.2.1. The *Lisp Language...................................223
    7.2.2. The C* Language 227
    7.2.3. The Paris Language 233
7.3. Applications 236
    7.3.1. Physics 237
        Numerical Computation of Electromagnetic
        Scattering on the Connection Machine Using
        the Method of Moments..............................237

	7.3.2.	Parallel Implicit Methods for Numerical Physics Using the Connection Machine 241
		Operations Research 250
		Nonlinear Network Optimization on a Massively Parallel Connection Machine 250
		Massively Parallel Implementation of Some Common 0/1 Knapsack Approximation Algorithms 264
	7.3.3.	Image Processisng 268
		Automatic Target Detection on the Connection Machine 268
	7.3.4.	Miscellaneous 273
		Parallel Free-Text Search 274
		Neural Network Implementation Approaches for the Connection Machine 279
		Orbit Collision Problem Benchmark Study 291
References 298		

Chapter 8
Research SIMD Computers 301
8.1. The GAM II Pyramid 301
    8.1.1. Introduction 301
    8.1.2. The Pyramid Structure 301
    8.1.3. The Processing Element 302
    8.1.4. The Daughter Card 302
    8.1.5. The Back Plane 302
    8.1.6. The Adder Network 303
    8.1.7. The Sequencer 303
    8.1.8. The Program Flow Sub-Unit 304
    8.1.9. The Next Address Generator 304
    8.1.10. Micro Memory 305
    8.1.11. Pipeline Register 305
    8.1.12. Condition Code Register 305
    8.1.13. The Data Executive Sub-Unit 305
    8.1.14. The General Purpose Registers 305
    8.1.15. Special Purpose Registers 305
    8.1.16. Data Memory 306
    8.1.17. Array Instruction Generation 306
    8.1.18. The Input and Output Unit 306
    8.1.19. The Host System 306
    8.1.20. Future Control System Expansion 307
8.2. CLIP4 308
    8.2.1. Introduction 308
    8.2.2. Array Processing 309
    8.2.3. Processor Design General Considerations 311

8.2.4.		The CLIP4 Processor	312
	8.2.4.1.	Boolean Functions of Two Binary Images	313
	8.2.4.2.	Shifting Binary Images	315
	8.2.4.3.	Local Neighborhood Operations	316
	8.2.4.4.	Labeled Propagation Operations	318
	8.2.4.5.	Arithmetic Operations	318
8.2.5.		Further CLIP4 Functions	323
	8.2.5.1.	Counting	323
	8.2.5.2.	The INSERT Instruction	324
	8.2.5.3.	Input and Output of Images	324
	8.2.5.4.	Serial Computer Functions and Register Operations	325
	8.2.5.5.	Array Tests	325
8.2.6.		The Integrated Circuit	325
8.2.7.		Operating Speeds	326

Chapter 9
Commentary 329

Appendix
Fault Tolerance in Highly Parallel Mesh Connected Processors 347

Introduction 347
Mesh Connected Parallel Processors 348
A VLSI PE Assay Organization 351
PE Fault Tolerance 353
MIC Mesh Node Fault Tolerance 356
Module Fault Tolerance 359
Chip Level Fault Tolerance 361
Cost of MIC Fault Tolerance 362
Conclusion 362
References 363

Acknowledgments 365

Index 367

Eight men dominate the history of SIMD computer architectures. Their names are listed here to acknowledge their pioneering efforts:

S. H. Unger	First to propose spatially organized architectures (1958)
Daniel Slotnick	SOLOMON and Illiac IV
David Schaefer	TZE, MPP, GAM
Kenneth Batcher	MPP
W. Holsztynski	GAPP
Dennis Parkinson	DAP
Michael J. B. Duff	CLIP4
Danny Hillis	Connection Machine

1 INTRODUCTION

As demands emerge for ever greater computer processing speeds and capacities, traditional serial processors have begun to encounter physical laws that prevent further speed increases. One impediment is the speed of light. Signals cannot propagate faster than about a foot in a nanosecond. Hence, computer components commanded by another component 10 ft away cannot respond in less than 10 ns. To defeat this limit, designers try to make computers smaller. This effort encounters limits on the allowed smallness of chip feature sizes and the need to dissipate heat.

The most promising strategy to date for overcoming these limits is the abandonment of serial processing in favor of parallel processing. Parallel processing is the use of multiple processors simultaneously working on one problem. The hope is that if a single processor can generate X floating point operations per second (FLOPS), then 10 of these may be able to produce 10X FLOPS, and 10000 processors may produce 10000X FLOPS.

Problems of obvious interest for parallel processing because of their computational intensity include

- Matrix Inversion
- Artificial Vision
- Data Base Searches
- Finite Element Analyses
- Computational Fluid Dynamics
- Simulation
- Optical Ray Trace
- Signal Processing
- Optimization

However, the range of applications for parallel processing has proven to be much broader than expected. This volume will examine one type of parallel processing termed SIMD (Single Instruction Multiple Data) and describe by example the wide variety of application areas that have shown themselves to be well suited to parallel processing.

Parallel processing, or concurrent computing as it is sometimes termed, is not conceptually new. For as long as there have been jobs that can be broken up into multiple tasks which in turn can be handed out to individual workers for simultaneous performance, team projects have been an effective way to achieve schedule speedup. In the realm of computation, one recalls the WPA projects of the 1930s to generate trigonometric and logarithm tables, that employed hundreds of mathematicians each calculating a small portion of the total work. Lenses were designed the same way, with each optical engineer tracing one ray through a candidate design.

The recent excitement for parallel computer architectures results from the rising demand for supercomputer performance and the simultaneous maturing of constituent computer technologies that make parallel processing supercomputers a viable possibility.

The term supercomputer enjoys an evolving definition. It has been facetiously defined as those computers that exhibit throughput rates 50% greater than the highest rate currently available. The advent of the term occurred in the 1975 time frame when it was variously applied to the CDC-7600, the Illiac IV and other high performance machines of the day. Upon the arrival of the CRAY-1, the usage became firmly established. Today with the need for high performance computing greater than ever, the supercomputer identifier is commonplace. For the purposes of this book the term supercomputer means that class of computers that share the features of 80% of the highest speed and 80% of the largest capacity available at any given time. Both elements are important; high speed on small problems is insufficient. With this definition supercomputers over time cease being supercomputers and retire to the category "former supercomputers".

Parallel processing supercomputers haven't always been technically feasible. They require interprocessor communication, as one example of a maturing constituent technology, to perform sufficiently well that multiple processors can execute an application more quickly than a single processor acting alone can execute that application. Even today we see cases where 32 processors are slower than 16 processors working the same problem, not because the problem is insufficiently parallel, but because the interprocessor communications overhead is too high. These cases are becoming less common.

Another maturity issue making parallel processing supercomputers feasible today is that of implementation cost. VLSI chip technology is revolutionizing the cost-performance characteristics of recent systems. In the July 1987 *IEEE Spectrum* magazine a scatterdiagram appeared in which dozens of fast computers were plotted in terms of speed in megaflops versus base price in thousands of dollars. Some computers were rated at over $1M per megaflops. Most fell into the range of $100K to $1M per megaflops, with a few just under the $100K per megaflops line. Since the leverage of reduced implementation costs is so effective for the SIMD architectures discussed in this book, partly due to the economies of scale associated with the massive replication of simple components, the cost per megaflops for SIMD machines is close to or even under $10K.

SIMD computers have been evolving for 25 years. Today they have become an essential and undeniable force in large scale computing. This book explores their design, their history, their programming languages, and a selection of their applications in some depth.

2 BACKGROUND

2.1. SIMD VS. MIMD

Parallel processors fall mainly into two general classes as described in Table 1 (see following page).

The fundamental distinction between the two classes is that one class is SIMD (single instruction, multiple data) while the other is MIMD (multiple instruction, multiple data). This distinction can be summarized as follows:

Single Instruction Multiple Data
  All processors are given the same instruction
  Each processor operates on different data
  Processors may "sit out" a sequence of instructions

Multiple Instruction Multiple Data
  Each processor runs its own instruction sequence
  Each processor works on a different part of problem
  Each processor communicates data to other parts
  Processors may have to wait for other processes or for access to data

In SIMD architectures, a single control unit (CU) fetches and decodes instructions. Then the instruction is executed either in the CU itself (e.g., a jump instruction) or it is broadcast to a collection of processing elements (PEs). These PEs operate synchronously but their local memories have different contents. Depending on the complexity of the CU, the processing power and addressing method of the PEs, and the interconnection facilities between the PEs, we can distinguish between pipeline, or vector, processors, array processors, processing ensembles and associative processors.

Pipeline processing is analogous to an assembly-line organization. The computational power is segmented into consecutive stations. Processes are decomposed into subprocesses which have to pass through each station or stage.

In array processors the CU has limited capabilities. PEs communicate with

TABLE 1
Two General Classes of Parallel Processing

Few powerful processors (<1000)	Many elementary processors (>1000)

Control level parallelism
  Assign a processor to a unit of code
	Data level parallelism
	  Assign a processor to a data unit

Coarse grain parallelism	Fine grain parallelism

Typically shared memory
  Memory contention problems
  Memory speed limiting
	Typically distributed memory
	  Data communication problem
	  Communication speed limiting

Typically MIMD	Typically SIMD

their neighbors through a network. The array processors are therefore well suited for problems involving vector processing and for grid problems. We will see that array processors are useful as general purpose machines. Array processors are the type of SIMD computers we focus on throughout this book.

In processing ensembles the CU is a complete computer and in order to communicate the PEs have to pass their messages through the CU. In general, each PE will operate on an associative memory. This is mandatory in the case of associative processors for which the associative memories are larger and interconnections are more extensive. Because the concept of associative processing requires a large amount of hardware it has not yet been proven cost-effective except for very specific functions. Thus, processing ensembles and associative processors are special purpose machines.

In MIMD architectures, several processors operate in parallel in an asynchronous manner and share access to a common memory. Two features are of interest to differentiate among designs: the coupling of processor units and memories, and the homogeneity of the processing units.

In tightly coupled MIMD multiprocessors, the number of processing units is fixed and they operate under the supervision of a strict control scheme. Generally, the controller is a hardware unit. (Note that this tight coupling is also present in array processors and pipeline computers.) Most of the hardware controlled tightly-coupled multiprocessors are heterogeneous in the sense that they consist of specialized functional units (e.g., adders, multipliers) supervised by an instruction unit which decodes instructions, fetches operands and dispatches orders to the functional units.

2.2. FINE GRAIN VS. COARSE GRAIN

There is some confusion about the terms fine grain and coarse grain as applied to parallel computers. Some use the terms to characterize the power of the individual processing elements. In this sense a fine grain processor is rather elemental, perhaps operating on a 1-bit word and having very few registers. This is contrasted with powerful processing elements in coarse grain computers, each a computer in its own right with a 16-, 32-, or 64-bit word size.

Others use the terms to differentiate between computers with a small number of processors and those with a large number. Of late the dividing line between these classes has been set at 1000 so that computers with more than 1000 processors are termed fine grain.

2.3. CONNECTIVITY

Another distinction between the two classes is whether memory is shared or distributed, i.e., do all processors have access to all memory banks or does each processor have access only to its own local memory? These are called shared memory architectures and distributed memory architectures.

The final major distinction is the topology of the interconnects, i.e., which processors have a direct interconnection with which other processors. Diagrams of various examples are shown in Figures 1 and 2. Figure 1 shows variations in connectivity for shared memory architecture. Figure 2 illustrates five kinds of connectivity for distributed memory architecture.

2.4. EARLY MACHINES

2.4.1. PEPE (Parallel Element Processing Ensemble)

PEPE was a special-purpose "attachment" to a general-purpose computer called the host. It was housed in the Ballistic Missile Defense Agency quarters in Huntsville (Alabama) in the 1970s. PEPE's main application was radar processing. A block diagram of the system is shown in Figures 3, 4, and 5. The host is a CDC 7600 and the test and maintenance station is controlled by a Burroughs B1700. PEPE itself consists of a control system and up to 288 processing elements (PEs) (see Table 2, p. 12). PEs can be added or disconnected without impeding the normal operation of the system. One reason for this flexibility is that, unlike other SIMD systems, there is no direct connection between PEs. If data has to be transferred between PEs it will have to be routed through the host.

A PE consists of three units: the arithmetic unit (AU), correlation unit (CU) and associative output unit (AOU) sharing an element memory (EM). The three types of units are under the global control of three control units: arithmetic control unit (ACU), correlation control unit (CCU) and associative output control unit (AOCU) which are part of the control system. Not shown in the figure are units in the control system for output data control and element memory control to resolve conflicts in data access, the interconnection logic between ACU, CCU and AOCU, and an I/O unit to connect to the host.

The PEs perform most of the computational work. For example, in a radar processing application each PE would have the responsibility of an object in

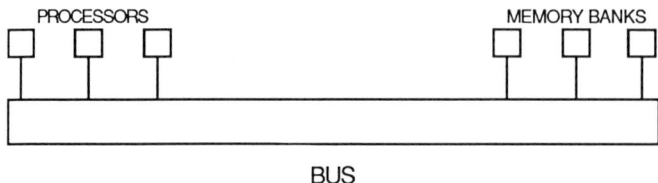

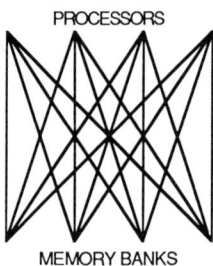

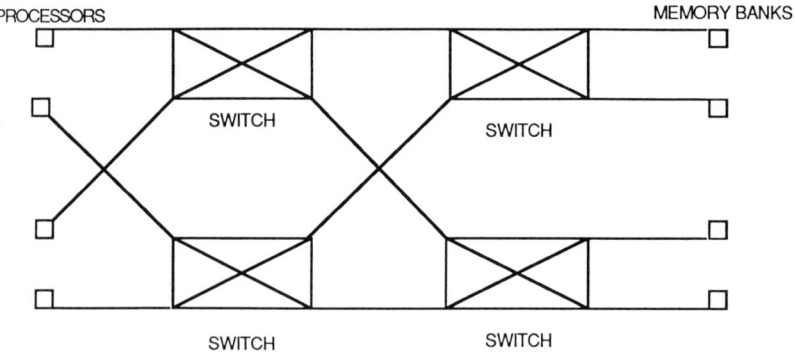

FIGURE 1. Shared memory architecture.

the sky, maintaining a data base for this object in its EM. Arithmetic computations such as track updating and prediction can be performed, in parallel, in each AU. Each PE can be enabled/disabled by setting a control flip-flop. The AUs execute under control of the ACU which sends microprogrammed sequences to them. Individual AUs do not have a stored program of their own.

Inputs to the PEs are controlled by the CUs under the supervision of the CCU. For example, information on a new object can be broadcast to all CUs at the same time. Each CU will correlate the new coordinates with predictions performed by the AU in the same element. The new information can then be

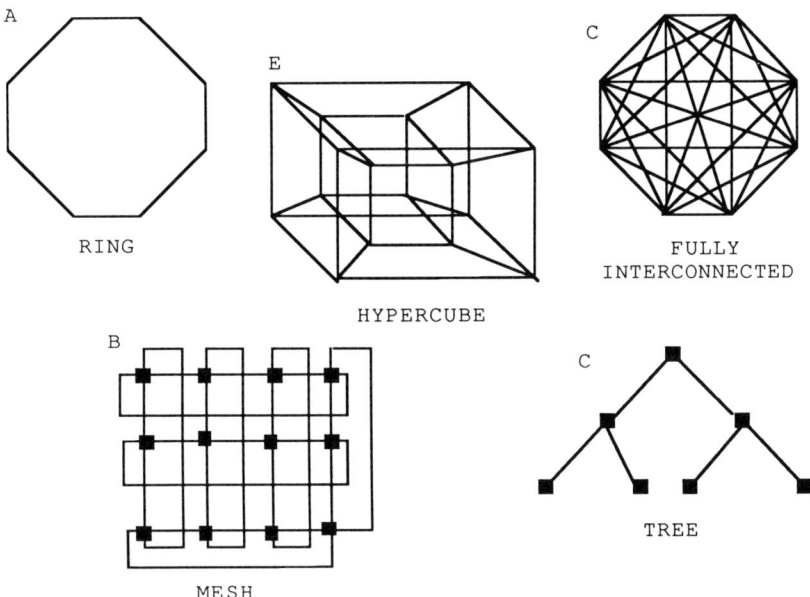

FIGURE 2. Distributed memory architecture. (A) ring; (B) mesh; (C) tree; (D) fully interconnected; (E) hypercube.

input to the CU whose data correlates, or to the first empty element if there were no correlation. It is this type of processing which gives to PEPE its associative label since the broadcast data can be viewed as an argument to be matched by the CUs acting as memory cells.

The AOUs, under control of the AOCU in an enable/disable fashion similar to the ACUs, provide data for the radar connected to the host. The data has to be ordered on an object by object basis and this ordering is performed through an associative maximum-minimum search. The three units in a PE can operate concurrently. Since PEPE can have 288 elements, up to 864 operations can be in execution at the same time. In order to alleviate the loads on the input and output buses connecting the global control units and the PEs, most programs for the latter are loaded at initialization time with very few parameter modifications during execution.

The three global control units have the responsibility of the control flow. Each control unit has its own program and data memories. They can communicate with each other. A program is a mix of instructions executed either in the global control unit or broadcast to all PEs.

This architecture is well suited for parallel tasks with low intertask communications requirements. The associative processing and the fact that the system can be in operation with a variable number of PEs are interesting and original features.

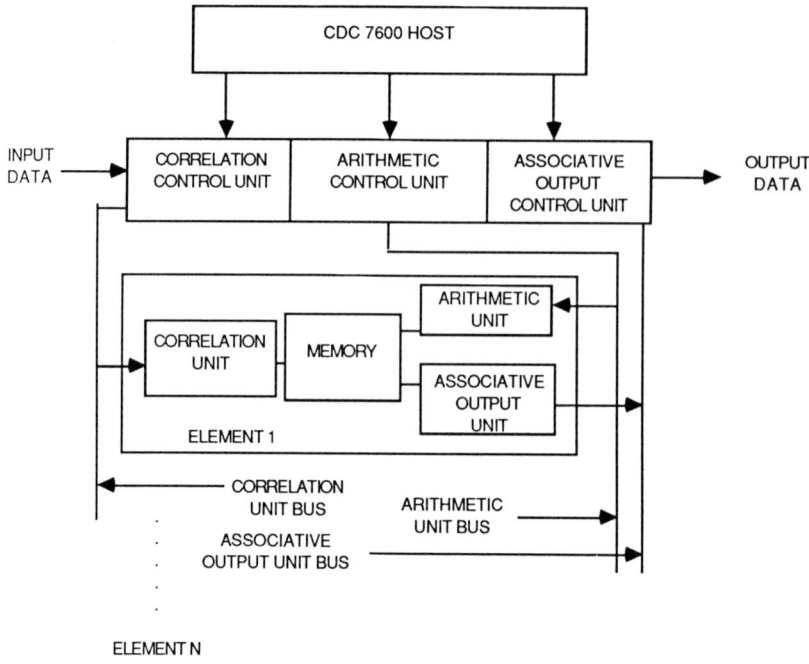

FIGURE 3. PEPE architecture.

2.4.2. STARAN

During the seventies, Goodyear Aerospace Corporation produced an associative processor called STARAN. The last model, Series E, had the same general architecture as its predecessors. The main improvement was an expanded memory capacity. The salient features of STARAN were that it could simultaneously perform search, arithmetic and logical operations on either all or selected words of its memory. The memory could be accessed on a word by word basis or in a bit-slice manner. A processing element was associated with each memory word of 256 bits. The words were grouped into arrays of 256 words. STARAN could have from 1 to 32 array elements.

Because STARAN's designers felt that each system should be custom built for its users, the overall configuration has the form of Figure 6 with STARAN interconnected through a custom interface unit to peripherals and other computers. The parallel I/O channel, PIO, can transfer in parallel up to 32×256 = 8192 words, i.e., it has a direct connection to all memory words. The STARAN system is shown in more detail in Figure 7.

An array element is shown in Figure 8. It consists of a 256×256 multidimensional access memory (in STARAN E, this was expanded to several planes of 256×256 bits), 256 processing elements (one per memory word) and a flip

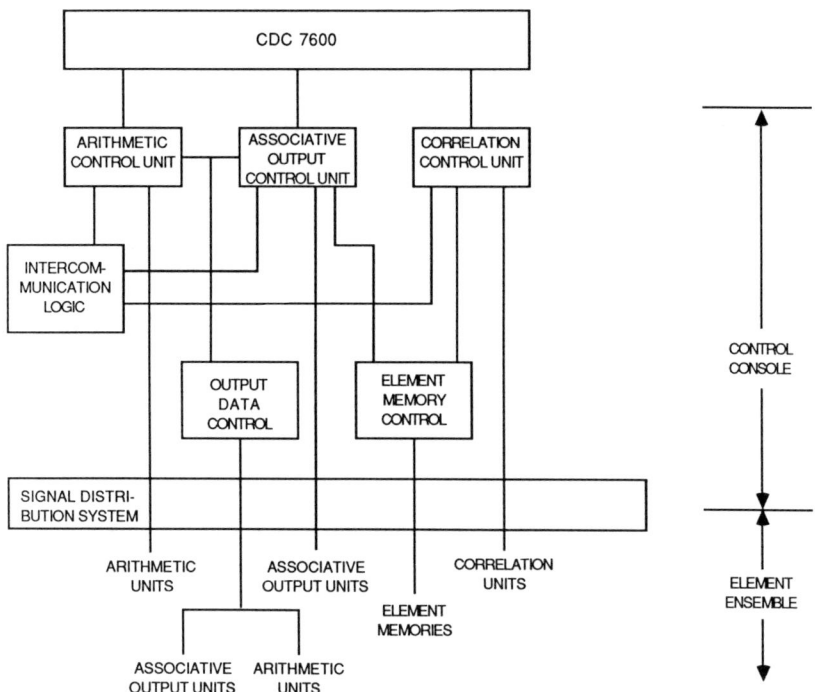

FIGURE 4. PEPE control console.

(also called scramble/unscramble) interconnection network between the memory and the processing elements. The flip network was designed in such a way that words and bit-slices, as well as other templates, could be implemented using conventional RAM chips.

STARAN is well suited for applications which require parallel processing at the bit level (e.g., data base management, air traffic control) and word (or bit-group) level (e.g., arithmetic operations). Furthermore, the inclusion of the permutation network facilitates the efficient programming of functions such as the fast Fourier transform. STARAN cannot be used as a stand-alone facility but must be part of a complex where it only performs parallel tasks.

2.4.3. SOLOMON I

A third early machine, SOLOMON I, built at Westinghouse by a team led by Daniel Slotnick, is generally considered the forerunner of the Illiac IV. A diagram of a processing element is shown in Figure 9. A partial block diagram of the control unit of the SOLOMON I is shown in Figure 10.

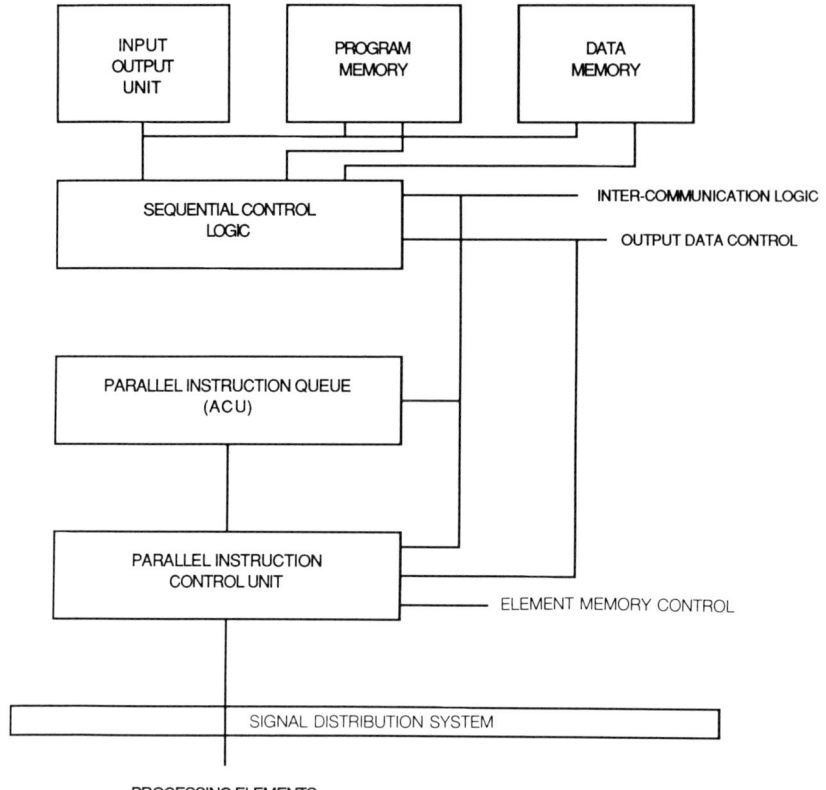

FIGURE 5. PEPE arithmetic control unit.

TABLE 2
PEPE Characteristics

Class
    Single instruction stream, multiple data stream
    Parallel processing
    Associative data match for input
    Associative search for output
    Simultaneous input, output, compute
    Conventional floating point, integer, logical instructions plus associative match
    and search instructions

Data processing speed
    Arithmetic control unit   1 MIP average times number of active elements
    Correlation control unit   5 MIP average times number of active elements
    Associative output control unit: 5 MIP average times number of active elements

Background 13

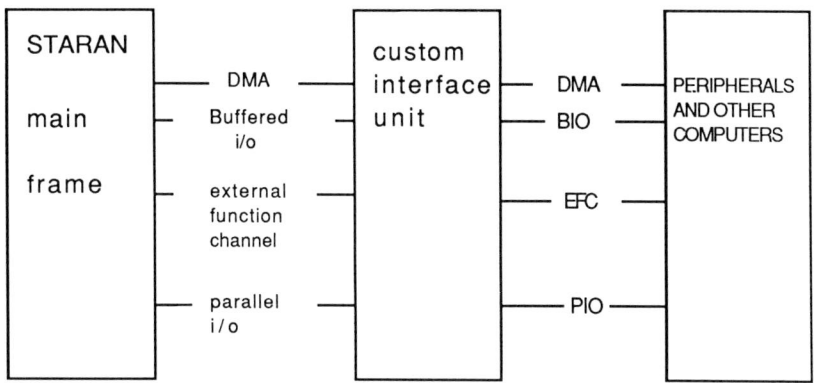

FIGURE 6. STARAN system configuration.

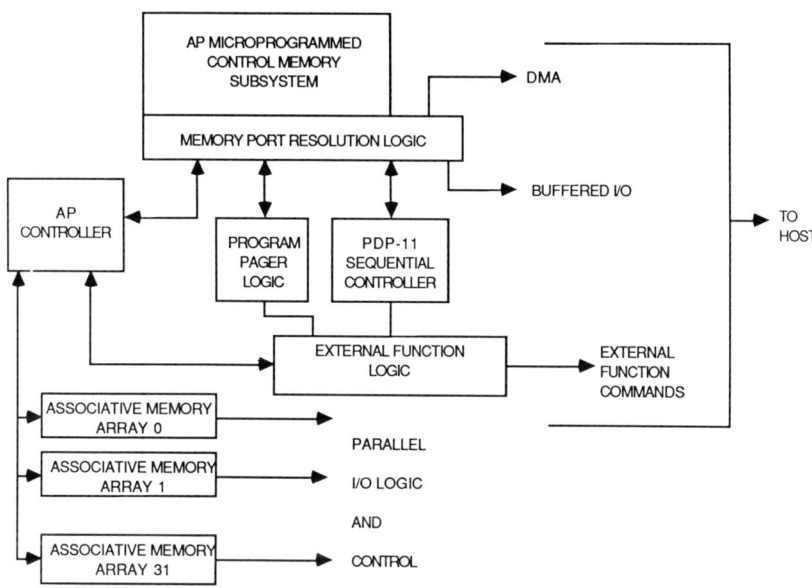

FIGURE 7. STARAN system.

14 *Parallel Supercomputing in SIMD Architectures*

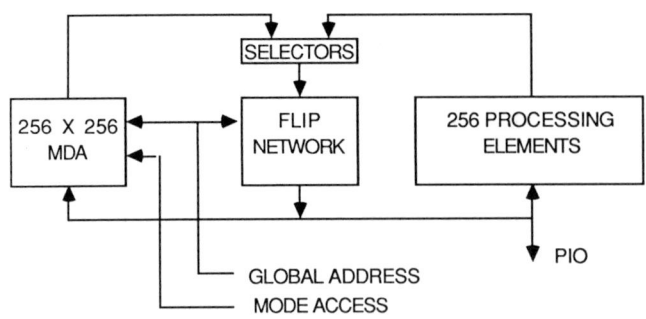

FIGURE 8. STARAN's array element.

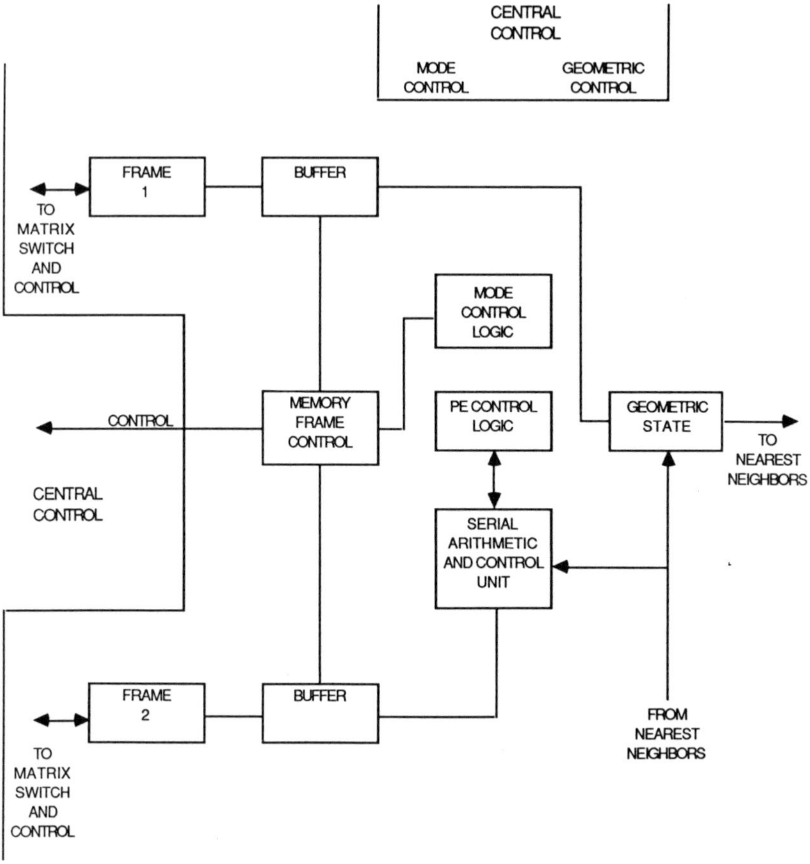

FIGURE 9. SOLOMON I PE.

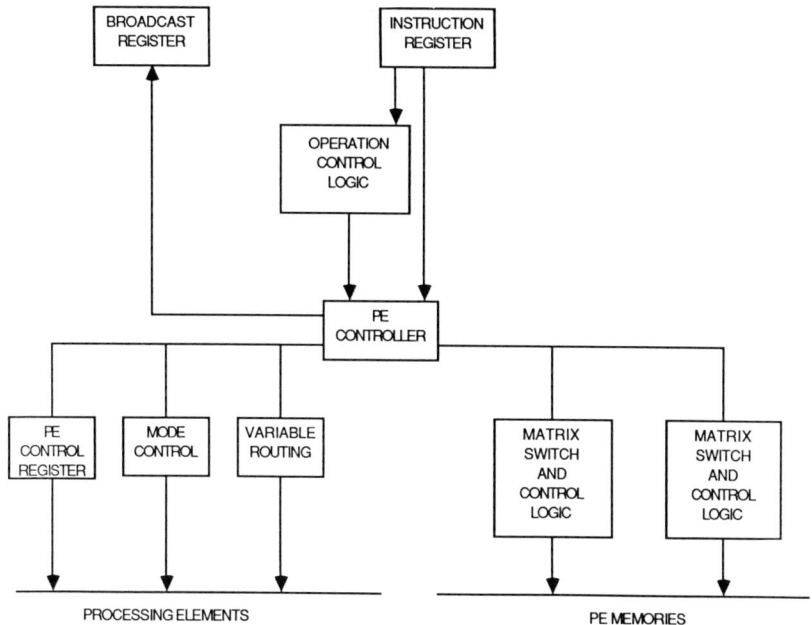

FIGURE 10. SOLOMON I sequencer (partial block diagram of SOLOMON I control unit.

3 ILLIAC IV — THE FIRST SIMD SUPERCOMPUTER

The Illiac IV was the first large scale array computer. As the forerunner of today's advanced computers, it brought whole classes of scientific computations into the realm of practicality. Conceived initially as a grand experiment in computer science, the revolutionary architecture incorporated both a high level of parallelism and pipelining.

After a difficult gestation, the Illiac IV became operational in 1975.

The Illiac IV consisted of a single control unit that broadcast instructions to 64 processing elements operating in lock step. Each of these processing elements had a working memory of 2K 64-bit words. The main memory of the Illiac was implemented in disk with a capacity of 8 million words and with a transfer rate of 500 megabits per second. Arithmetic can be performed in 64-, 32-, or 8-bit mode. In 32-bit mode, on algorithms well suited to the parallel architecture, the Illiac performed at a rate of 300 million instructions per second. Although it used electronics from the late 1960s, for certain classes of important problems, the Illiac was the fastest computer of its time.

3.1. HISTORY

The Illiac IV story begins in the mid-1960s. Then, as now, the computational community had requirements for machines much faster and with more capacity than were available. Large classes of important calculational problems were outside the realm of practicality because the most powerful machines of the day were too slow by orders of magnitude to execute the programs in plausible time. These applications included ballistic missile defense analyses, reactor design calculations, climate modelling, large linear programming, hydrodynamic simulations, seismic data processing, and a host of others.

This demand for higher speed computation began in this time frame to encounter the ultimate limitation on the computing speed theoretically achievable with sequential machines. This limitation is the speed at which a signal can be propagated through an electrical conductor. This upper limit is somewhat less than the speed of light, 186,000 miles per second. At this speed the signal travels less than a foot in a nanosecond. Through miniaturization the length of the interconnecting conductors had already been reduced substantially. Integrated circuits containing transistors packed to a density of several thousand per square inch helped greatly. But the law of diminishing returns had set in.

Designers realized that new kinds of logical organization were needed to break through the speed of light barrier to sequential computers. The response to this need was the parallel architecture. It was not the only response. Another architectural approach that met with some success is overlapping or pipelining wherein an assembly line process is set up for performing sequential operations at different stations within the computer in much the way an automobile is fabricated. The Illiac IV incorporates both of these architectural features.

3.1.1. The Design Concept

The Illiac IV computer is the fourth of a series of advanced computers designed and developed at the University of Illinois, and this accounts for the origin of its name. Its predecessors include a vacuum tube machine completed in 1952 (11,000 operations per second), a transistor machine completed in 1963 (500,000 operations per second) and a 1966 machine designed for automatic scanning of large quantities of visual data. The Illiac IV is a parallel processor in which 64 separate computers work in tandem on the same problem. This parallel approach to computation allows the Illiac IV to achieve up to 300 million operations per second.

Conventional computers solve problems by a series of sequential steps in much the same way an individual mathematician would solve the same problem. In a parallel processor, however, many computations can be performed simultaneously; on the Illiac IV, for example, 64 calculations are done at once.

If the problem at hand is to calculate the price earnings ratio for the stock of a corporation, parallelism is of no advantage since the problem cannot be broken into pieces that the separate processors can address independently. Hence 64 mathematicians can solve the problem no faster than one mathematician. If, on the other hand, the problem is to calculate the average price earnings ratio for all of the stocks listed on the New York Stock Exchange, then by assigning the calculation of the different ratios to different mathematicians, a productive division of labor is achieved and the result is obtained more quickly than one mathematician could obtain it sequentially.

Fortunately, a very large fraction of the world's scientific computational problems satisfies this parallelism requirement. For these problems that are suitable for implementation on the Illiac, very handsome run-time reduction factors have been achieved.

The father of the Illiac IV was Professor Daniel Slotnick who conceived that machine in the mid-1960s. The development was sponsored by the Defense

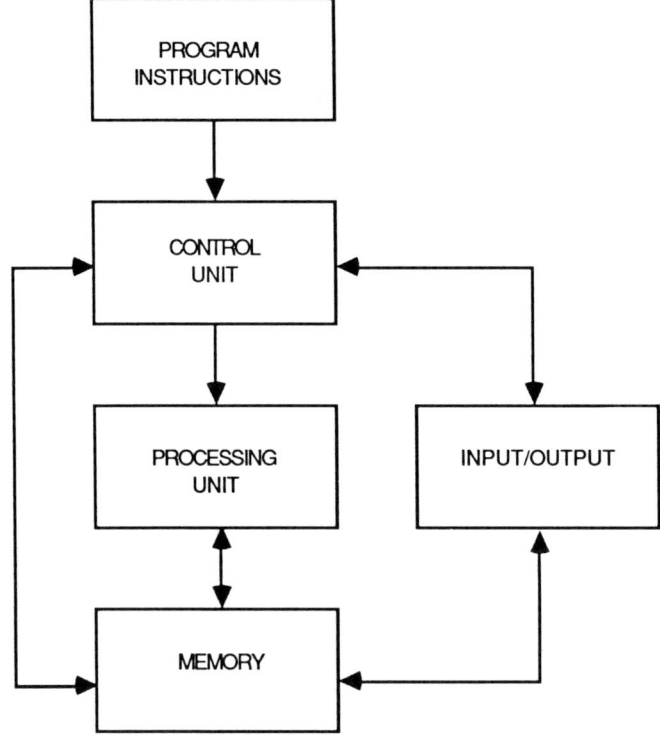

FIGURE 11. Conventional computer architecture.

Advanced Research Projects Agency. Subsystems for the Illiac were manufactured in a number of facilities throughout the U.S. These subsystems were then shipped to the Burroughs Corporation in Paoli, Pennsylvania for final assembly. The Illiac was delivered to the NASA Ames Research Center south of San Francisco in 1971.

The logical design of the Illiac IV is patterned after the Solomon computers. Prototypes of these were built in the early 1960s by the Westinghouse Electric Company. In this design there is a single control processor which sends instructions broadcast style to a multitude of replicated processing units termed elements. Each of these processing elements has an individual memory unit; the control unit transmits addresses to these processing element memories. The processing elements execute the same instruction simultaneously on data that differs in each processing element memory.

For comparison, the logical structure of a conventional sequential computer is illustrated in Figure 11, while Figure 12 shows the architecture of the SIMD machine.

In the particular case of the Illiac IV, each of the processing element

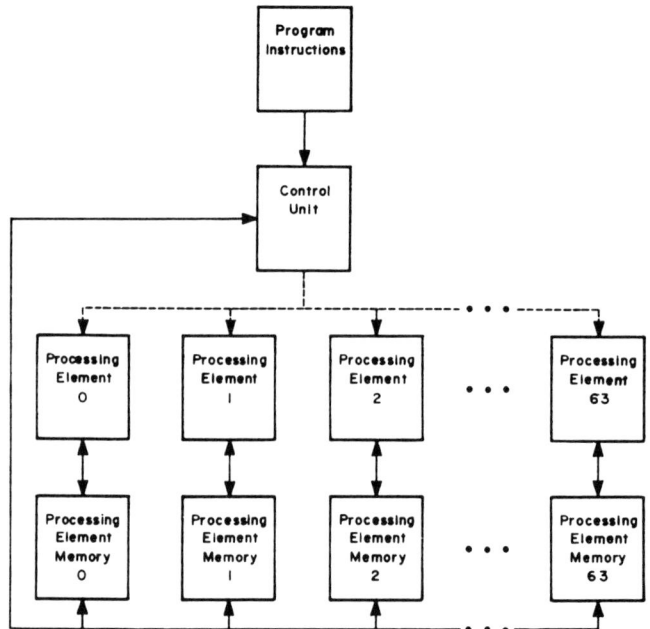

FIGURE 12. Parallel organization of a SIMD computer.

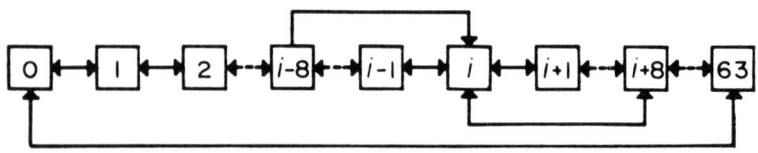

FIGURE 13. Illiac IV routing paths.

memories has a capacity of 2048 words of 64-bit length. In aggregate, the processing element memories provide a megabyte of storage. The time required to fetch a number from this memory is 188 ns, but because additional logical circuitry is needed to resolve contention when two sections of the Illiac IV access memory simultaneously, the minimum time between successive operations is somewhat longer.

In the execution of a program it is often necessary to move data or intermediate results from one processor to another. Routing paths for this purpose are provided as shown in Figure 13. One way of regarding this interconnection pattern is to consider the processing elements as a linear string numbered from 0 to 63. Each processor is provided a direct data path to four other processors, its immediate right and left neighbors and the neighbors spaced eight elements away. So, for example, processor 10 is directly connected to processors 9, 11, 2 and 18. This interconnection structure is wrapped around, so processor 63 is directly connected to processor 0. To transfer values among processors not directly connected, multiple routing steps are required. For example, to move a number from processor 9 to processor 18 it must first be moved to processor 17 and then to processor 18.

The other major control feature that characterizes the Illiac IV is the enable/disable function. While it's true that the 64 processing elements are under centralized control, each of the processing elements has some degree of individual control. This individual control is provided by a mode value. This mode value for a given processor is either 1 or 0, corresponding to the processor being enabled ("on"), or disabled ("off"). The 64 mode values can be set independently under program control, depending on the different data values unique to each processing element. Enabled processors respond to commands from the control unit; disabled elements respond only to a command to change mode. Mode values can be set on specific conditions encountered during program execution. For example, the contents of two registers can be compared and the mode value can be set on the outcome of the comparison. Hence iterative calculations can be terminated in some processors while the iteration continues in others when, say, a quantity exceeded a specified numerical limit.

In addition to the megabyte of processor element memory, the Illiac IV has a main memory with a 16 million word capacity. This main memory is implemented in magnetic rotating disks. Thirteen fixed head disks in synchronized rotation are organized into 52 bands of 300 pages each (an Illiac page is 1024 words). This billion-bit storage subsystem is termed the Illiac IV Disk Memory or I4DM. The access time is determined by the rotation rate of the disks. Each disk rotates once in 40 ms so the average access time is 20 ms. This latency makes the access time about 100,000 times longer than the access time for processor element memory. The transfer rate, however, is 500 million bits per second.

This memory subsystem, the input/output peripherals and the management of the other parts of the system are under the direction of a Digital Equipment Corporation PDP-10 conventional computer. A Burroughs B-6700 computer compiles the programs submitted to the Illiac into machine language.

This design concept came to fruition in November 1975 when the Illiac IV was pronounced operational.

3.1.2. Implementation Difficulties

It was during the firebombing and rioting that shook the University of Illinois campus in the spring of 1970 that the Illiac IV computer project reached its climax. Illiac IV was the culmination of a brilliant parallel computation idea, doggedly pursued by Daniel Slotnick for nearly two decades, from its conception when he was a graduate student to its realization in the form of a massive supercomputer. Conceived as a machine to perform a billion operations per second, a speed it was never to achieve, Illiac IV ultimately included more than a million logic gates — by far the largest assemblage of hardware ever in a single machine.

Until 1970, Illiac IV had been a research and development project, whose controversy was limited to the precise debates of computer scientists, the agonizing of system and hardware designers, and the questioning of budget managers. Afterward, the giant machine was to become a more or less practical computational tool, whose disposition would be a matter of achieving the best return on a government investment of more than $31 million.

This article will discuss the success and failures that have made Illiac IV significant in the development of computer technology, but first let us return to the campus in Urbana-Champaign, IL, in 1970, when Illiac IV was at the center of boiling passions over the relationships between the university, government, and private industry.

Illiac was funded by the U.S. Department of Defense Advanced Research Projects Agency (ARPA) through the U.S. Air Force Rome Air Defense Center. However, the entire project was not only conceived, but to a large extent managed, by academicians at the University of Illinois. Finally, the system hardware was actually designed and built by manufacturing firms — Burroughs acted as the overall system contractor; key subcontractors included Texas Instruments and Fairchild Semiconductor.

When headlines in the Daily Illini, January 6, 1970, proclaimed "Department of Defense to employ UI computer for nuclear weaponry", tensions rapidly escalated — not only between University of Illinois students and the faculty and school administration, but also between the parties directly involved in the Illiac IV project itself. Out of the campus cauldron bubbled heated phrases; some directed at Government's "dangerous fools", others at industry's "questionable business practices", and still others at the university's "volatile visionaries".

As a university-based project supported by military funds, Illiac IV was large, but by no means unique. Such funds had long been flowing into graduate schools and laboratories, and had always been accompanied by strain and contradiction. On the one hand, there was the university's need to train students and advance basic knowledge; on the other, there was the Department of Defense (DOD) need for new military technology. With the prodding of the Military Procurement Act of 1970, signed into law by President Richard Nixon on November 19, 1969, DOD funding agencies were under increased pressure to demonstrate the military value of all the research and development projects they supported. David Packard, then Deputy Secretary of Defense, was publicly reiterating DOD determination to support only work that had a "direct, apparent, and clearly documented relationship" to military functions and operations. Meanwhile, on the campus, there were antiwar sit-ins, demonstrations, and rising feelings — extending beyond the students to the faculty — that military projects did not belong at the University of Illinois. Confrontation over the military R & D issue was imminent in 1970, and the news of military uses for Illiac IV was explosive.

When the dust settled, what remained for those in industry who had been observing the Illiac IV project was an impression that universities might be bold initiators of new ideas but were not equipped to manage large engineering projects. For those in government, there was a hardening determination to keep Illiac in a protected, secure environment away from any campus. For university administrators and faculty, there was a growing conviction that military R & D support was a very mixed blessing, and one that in many cases might not be worth pursuing.

Despite misgivings, the university prepared itself to receive the giant computer — in a new building specially designed for the machine — but the move from the Burroughs plant in Paoli, Pa. to Illinois was never to occur. Instead, Illiac IV would find its permanent home at a NASA facility in California.

Lawrence Roberts, then director of ARPA Information Processing Techniques, recalls the decision not to place Illiac IV at Illinois as mainly a question of finding the best possible operational managers for the machine: "University people who might run it. . .are unwilling to look at some kinds of problems; maybe the classified ones, maybe just sensitive ones. . .Was the university the right organization to manage a large operational undertaking?...The answer was generally no."

Just as the story of Illiac IV can be divided into the periods before and after the campus turmoil of 1970, so the successes and failures of the project can be measured in two quite separate senses. For the Illiac IV balance sheet, there are the achievements and shortcomings of an R&D project, deliberately designed to press computer architecture and design forward as far and fast as possible. There are also the more practical considerations surrounding a multimillion-dollar conglomeration of hardware that is expected to prove its worth by performing day-to-day computational tasks.

This research-operational ambivalence in the Illiac IV project is reflected in the divided feelings expressed by those involved. For example, Daniel Slotnick says: "I'm bitterly disappointed, and very pleased...delighted and dismayed. Delighted that the overall objectives came out well in the end. Dismayed that it cost too much, took too long, doesn't do enough, and not enough people are using it."

Perhaps the greatest strength of Illiac IV, as an R & D project, was in the pressures it mounted to move the computer state of the art forward. There was a conscious decision on the part of all the technical people involved to press the then-existing limits of technology. Dr. Slotnick, who was the guiding spirit of the project, made it clear to his co-workers that the glamour and publicity attendant to building the fastest and biggest machine in the world were necessary to successfully complete what they had started.

The end results this pioneering urge had on computer hardware were impressive: Illiac IV was one of the first computers to use all semiconductor main memories; the project also helped to make faster and more highly integrated bipolar logic circuits available (a boon to the semiconductor and computer industries, this development actually proved a disaster for Illiac IV — more on this subject later); in a negative but decisive sense, Illiac IV gave a death blow to thin-film memories; the physical design, using large, 15-layer printed circuit boards, challenged the capabilities of automated design techniques.

As it began to take shape in 1965 and 1966, Illiac IV seemed so exciting that engineers, physicists, and computer scientists pressed to be assigned to the project. Its overall architecture — using many separate processing units all operating simultaneously — was underway.

On the software side, the Illiac IV programming work at the University of Illinois spawned a whole new generation of experts in parallel and high-speed computation. David Kuck and his students at the university stopped full-time work on Illiac IV in 1968, but the impact on software and applications thinking was a lasting one. Students who wrote their master's theses at Illinois on Illiac IV have gone on to promising careers in the field. For example, Muraoka became manager of computer architecture at NTT Laboratories in Japan. According to Dr. Kuck, work on extracting the ultimate computation speed from programs, in organizing algorithms for ultimate speed, has been greatly stimulated by experience with the Illiac IV project. He points out that people at other schools, such as Stanford, the Massachusetts Institute of Technology, and Carnegie-Mellon, are now doing theses and research that have been influenced, however indirectly, by Illiac IV.

In terms of hardware, deficiencies in Illiac IV's bipolar logic circuits set off a series of design changes that ultimately delayed by years the completion of the machine, while they also ushered in dramatic changes in memory technology.

Initial plans for Illiac IV circuitry envisioned bipolar emitter-coupled logic (ECL) gates capable of speeds of the order of 2 to 3 ns. The ECL circuits were to be packaged with 20 gates per chip — a level of complexity that later would be called medium-scale integration. Chosen as the subcontractor for these circuits, Texas Instruments seemed eager to do the job and sincere in the belief that it could produce the expected circuits.

As the development process moved ahead, it became evident that the 20-gate chips were not functioning properly. Noise margins for these circuits were inadequate. The power distribution design inside the circuit packages was such that crosstalk was excessive. At the root of such problems was an inability to produce multilevel circuit substrates that could meet the necessary precision requirements for lead definition, resistivity, and level-to-level registration. TI asked for an added year of development time to produce the original circuits. Instead, the decision was to go to a simpler integrated circuit — with only seven gates per chip — while maintaining substantially the same circuit speeds.

Although the initial ECL development effort for Illiac was a failure, the millions of dollars of government money that were invested in that effort played a substantial role in advancing the ECL integrated circuit art, so that within about a year TI was able to solve the substrate problems and to offer commercial medium-scale integrated ECL circuits similar to those the Illiac IV project had hoped to use.

But for Illiac IV, problems with ECL circuits were just beginning. The shift to smaller circuit packages was to have a pervasive impact on other portions of the hardware, such as processing element memories, printed circuit boards, and cabling — and overall system design and capabilities would be drastically affected as well. But even the smaller circuit packages themselves proved to be a continuing source of trouble. The plastic encapsulation for these circuits proved to be very sensitive to the operating environment, particularly to the ambient humidity. This required an unusual effort to provide stable humidity in the final Illiac IV installation at the NASA Ames Research Center at Moffet Field, California. Internal short circuits between leads to external circuit pins provided a second major problem — and one that was more subtle since it developed only over a period of time. Test procedures were devised to adjust power-supply voltages to maximum and minimum marginal values in an attempt to show up potential short circuits. Dynamic impedance between leads was also checked, using a variable-current supply source while monitoring voltage output. For the design and production schedule of the overall Illiac IV system, the shift from medium-scale to small-scale ECL chips was a disaster that led to delays probably totaling about 2 years.

Illiac IV initial specifications called for a 2048-word, 64-bits-per-word, 240-ns cycle-time memory for each of its processing elements. In 1966, when the initial design study for the system was underway, the only technology that seemed to be available to meet these requirements was the thin-film memory. At that time, a few developmental semiconductor memory chips were being studied, but no computer manufacturer would yet consider them seriously for main memory use.

Fortunately, Burroughs, the Illiac IV system contractor, had already developed thin-film memories for its B8501 computer. Two years and about a million dollars later, the memory design had been modified to meet initial Illiac IV requirements and prototype memories were in operation.

The change to smaller ECL circuit chips proved to be a death blow to the thin-film memory. When the smaller chips' requirements for added space on circuit boards and interconnections were taken into account, it turned out that there was not enough room for the smallest feasible thin-film memory configuration. Attempts to increase the overall size of the processing elements were frustrated by limitations on propagation time through system interconnections and cables. Even when use of the small-scale ECL circuits forced the designers to drop the system clock rate from 25 MHz down to 16 MHz, the thin-film memory still could not be made to fit. Not only was the thin-film development money wasted, but thin-film memory technology received what has since proved to be a fatal blow — at least as far as its use in computer main memories is concerned.

Strangely, the failures and disappointments of the ECL circuits and thin-film memories also set the stage for a brilliant hardware success; Illiac IV was to be one of the first computers to use all-semiconductor main memories. While interviewing EE students at the University of Illinois for jobs at Fairchild's Semiconductor Division, Rex Rice also happened to meet an old friend and former co-worker, Daniel Slotnick. When the conversation turned to the computer memory art, Rice, who was managing advanced development projects at Fairchild, described, in confident and optimistic terms, the work then underway on bipolar semiconductor memories. The conversation may have been just interesting shoptalk at the time, but the idea that high-speed semiconductor memories had become a feasible alternative was to play a key role in Illiac IV developments.

When it became clear that thin-film memories could not be used without drastically slowing down the entire system, the stage was set for semiconductor memories. Proposals were taken from Texas Instruments, Motorola, and Fairchild for the development and production of memories that would meet Illiac IV specifications. Over the contrary advice of some of the engineers working on the project, Slotnick chose Fairchild as the semiconductor memory subcontractor.

Called for were 2048 words (64 bits/word) of memory for each of the 64 Illiac processing elements, a total of 131,072 bits per processing element. And the memory was to operate with a cycle time of 240 ns and an access time of 120 ns. A complication, in the packaging of the memories, was the need to provide access to each memory not only from its own processing element but from the overall system control unit and the system input-output connections as well. Meeting these requirements meant some extension of the semiconductor art, as well as overcoming a host of design and production problems.

When Fairchild was awarded the contract, its facilities for the project consisted of an empty room, a naked facility that was to be converted for development and production of new devices. Within a few months, with an all-out effort, the company would churn out some 33,000 memory chips (256 bits per chip).

Slotnick recalls that development proudly: "I was the first user of semiconductor memories, and I took a lot of criticism for thinking that we'd have them on time and within specifications. Illiac IV was the first machine to have all-semiconductor memories. Fairchild did a magnificent job of pulling our chestnuts out of the fire. The Fairchild memories were superb and their reliability to this day is just incredibly good." For the semiconductor industry, this dramatic demonstration of memory capabilities had a decisive effect. It put Fairchild firmly into the memory business and, together with IBM's announcement of 64-bit bipolar memory chips for its 360-85 system, the effect was to speed up the pace toward the widespread acceptance that semiconductor memories now enjoy in computers and related systems.

One of the most formidable problems faced by the Illiac IV designers was that of packaging and interconnecting the control unit and the 64 processing elements. Speed was a prime objective of the design, and in the early stages there was no indication that the project would be moving into massive cost overruns, so guaranteeing 25-MHz operation appeared to be an unconditional design criterion. Optimization was to be strictly on performance, not cost.

Configuration studies revealed that the principal packaging problems were to minimize the volume of the equipment and the length of the interconnections so as to reduce propagation delays. Because of the tight system control requirements and the limited space available for interconnections, the designers felt forced into the use of multiple-layer printed circuit boards. For the control unit, four signal layers were needed to make connections between the 165 circuit package positions accommodated by each board (the final control unit boards averaged about 140 to 150 circuits each).

Because of impedance problems, ground layers had to be spaced between the signal layers and the board designs grew until they included 15 different layers. They were expensive and extremely difficult to produce. Furthermore, the designs turned out to be so complex that board layout by human beings was virtually impossible. Initially, a number of wiring patterns were attempted by designers and draftsmen, but these proved to contain so many errors that they were unusable. In addition to the 15-layer complexity, wiring rules were complicated by the use of 50-ohm transmission lines loaded with 100-ohm stubs throughout the design. There were limitations on how close, and how far, loads could be placed from sources — because of the problem of transmission reflections. The human designers simply could not cope with all the rules and requirements.

Fortunately, computer-based design automation techniques were available at the time the Illiac control unit boards were being designed. At first, a printed-wiring routing program supplied by a subcontractor proved inadequate, but with the help of the University of Illinois faculty and students, as well as the Burroughs design team, a satisfactory routing program was finally developed. The boards were designed and produced — a minor triumph for the design automation art.

That was not the end of the printed circuit board story, however. In its final incarnation at NASA Ames, Illiac IV continued to be plagued by board problems, and faults such as small cracks in the printed circuit connections were uncovered in the process of bringing the computer into regular daily operation.

In looking back at the history of the Illiac IV project, Lawrence Roberts, former director of ARPA Information Processing Techniques, decided that Illiac's strongest virtue — its pioneering role in pressing forward the computer state of the art — became in the end its greatest weakness. Dr. Roberts felt that the best course would have been to build the machine using transistor-transistor logic (TTL) rather than ECL circuits. TTL logic was, in the late 1960s, a straightforward, widely employed technology, and its use could have considerably reduced the cost and duration of the project. Said Roberts: "I feel it is absolutely clear that it should have been done with older technology. I've used that lesson many times since then. People complain bitterly but it has always worked out better."

When Illiac IV was delivered to its final home at the NASA Ames Research Center in California in the spring of 1972, the question in the mind of Dr. Pirtle, former Director of NASA's Institute for Advanced Computation, was whether or not the machine could actually be made to perform useful work. By the following summer, the educated outlook was positive. Illiac was then operating at reduced speed, but it would almost always execute its control sequences correctly, and, occasionally, it would actually deliver correct results. At that time, the machine was made available to a few users, just to demonstrate that useful programming codes could be made to run — but knowing full well that most of the computed results would be erroneous or inaccurate.

Then, in June 1975, a concerted effort began to check out Illiac fully and make it operational. Over the next 4 months, thousands of manufacturing faults were uncovered in printed circuit boards and connectors; 110,000 low-reliability terminator resistors, wire-wrapped to backplanes, were replaced by circuit-board-mounted resistors; and logic design faults — principally involving signal-propagation times — were corrected, as were improper line terminations and inadequate power-supply filtering in the disk controllers.

The system eventually operated from Monday morning to Friday afternoon, including 60 to 80 h of good, verified up-time for the users, along with 44 h of maintenance and downtime.

Above all, speed was to be the most crucial characteristic of Illiac IV. A billion instructions per second was Slotnick's initial goal. As the system design took shape, that target was expressed more specifically as 256 parallel processing elements that would each perform a 64-bit floating point addition in 240 ns.

Then, when the size of the machine had to be dropped from 256 to 64 processing elements, this goal faded from sight, retreating even further as the clock speed was dropped from 25 to 16, and finally to 13 MHz. Still even in 1970, after major hardware disappointments with the available circuitry had been absorbed into the systems design, the system was still believed by its creator to be capable of performing 200 million instructions per second.

When the University of Illinois trustees signed the initial Illiac IV contract with the United States Air Force in February 1966, the cost of the project was estimated at just over $8 million, a number that was remarkably close to the gate of the Clay-Liston world championship heavyweight title bout held just months earlier. By January 1970, funding for the project had grown far beyond the dimensions of a prizefight, to over $24 million, and by April 1972, when the huge computer had been delivered to its permanent site in California, its estimated cost had reached $31 million.

Clearly, inflation played a role in these escalating costs, as did the millions that were spent for development of key components such as the ECL circuits, and for components that were discarded, such as the thin-film memories.

At the same time, Illiac was originally planned to include 256 processing elements. As it became evident that costs were rapidly rising, the number of processing elements was cut back to 64 — so the machine ended up at one-fourth its original size, although costing about four times as much as initially estimated.

University-based project managers apparently had no clear idea of the costs of developing and manufacturing in an industry environment. Slotnick felt that the primary source of the cost overruns was at Burroughs where the cost-plus-fixed-fee environment in the company's defense-space operations set it up to jump on the Illiac IV contract "with both feet". From the Burroughs viewpoint, it was a "hairy" project; their aim was to avoid losing money. Actually, Burroughs management consistently underestimated the man-hour costs of the project.

It wasn't until 1971 that those costs came under more accurate control. At the time, a group was set up at DOD's Advanced Research Projects Agency to review the Illiac IV situation every few months and make estimates of costs to completion. Their figures proved to be accurate, probably because they were from a relatively uninvolved source.

From the system software standpoint, Illiac IV is quite rudimentary. There is almost no operating system. A user takes hold of the machine, runs his problem, and then lets it go; the next user does the same. No shared use of Illiac's 64 processing elements is provided. In smaller computers that surround Illiac's control unit and processing elements, there is more complex software that forms a queue of users waiting to get at the big machine and allows them to perform nonarithmetic "companion" processes. But the actual Illiac operating software itself is very simple, capable of such basic operations as monitoring input/output and loading data into the processing element memories. An operating system, along with two Illiac IV languages called TRANQUIL and GLYPNIR, was written at the University of Illinois beginning in 1966. This effort amounted to perhaps a dozen man-years of programming. Later, when the system was moved to California and connected to the ARPA network, it was decided that entirely new system software was needed, since PDP-10 and PDP-11 computers were used — in place of the original B6500 machine — to connect Illiac IV to the outside world.

There, the NASA Ames users decided to write a new Illiac IV language, which would be called CFD, to efficiently communicate problems involving the solution of partial differential equations to the big machine. This was accomplished with approximately 2 man-years of programming effort.

These equations were important to the NASA Ames users, who took up about 20% of Illiac IV operating time solving aerodynamic flow equations.

The remaining 80% of Illiac IV time was taken up by a diverse, and often anonymous, group of users, many of whom used the GLYPNIR language.

A giant computer should be useful for tackling giant computing problems, and that is pretty much the story of Illiac IV applications programs. Beyond the NASA Ames aerodynamic flow problems, users of the big computer have been running several small weather-prediction and climate models with improved and larger models still under development.

Several types of signal processing computations, including fast Fourier transforms, were a regular part of Illiac IV's diet, and a large-scale experiment with real-time data is now underway. Other applications problems that have actually found their way to Illiac IV include beam forming and convolution, seismic research, radiation transport for fission reactors, and linear programming software.

As Slotnick saw it, applications have gone just about as he thought they would — "No huge new computational areas have succumbed to Illiac, but nothing we thought would work has not worked."

3.2. I4 SYSTEM

The architecture of the Illiac is shown at a conceptual level in Figure 14.

Illiac is a parallel processor. It consists of a control unit (CU), 64 processing elements (PE), 131,072 words of core memory, and 15,974,400 words of disk memory. The control unit has access to all of core memory. Its basic cycle time is 60 ns. However, greater processing power is achieved through the simultaneous execution of an instruction in each of the 64 processing elements.

The control unit fetches and decodes all instructions. After decoding, some instructions are broadcast for execution in the processing elements while others are executed in the control unit. The arithmetic capability of the control unit is limited to 24-bit two's complement addition and subtraction, masking, and comparison for use in branching. The control unit has no floating point capability. One operand at a time is processed by the control unit. The control unit also initiates data transfers between core and Illiac disk.

The processing power of Illiac resides in 64 identical processing elements. Each PE executes instructions broadcast from the CU. Though each PE has its own index registers and memory to operate upon, all 64 PEs always execute identical instructions in lockstep. Each PE has direct access to 2048 words of core memory.

There are three data paths available for communication among PEs and between the PEs and the control unit (CU). First, the CU can access all of core, so it can load a word from one processing element memory (PEM) and either use it or store in another PEM. This method of communication is both simple and flexible, allowing for any data movement desired, but, since only one word at a time is transferred, it is relatively slow compared to the two other methods available.

Second, the CU can communicate with all PEs by broadcasting the same word to all PEs simultaneously. This method is faster than the first since 64 words are transmitted at once, but provides only a limited form of communication.

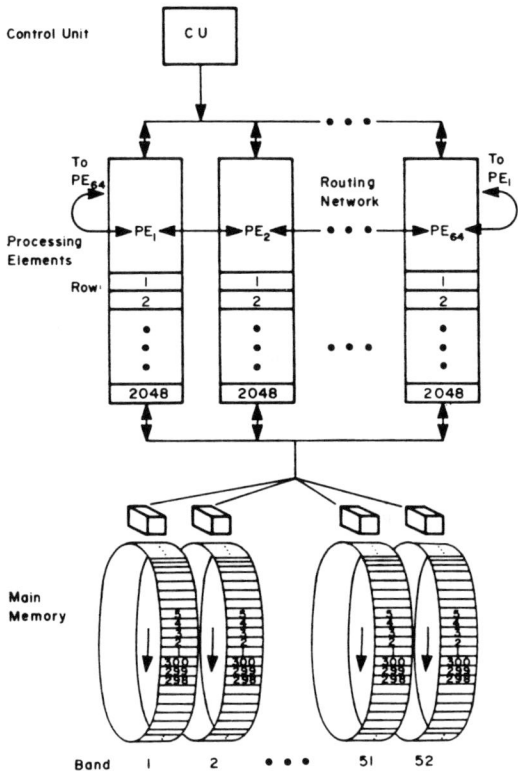

FIGURE 14. Conceptual architecture of Illiac IV.

Third, the PEs can communicate with each other via the ROUTE instruction which transfers the contents of a register in each PE to the PE determined by the following scheme: If PEN (processing element number) is the number of the source PE and R is the route amount supplied with the instruction, identical in all PEs, the number of the destination PE is MOD_{64} (PEN+R). If the PEs are thought of as arranged in a circle with PE 63 adjacent to PE 0, the ROUTE instruction consists of loading the data, rotating the circle, and storing the data. This data transfer is very fast since 64 words are transferred simultaneously. It is general in that all 64 words can be different but the pattern set by the fact that the routing distance is the same for all PEs is restrictive. It does not transfer 64 words randomly distributed in core to 64 different locations simultaneously.

The primary memory used by Illiac is a disk memory with capacity approximately 100 times that of core memory. One page (1024 words) of memory is the minimum amount of data that can be transferred between core and disk.

Although the bandwidth between core and disk is $5. \times 10^8$ bits/s, the average access time to a particular spot on the disk is 20 ms. This relatively long access time (compared to an 80-ns clock time in 64 parallel processors) necessitates careful planning of disk usage. The number of disk transfers must be kept to a minimum to avoid waiting for disk accesses.

Since the most important feature of Illiac is its computing power one of the prime objectives in the design of any Illiac program is to minimize execution time. The best possible result is a running time $1/_{64}$ that possible with only one PE, but due to the architecture of the machine the degree to which this is achieved is dependent upon the design of the algorithm. First, suppose that it is necessary to code the trigonometric SIN function for Illiac. If the particular usage makes it possible to always compute 64 functions simultaneously, one simply has the same SIN routine running in all PEs on different data, and a speedup by a factor of 64 is very nearly achieved. (Some time is lost if there is conditional branching in the original SIN routine which is changed to enabling and disabling of PEs.) A second approach is to devise a method for utilizing all 64 processors to compute one function value. No method has been devised for doing this 64 or even 10 times faster than is possible with one processor. The first approach is both faster and simpler, but certain algorithms may preclude calculation of more than one value of SIN simultaneously or may require significant overhead elsewhere in order to do so.

One misconception is that if all of the PEs are kept "busy" the machine is running at maximum efficiency. In fact this statement is not true and one must be very careful in relating the word efficiency to the use of Illiac. For example, consider the problem of summing groups of numbers. If it is desired to sum 64 pairs of numbers, keeping 64 different results, each PE forms one sum and the work is done 64 times faster than could be done by one processor. If however, it is desired to find the sum of one group of 64 numbers, a more complicated method must be used. In order to simplify the explanation somewhat, consider an eight PE machine and the summation of eight numbers, one in each PE. Figure 15 depicts a method whereby this can be done in three routes and three additions. Since the routes require roughly equivalent CPU time as the register loads necessary before any operation, the time taken for an 8 PE machine to sum eight numbers is equal to the time taken for three additions. If this algorithm is extended to the summation of 64 numbers within a 64-PE machine it takes six additions to form the sum. Given that 1 PE requires 63 additions to sum 64 numbers, the 64-PE machine is 63/6 or 10.5 times faster. Note that although none of the PEs are ever disabled and all are forming the sum, this algorithm does not achieve the factor of 64 speedup. However, the factor of ten speedup that is achieved makes this algorithm usable if data organization requires its use.

The choice of which design approach to take for a particular problem is

STATE OF REGISTERS

OPERATION	PE_0	PE_1	PE_2	PE_3	PE_4	PE_5	PE_6	PE_7
Initial Conditions	$\$A=I_0$ $\$R=0$	$\$A=I_1$ $\$R=0$	$\$A=I_2$ $\$R=0$	$\$A=I_3$ $\$R=0$	$\$A=I_4$ $\$R=0$	$\$A=I_5$ $\$R=0$	$\$A=I_6$ $\$R=0$	$\$A=I_7$ $\$R=0$
1. Route Contents of $\$A_N$ to $\$R_{N+1}$.	$\$A=I_0$ $\$R=I_7$	$\$A=I_1$ $\$R=I_0$	$\$A=I_2$ $\$R=I_1$	$\$A=I_3$ $\$R=I_2$	$\$A=I_4$ $\$R=I_3$	$\$A=I_5$ $\$R=I_4$	$\$A=I_6$ $\$R=I_5$	$\$A=I_7$ $\$R=I_6$
2. Add $\$A to $\$R and leave result in $\$A.	$\$A=I_0+I_7$ $\$R=I_7$	$\$A=I_1+I_0$ $\$R=I_0$	$\$A=I_2+I_1$ $\$R=I_1$	$\$A=I_3+I_2$ $\$R=I_2$	$\$A=I_4+I_3$ $\$R=I_3$	$\$A=I_5+I_4$ $\$R=I_4$	$\$A=I_6+I_5$ $\$R=I_5$	$\$A=I_7+I_6$ $\$R=I_6$
3. Route Contents of $\$A_N$ to $\$R_{N+2}$.	$\$A=I_0+I_7$ $\$R=I_6+I_5$	$\$A=I_1+I_0$ $\$R=I_7+I_6$	$\$A=I_2+I_1$ $\$R=I_0+I_7$	$\$A=I_3+I_2$ $\$R=I_1+I_0$	$\$A=I_4+I_3$ $\$R=I_2+I_1$	$\$A=I_5+I_4$ $\$R=I_3+I_2$	$\$A=I_6+I_5$ $\$R=I_4+I_3$	$\$A=I_7+I_6$ $\$R=I_5+I_4$
4. Add $\$A to $\$R and leave result in $\$A.	$\$A=I_0+I_7+I_6+I_5$ $\$R=I_6+I_5$	$\$A=I_1+I_0+I_7+I_6$ $\$R=I_7+I_6$	$\$A=I_2+I_1+I_0+I_7$ $\$R=I_0+I_7$	$\$A=I_3+I_2+I_1+I_0$ $\$R=I_1+I_0$	$\$A=I_4+I_3+I_2+I_1$ $\$R=I_2+I_1$	$\$A=I_5+I_4+I_3+I_2$ $\$R=I_3+I_2$	$\$A=I_6+I_5+I_4+I_3$ $\$R=I_4+I_3$	$\$A=I_7+I_6+I_5+I_4$ $\$R=I_5+I_4$
5. Route Contents of $\$A_N$ to $\$R_{N+4}$.	$\$A=I_0+I_7+I_6+I_5$ $\$R=I_4+I_3+I_2+I_1$	$\$A=I_1+I_0+I_7+I_6$ $\$R=I_5+I_4+I_3+I_2$	$\$A=I_2+I_1+I_0+I_7$ $\$R=I_6+I_5+I_4+I_3$	$\$A=I_3+I_2+I_1+I_0$ $\$R=I_7+I_6+I_5+I_4$	$\$A=I_4+I_3+I_2+I_1$ $\$R=I_0+I_7+I_6+I_5$	$\$A=I_5+I_4+I_3+I_2$ $\$R=I_1+I_0+I_7+I_6$	$\$A=I_6+I_5+I_4+I_3$ $\$R=I_2+I_1+I_0+I_7$	$\$A=I_7+I_6+I_5+I_4$ $\$R=I_3+I_2+I_1+I_0$
6. Add $\$A to $\$R and leave result in $\$A.	$\$A=I_0+I_7+I_6+I_5+I_4+I_3+I_2+I_1$ $\$R=I_4+I_3+I_2+I_1$	$\$A=I_1+I_0+I_7+I_6+I_5+I_4+I_3+I_2$ $\$R=I_5+I_4+I_3+I_2$	$\$A=I_2+I_1+I_0+I_7+I_6+I_5+I_4+I_3$ $\$R=I_6+I_5+I_4+I_3$	$\$A=I_3+I_2+I_1+I_0+I_7+I_6+I_5+I_4$ $\$R=I_7+I_6+I_5+I_4$	$\$A=I_4+I_3+I_2+I_1+I_0+I_7+I_6+I_5$ $\$R=I_0+I_7+I_6+I_5$	$\$A=I_5+I_4+I_3+I_2+I_1+I_0+I_7+I_6$ $\$R=I_1+I_0+I_7+I_6$	$\$A=I_6+I_5+I_4+I_3+I_2+I_1+I_0+I_7$ $\$R=I_2+I_1+I_0+I_7$	$\$A=I_7+I_6+I_5+I_4+I_3+I_2+I_1+I_0$ $\$R=I_3+I_2+I_1+I_0$

FIGURE 15. Detailed view of Rowsum operation.

dependent upon data organization which is much faster than any other. It must be decided whether the overhead and execution time involved in data transportation is compensated by decrease overhead and execution time elsewhere in the algorithm.

The original Illiac IV design included two methods for achieving high execution rates: parallelism (the concept of multiple processing units acting in concert) and overlap (the ability of the computer to begin the execution of new instructions before completing the execution of previous ones). In order to begin running actual computer programs on Illiac IV as soon as possible, NASA-Ames decided in 1971 to disable the computer's overlap capability and concentrate upon making the Illiac IV execute instructions reliably and correctly. Though limited usage was delivered to users prior to 1975, the Illiac IV was brought to full operation in November of that year, but without the overlap capability. Following this achievement, the hardware development team focused upon implementing the circuitry to support overlap.

This discussion details how the Illiac IV (I4), the control unit (CU), the processing elements (PEs), the processing element memory (PEM), and other hardware interact under overlap to increase throughput. Programs execute more rapidly under overlap mode, even if the codes are not designed to exploit the overlap feature, but designing code with overlap in mind produces greater speed improvement.

Two types of overlap are considered here: CU overlap and FINST (Final Station) overlap. The term overlap is sometimes associated with input/output operation, but this is not of concern here.

CU overlap is concerned with processing CU and PE instructions as fast as possible. To this end, the CU is divided into several units using the design concept that each unit should be capable of independent operation with small, well-defined interfaces between units. This design philosophy allows for different operations to go on at the same time in each independent unit. The term CU overlap describes the ability of these independent units to function in this manner.

FINST overlap refers to the capability of simultaneously processing more than one PE instruction by FINST. Operations such as getting an operand ready for the next instruction while the present operation is in progress are typical of this multiprocessing capability.

Both of the overlap processes listed above can be turned off. The mode of operation when both are disabled is called single instruction mode or nonoverlap. It has the action of processing only one operation throughout the CU at a time.

Four of the five units in the CU are involved with overlap. The fifth unit's operation, the Test and Maintenance Unit (TMU), does not differ in the two modes of operation, overlap and nonoverlap, so it is not discussed.

Starting from the bottom up, FINST is the unit that sends the instructions in the form of microsequences to the PE. By design, the PEs are void of control and depend on these microsequences to manipulate the data in the proper fashion to obtain the desired results. The PEs are driven by the CU, in particular, FINST. The task for FINST is decoding individual instructions, deciding proper action for the instructions, and issuing microsequences to accomplish the correct actions.

The Advanced Station's (ADVAST) primary task is to differentiate between two types of instructions. The first type are those destined for the PEs. The second type are those that will operate within the CU. A third type could be considered those instructions that do considerable action in the CU and also reference the PEs. Basically ADVAST either processes the instructions or passes it on to FINST for processing by the PEs.

Instruction Look Ahead (ILA) has as its function the prefetching of instructions. In an attempt to optimize the instruction processing speed, the unit ILA was established to maintain a significant number of instructions in the CU, thus negating the need to go to PEM for each new instruction.

The Memory Service Unit (MSU) correlates and processes memory requests. There are five different requests that can be made of memory and MSU must arbitrate the requests and give them all proper service.

Important to understanding overlap is understanding the interfaces between the units involved. The following description is not intended to completely describe the interface but merely to give the reader an introduction to aid in more fully understanding overlap.

ADVAST and ILA share an interface that primarily deals with the obtaining of instructions. ILA is the station where the instructions are stored locally and ADVAST is the station where the individual instructions will first be examined, so their interface is quite simple. A simple handshaking is all that goes on, with ADVAST notifying ILA that it is ready for another instruction, and ILA notifying ADVAST when the instruction is ready. Since ADVAST has the only connection with all 64 PEMs in the CU, by means of the Control Unit Buffer (CUB), ADVAST also participates in the block fetching of instructions by ILA in its storing multiple instructions in the CU.

The interface acts differently during overlap operations. ADVAST monitors conditions throughout the CU. These conditions determine when ADVAST goes to ILA for another instruction. In overlap mode this request comes more quickly on the heels of the previous request than in the nonoverlap mode. ADVAST no longer waits for the rest of the machine to reach an idle state as it did in the nonoverlap mode. Thus, as soon as ADVAST decided that it can handle another instruction it requests one from ILA.

Another important interface is the one between ADVAST and FINST. An eight-position queue exists as a buffer between the two units. Its purpose is to allow ADVAST to deposit instructions destined for the PEs and return for another instruction to ILA. Meanwhile FINST is free to remove an instruction from the queue as soon as one is available and FINST is ready to process another. Some important control states of the queue are Queue Full, Queue Empty, and Queue Not Full.

In the nonoverlap or single instruction mode, the purpose of the queue is defeated because only one instruction is available for processing throughout the CU. ADVAST will not request another instruction from ILA until FINST is finished with the last instruction. Hence, there is no possibility that more than one instruction can exist in the queue at any time. In overlap, however, ADVAST only need deliver the instruction to the queue and it is free to return to ILA for another instruction. Of course, the condition of the queue is important, and ADVAST cannot deliver an instruction to a queue that is already full. Similarly, FINST cannot remove an instruction from a queue that is empty. The purpose of the queue is how to be a buffer for instruction between the two units allowing for more independent operation in the overlap mode.

FINST, as noted before, is the section that sends the microsequences to the PEs. The main objective in the design of the FINST/PE interface is to keep the PEs busy. Every attempt was made to make sure the most vital resource, the array, does not sit idle. The result of this design objective is called FINST overlap. Basically, this means the starting of one instruction before the conclusion of the previous instruction.

Every instruction reaching FINST by way of ADVAST is considered to have two parts. The first part is referred to as the overlap portion. The second section is referred to as the execution portion. The combination of the overlap and execution sections can vary. That is, an instruction can have 1 clock of overlap and 70 of execution, or 10 of overlap and 1 of execution. Instructions may also be completed in overlap and have no execution, or contain no overlap and only execution. The determination is fixed and will be discussed later. One single clock contains a portion of the execution of one instruction and the decoding of another instruction. This is the basis for FINST overlap: the simultaneous processing of more than one instruction.

The description of the FINST/PE instruction in the previous section led the designers to the configuration of FINST hardware. The FINST hardware is divided into two sections, one dedicated to the processing of the overlap portion of the instruction, and the other dedicated to the execution portion of the instruction. There is no duplication of hardware; the execution portion of the instruction cannot be accomplished from the overlap section and vice versa. Each section is dedicated to its portion of the instruction. The flow of instructions through FINST is also fixed. An instruction must always appear in the overlap station before moving on to the execution station.

In more detail, the instructions are deposited in the queue by ADVAST. On a first-in first-out basis, an instruction is removed from the queue and placed in the instruction register on the overlap station. The instruction is examined for the type of overlap to be accomplished, if any. At the precise time it is determined that it can proceed with the overlapping action, the instruction is simultaneously transferred to the Read-Only-Memory (ROM) address register for the overlap portion of the ROM and sent on to the instruction register for the execution section. The ROM address register will then select the proper word from the ROM to accomplish the desired action for overlap. At the next clock period the enables necessary to accomplish the desired action in the PE will appear in the FINST command register. A copy of the command register appears in the PE each clock. Meanwhile the instruction register of the execution station is decoding the instruction. The next clock will select the address of the word or words in the ROM that are dedicated to the particular instruction decoded. And as in the overlap section, the next clock will load the particular set of enables needed by the PE to perform the instruction.

In examining how FINST overlap works with respect to instruction flow, it is seen that as soon as the instruction overlap is decoded and sent to the ROM address register, the overlap station is ready to get another instruction from the queue. In the case of instructions with short overlap portions, this leads to a pipelining effect with instructions in the overlapping station and the execution station.

The following are some considerations taken into account at the time of design. The decisions made will not be justified, but their impact on the operation of FINST with respect to overlap will be discussed in an attempt to give the reader a better understanding of FINST.

The decision was made to limit the scope of operations that could take place in the overlap station. There are only several types of operations, such as memory references, register transfers, literal transfers and shift count modifications that go on in overlap. With minor exception, overlap is used to get the second operand of an instruction in place before the instruction is executed. Because the action of getting the second operand in place in many cases is similar to other FINST/PE instructions, in most cases those other instructions are also executed in the overlap station.

The portion of any instruction in FINST done in overlap and the portion in the execution station is predetermined. An instruction does not move from station to station because one station becomes available, it moves to the next station only after it has performed all the tasks it was designated to perform while in that station. A memory reference cannot start in the overlap station and complete in the execution station. It must wait in the overlap station until the operand has returned from memory, and then is permitted to move to the execution station.

The ROM is conceptually divided into two sections, one addressed from the execution station and one addressed from the overlap station. No cross-addressing is allowed.

When an instruction enters the execution station, is decoded and starts addressing the ROM, a state of ROM busy is set up. This state precludes any other instructions from being executed from the ROM, but does not prohibit noninterfering instructions from generating PE enables from the overlap station. The method for determining if the overlapping instruction is noninterfering will be discussed shortly. An instruction may continue to address the ROM for a considerable time. A divide, for instance, can use 69 clocks in order to accomplish its task. Any following instructions that have completed their overlap portion will wait in the execution until the resources needed to accomplish its desired action are available. Even further overlap at some point will be stopped until the long instruction processing in the execution station is completed. Resources do not exist in the PE to store many operands so it makes little sense to get too far ahead in fetching operands.

So far, only the processing of the instruction has been examined. Note also that there are a series of registers in FINST that allow the data associated with the instruction to keep in step with the instruction. Therefore, when the enables appear at the PE, the data, if any, associated with these enables will be on the common data bus.

The mechanism alluded to in the previous section which determines when overlap is allowed to proceed is referred to as the busy bits. This mechanism must be understood in order to get the most advantage out of programming for the overlap mode. For the sake of this mechanism, the PE has been divided up into seven areas, each labeled with its own busy bit. Registers A, B, and R all have a busy bit. The Mode Register has a busy bit M. The Address Adder has the busy bit Z. The Operand Select Gates, a very important resource in the PE, has the busy bit D. Finally, the Logic Unit (LOG) and the Barrel Switch have the busy bit L. These busy bits were selected by careful examination of all PE instructions and the design considerations of FINST and overlap.

Keep in mind there is an instruction processing in the overlap section that requires a portion of the hardware in the PE, and there is an instruction processing in the execution station that requires a portion of the PE hardware. When both sections require the same hardware, there is a conflict and it is up to the busy bit hardware to resolve it. Solving conflicts is not difficult; the execution station always has priority. If the execution station requires the use of the R register and the overlap station has an identical requirement, the execution gets first use. The overlapping instruction must wait until the executing instruction is furnished with the register.

This method of arbitration has the effect of keeping instruction sequences in order and still remaining quite simple. For example, a simple instruction sequence is that of adding registers A and B and placing the result in memory. The ADD instruction will obtain the two operands in A and B and proceed to add them together. The result of the addition, which takes place in the execution station, will be deposited in the A register. Both A and B will be unavailable to the overlapping station until the results are in A. The next instruction, loading A into memory, takes place in the overlap station. Had the A register not been off limits to the overlapping instruction, some intermediate results not desired would have been placed in memory instead of the final sum which is correct.

The busy bit hardware is then a mechanism by which the executing instruction notifies the overlapping instruction of those parts of the PE hardware it intends to use during the execution of its instruction. The overlapping instruction then observes the portions of the PE the executing instruction requires, and proceeds with overlap only after determining that all the hardware necessary for completing the overlap sequence is available. The executing instruction will record in the busy bit register those portions of the PEs it will use during the execution of its instructions. The mechanism is available to release those portions of the PE as soon as they are no longer needed by the instruction. The overlapping instruction awaits all hardware necessary for the successful conclusion of the overlap portion. If, for example, the overlapping instruction has accounted for all hardware necessary to complete overlap except the R register, and the execution station is processing a divide, the overlapping instruction will begin its overlap as soon as the divide releases the R register, which will come before the instruction in the execution stage is completed. That is considered FINST overlap.

Each I4 instruction takes the same time to execute as in nonoverlap mode (the I4 continues to operate at a "clock rate" of 12.5 MHz) but instruction sequences can be executed more quickly with overlap. During testing of overlap, I4 executed program sequences up to five times faster. Many users' codes were executed at more than twice their previous rate. Clearly the time required to execute a highly I/O bound program would not be significantly affected by overlap since I/O mechanism is unaltered.

A number of arithmetic expressions for the Illiac IV have been coded in assembly language in order to analyze the performance of the machine in overlap mode. The following results were produced using the two most common measures of computer performance — millions of operations per second (mips) and millions of floating-point operations per second (megaflops).

Precision	Mips	Megaflops
64 bits	140 — 195	40 — 55
32 bits	250 — 310	70 — 90

These results were for vector lengths that were multiples of 64, and for floating-point operations that were rounded and normalized. No routing was involved, and all arithmetic operations were done from memory to memory. The sample problems were carefully coded to take advantage of their inherent parallelism.

3.3. THE CFD LANGUAGE

To understand the evolution of CFD it is necessary to go back to 1970 and 1971 when the Computational Fluid Dynamics Branch of NASA Ames Research Center first learned that it would be able to use the Illiac IV. For a great many years this branch had been coding fluid flow problems in FORTRAN so that they could be run on the conventional serial machines of that period (IBM 360 and CDC 6000 series computers). Thus the advent of the Illiac IV forced the branch to determine how to run the next generation of these fluid flow problems on the Illiac. To do this the branch first looked closely at how the Illiac hardware performed.

They wanted to understand the Illiac hardware from the standpoint of how best to generate code for it. To do this, the branch looked at the four functional parts of the Illiac IV. Those parts are the control unit, the 64 processing elements, the processing element memories, and the Illiac main memory. (See Figure 16 for a diagram of the hardware described below.)

The control unit (CU) contains the instruction stack which interprets all instructions, some of which may be completely executed within the CU.

40 *Parallel Supercomputing in SIMD Architectures*

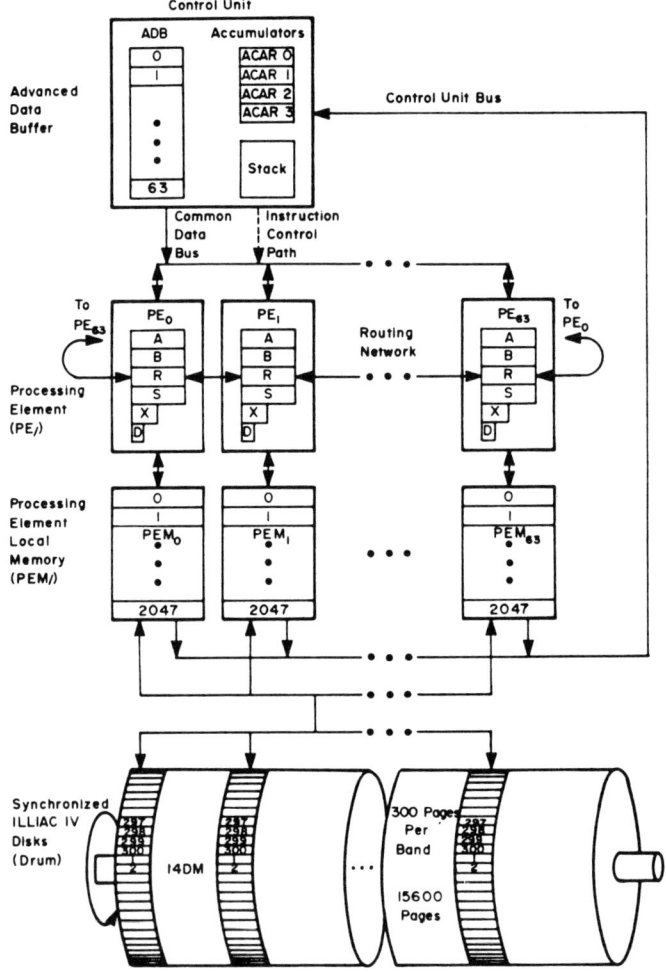

FIGURE 16. Conceptual architecture of Illiac IV with Registers detailed.

Instructions are partially executed and then broadcast to the 64 processing elements; there the execution is completed by all the processing elements in lock-step. In addition to managing the instruction stack, the CU may be thought of as a small, self-contained computer. If has four accumulators which are capable of a full set of shifting, bit-setting, and Boolean operations, as well as addition and subtraction. Furthermore, these accumulators may be used as index registers for fetching and storing in the processing elements. The CU also has 64 scratch registers called the Advanced Data Buffer (ADB).

A processing element (PE) has six programmable registers called RGA, RGB, RGS, RGR, RGX, and RGD. RGA is the accumulator and RGB is its extension; RGS is a scratch register. The remaining registers are somewhat peculiar to the Illiac architecture. RGR is used for inter-PE communication of data. Data may be rotated end-around (data from PE 1 going to PE 64) within the 64 RGRs. RGX acts as an index for intra-PE fetching. This register allows independent fetching depths in each of the PE memories. The RGD contains fault bits and test result bits for that PE. It also contains bits called mode bits which, when set, allow the PE to take part in instructions and, when reset, protect the PE memory as well as RGA, RGS, and RGX from change.

The processing element memories (PEMs) may be thought of in two ways: (1) collectively as 131,072 64-bit words of memory from the CUs point of view, and (2) as a 64×2048 matrix of 64-bit words from the point of view of the PEs. In the latter case, each PE is able to access its own column of 2048 words. (Note that the RGX indexing permits the PEs to fetch independently any word within their own column.)

The main memory of the Illiac is logically a 16-million word drum. The drum is divided into 52 bands (tracks) each of which contains 300 Illiac pages (an Illiac page is 1024 64-bit words). The drum may be mapped, i.e., data may be stored upon it in predetermined locations and accessed asynchronously. This enables the programmer to ensure that the data he wishes to fetch are coming under the read/write heads when he needs it. This allows the full billion-bit-per-second transfer rate to be realized during execution.

CFD may now be contrasted to FORTRAN, bearing in mind the hardware for which CFD must generate code. CFD statements are composed of CFD key words used in conjunction with the basic elements of the language (constants, variables, and expressions). These statements are written in card format similar to FORTRAN.

Types of Named Quantities

There are four classes of named quantities in CFD: (1) variables, (2) subprograms, (3) common blocks, and (4) disk areas. Variables may be divided into three subclasses: (1) scalars, (2) arrays, and (3) vector aligned arrays. An array may reside in either PE or CU memory and may be of any length, limited only by the memory size. These arrays, however, may not be used as vectors in vector operations, and may not have more than one subscript. Vector aligned arrays, on the other hand, must reside in PE memory, and may have one, two or three subscripts. The range of the first subscript of a vector aligned array is always 64. All vector aligned arrays have their first word in the first PE, hence the nomenclature "vector aligned".

There are five categories of CFD statements: (1) specification; (2) subprogram; (3) input/output; (4) control; and (5) assignment statements. Each statement category is discussed below.

Specification Statements

CFD supports the full range of FORTRAN specification statements; for example, IMPLICIT, DIMENSION, COMMON, EQUIVALENCE, and explicit statements. There are five types of variables in CFD: CU INTEGER, CU REAL, CU LOGICAL, PE REAL, and PE INTEGER. Note that the residence of the variable must be declared. Real and integer variables are similar to those in FORTRAN. However, CFD logical variables are quite different from FORTRAN logical variables. FORTRAN logical variables have one value (either .TRUE. or .FALSE.) while CFD logical variables always have 64 values, one in each bit of the 64-bit word. In this sense they are vectors, and when used to control the PEs, each PE receives one bit. The CFD variable MODE contains the current machine mode bit vector and is stored in the RGDs.

In the case of CU variables, a specific CU address must be assigned by use of an EQUIVALENCE statement. Because there is only one CU, these variables must be thought of as being in *common* to all subprograms. The following are examples of various types of CFD specification statements:

```
*   IMPLICIT CU LOGICAL (M)
*   CU REAL ALPH
*   PE INTEGER X(*)
*   DIMENSION RHO(*,64)
*   COMMON   /CONSV/   EO(*,64,2)
*   EQUIVALENCE   (1,I),   (2,MSK),   (3,MD),
    (4,ALPH)
```

In this example we have the integer variable I residing at CU location 1, while the logical variables MSK and MD reside in CU locations 2 and 3, respectively. X, RHO, and EO are vector aligned arrays of one, two and three dimensions respectively. The asterisk in column 6 is one of our concessions to ease of translator writing. All nonassignment statements must have an asterisk in column 6 to be valid Version-2.0-CFD statements.

Subprogram Statements

A subprogram may either be a FUNCTION, a SUBROUTINE, or a BLOCK DATA. The declaration statements for these subprograms are the same as in FORTRAN. A FUNCTION or a SUBROUTINE is referenced in the usual FORTRAN manner. The following is an example of a SUBROUTINE statement with one argument.

```
*   SUBROUTINE UPDATE(RHO)
```

Input/Output Statements

All CFD I/O is between Illiac main memory and the processing element memories — not printers and card readers — and this I/O is asynchronous to make use of the overlapping and mapping capabilities of the Illiac. Since this I/O is asynchronous, CFD also has a WAIT statement which will halt execution until a previously requested READ or WRITE is completed. For example:

```
*   DISK AREA EOSTAR(4)
*   READ(3,EO(1,1,2), EOSTAR(1),4)
*   WAIT 3
```

In this example the first statement declares that there is a previously mapped area on the Illiac main memory, called EOSTAR, which is 4 Illiac pages long. The second statement requests that four pages be read beginning with the first page of area EOSTAR into PEM beginning at EO(1,1,2). This second statement also gives this I/O request the identification number 3. The third statement will stop the program until the I/O request associated with identification number 3 has been completed. CFD WRITE statements have the same format as the READ statement.

Control Statements

There are two kinds of program control in Illiac IV: (1) branching, and (2) enabling or disabling PEs. These controls may be used separately or in combination. Branching is the type of control used in serial computers and determines which statements will be executed next. In Illiac, however, it is also necessary to specify which PEs will participate in the execution of a vector statement.

The following statements are implemented in CFD with their standard FORTRAN form and meaning.

```
GO TO (absolute, computed, and assigned)
ASSIGN
CONTINUE
RETURN
STOP
CALL
END
```

One of the most frequently used control statements is the DO statement, and in CFD it is slightly more general than in FORTRAN. The differences from FORTRAN are (1) the increment must be a constant, but may be negative, and (2) the starting and limit values may be a CU INTEGER variable plus or minus an integer constant. As in FORTRAN, the index must be greater than zero.

Logical IF statements are implemented in CFD, but arithmetic IF statements are not. IF statements are of two basic kinds: (1) scalar IFs having a single true/false result, and (2) vector IFs having 64 true/false results, one for each PE.

Scalar IFs determine the program flow, and vector IFs define the participating PEs. There are no single-result, logical variables in CFD, so the variety of scalar IFs is quite restricted. There are three basic forms: (1) those involving arithmetic tests between CU integer expressions using only addition and subtraction; (2) those involving quantified logical expressions; and (3) those testing for I/O request completion. A logical expression in CFD implies 64 true/false results, and "quantifying" reduces it to 1 true/false result. The logical quantifiers are .ANY., .ALL., .NOT ANY., and .NOT ALL.. The following are examples of scalar IF statements:

```
*  IF (INDEX .GT. LIMIT) RETURN
*  IF (.NOT ANY. ((A(*) .GT. EPSLON))) STOP
*  IF (.COM. 3) GO TO 123
```

The first statement is true if the CU INTEGER INDEX is greater that the CU INTEGER LIMIT. The second statement is true of all 64 A's are less than or equal to EPSLON. The third statement is true if the I/O request associated with the identification number 3 is completed.

The PEs are controlled in two ways: (1) the instruction stream in the CU determines the machine instruction to be executed; and (2) the enabling mode pattern in the PEs determines which PEs will perform the instruction and which will remain idle. At the CFD level, the enabling mode controls only vector arithmetic assignment statements and the evaluation of SUBROUTINE arguments that require scratch storage. Vector arithmetic statements do not alter variables in disabled PEs. The enabling mode patterns the logical variable MODE, a reserved symbol, at all times except when the vector assignment statement following a vector IF is executed. In that case, the enabling mode is the result of the vector IF. For example,

```
*  IF((A(*).LT.0.))  A(*) = -A(*)
```

is one way to replace A(*) by its absolute value. If the sequence

```
MODE = (-A(*).LT.0.)
A(*) = -A(*)
```

is used, A(*) is replaced by its absolute value before, but now the enabling mode has been set so that only the PEs in which A was negative will be active in statements following this sequence.

Assignment Statements

In a logical assignment statement a logical variable is assigned the value of a logical expression. The basic building block of a logical expression is the "base mode", which may be a logical variable, a logical constant (ON meaning all true and OFF meaning all false), a vector relation, or any of these preceded by .NOT., which implies logical negation. A vector relation consists of two vector arithmetic expressions separated by one of the following: .GT., .LT., .GE., .LE., .EQ. or .NE..

The logical expression may simply be a base mode, or it may contain operators having base modes as operands. There are three kinds of operators: (1) bit setting operators, (2) shifting and rotating operators, and (3) Boolean operators. The two kinds of bit setting operators are .TURN ON. and .TURN OFF. and are used to turn on (enable) or turn off (disable) discrete bits of the variable being defined. The bits themselves are specified in a list following the operator. For example:

```
MASK = ON .TURN OFF. 1,2,.LAST.2
```

This statement assigns false to the first two and the last two bits of MASK while assigning true to the remaining 60 bits. The list may indicate individual PEs or ranges of PEs as may be seen in the following CFD statements.

```
MODE = MODE .TURN ON. .FIRST. I-1
MASK = .NOT. MASK .TURN ON. MIN .TO. MAX
```

The two kinds of bit shifting operations are "end-off" shifts (.SHL. and .SHR. for left and right shifts, respectively) and "end-around" shifts (.RTL. and .RTR. for left and right); the end-around shifts are usually called "rotates" rather than shifts. In the end-off shifts, vacated bits are set to zero (false).

The three Boolean operators are .NOT., .AND., and .OR. all of which have their conventional meaning. The following are typical CFD logical assignment statements.

```
MASK = MODE .RTL. I+1
MODE = .NOT. MODE .AND. (A(*) .GT. 1.0)
```

There are three kinds of scalar arithmetic statements, all of which are specific and restricted. The limited vocabulary for CU arithmetic reflects the absence of the required hardware. The first kind of statement is an arithmetic assignment statement involving only CU INTEGER variables, integer constants, and the + and - operators. The second kind of statement involves the transfer of single words of data. No arithmetic is done, and the data may be REAL or INTEGER and have any residence (CU or PE). The third kind of statement has no FORTRAN equivalent and is required in Illiac to facilitate any necessary juggling between CU and PE memory due to the limited size of CU memory. The TRANSFER statement allows the programmer to move blocks of eight words between CU and PE memory (using special Illiac machine instructions). The following CFD statement causes variable I and the seven CU variables after it to be assigned the first eight values of the PE array TEMP.

```
*TRANSFER (8) I=A(1)
```

PE arithmetic is vector arithmetic, even when an expression involves only scalars. Expressions must be either REAL or INTEGER, and mixed type expressions are not allowed. The following standard FORTRAN operations are implemented: +; –; *; /; and **. The order of computation is the same as in FORTRAN. Exponents in CFD must be integer constants in the range 2 through 10 and may not be exponentiated themselves. The variable being defined in a PE arithmetic assignment must be vector aligned, and its first subscript must be * alone. This convention is followed because the enabling mode (MODE) then corresponds directly to the PEs of the defined variable.

When the first subscript contains *, the subscript possesses some non-FORTRAN qualities. Assume that all PEs are enabled, then the statement

```
A(*) = B(*-1)
```

is equivalent to the FORTRAN statements

```
A(1)=B(64), A(2)=B(1), A(3)=B(2), ...,
A(64)=B(63)
```

illustrated by the following diagram.

```
  A(1)  A(2)  A(3)  ...   A(63)  A(64)
 /B(1) /B(2) /B(3)  ...  /B(63) /B(64)
```

Note that the transfer of data is done in the RGRs and is end-around.

Suppose the central difference of the vector P(*) is needed. Its value, as given by

```
DIFT(*) = P(*+1) - P(*-1)
```

may have no meaning in PEs 1 and 64 unless P is in fact periodic. The difference would not be computed in these PEs if the statement above were preceded by

```
MODE = ON .TURN OFF. 1, 64
```

An * in the first subscript implies that the variable is a vector. When the first subscript contains no * the variable is used as a scalar, the same value being used in every PE.

When the first subscript contains an *, the second subscript, if present, may contain an integer vector. This allows each PE to refer to a different position in its memory. Suppose the variable X has been declared a PE INTEGER vector and has been assigned the values 1,0,1,0,...,0,1,0. Then, if RHO is a 64 × 64 matrix, RHO(*,X(*)+1) is the saw tooth pattern vector made up of the following variables:

```
RHO(1,2),  RHO(2,1),  RHO(3,2),  RHO(4,1) ...
RHO(64,1)
```

Note that this integer vector index is stored in and used from RGX.

There are two CFD translators in existence. One compiles CFD into relocatable machine code for the Illiac IV and the other translates CFD into standard serial FORTRAN.

Both these translators are written in FORTRAN which allows them to be easily brought up on a wide range of computers. These translators currently run on a PDP 10, an IBM 360/67, and IBM 360/91, and a CDC 7600.

CFD is clearly not a machine-independent language. It allows the programmer to use the power of the RGR, RGX, and RGD for intra-PE communication, independent PE indexing, and a wide range of mode control, respectively. It also restricts the user to simple scalar operations because complicated scalar operations are not possible on the Illiac without running at $1/_{64}$ its top speed. The machine-dependent nature of the CFD language forces the programmer to think parallel, leaving only bookkeeping chores to the compiler. This has allowed the Computational Fluid Dynamics Branch of Ames Research Center (and others) to develop a wide range of application programs which make efficient use of the Illiac IV parallel hardware; for this reason, CFD has met all of its goals.

Although the language can be said to be machine dependent, its dependence is not just on the Illiac IV. Rather, its dependence is on a machine which can execute vector as well as scalar instructions. To this end the Computational Fluid Dynamics Branch is developing a third CFD translator. This translator will translate CFD to CDC 7600 assembly language, which makes optimal use of all the pipelining and overlapping of which the 7600 is capable. Or, as pointed out in Feustel et al.,[5] the Branch will compile CFD for the "vector 7600", which runs from 1 to 5 times faster than the 7600 using FORTRAN.

Thus CFD appears to be a logical extension of FORTRAN which allows for the efficient use of the vector hardware of the Illiac IV and quite probably other parallel and vector machines.

3.4. LANGUAGE REVIEW

This review is a brief examination of some existing programming languages for the Illiac IV, namely GLYPNIR, CFD, and IVTRAN. A proposed language, APPLE, is also discussed briefly to contrast with the above languages. In this short overview, the comparison of the various programming languages is organized from three points of view.

The first point of view is that programming languages, for Illiac as for other computers, are tools of problem solving. High-level languages attempt to present the computational facilities of computers in terms more understandable to the user than machine language. The purpose is to help the user formulate a solution to his problem by providing him with computational "abstractions" that are "close" to his problem domain.

A second and closely related point of view is that programming languages are tools to implement solutions (programs). Over the past few years, people have begun to realize the high cost of software production. Interestingly, it has been found that the major part of this cost is not due to the initial design and programming efforts, but lies with program testing and debugging and with program maintenance. Even in these conditions, the reliability and the "quality" of large software systems is often questionable. Consequently, it is important to look at how languages can simplify testing and debugging and facilitate program maintenance.

The third point of view taken in the following sections is that Illiac IV is a unique architecture. Although the development of software for Illiac IV has similarities with the development of software for any other computer, programming the Illiac is much different from programming a classic sequential computer. Because the main advantage of Illiac IV is its speed, Illiac applications are usually applications that cannot be processed within a reasonable amount of time on most other machines. Thus, the major design and programming issues for Illiac IV result from a justified concern for efficiency. A first issue is to isolate, during the design phase, the parallelism inherent to the application, or to reformulate the problem to obtain some parallelism. The second issue is to map this parallelism onto the Illiac IV. The two major difficulties are (1) management of the two-level memory hierarchy, i.e., how to lay out the data on the disk memory to provide fast access to portions of the data sets needed at the same time for processing, and (2) management of the CU-PE ensemble, i.e., how to organize program data within this complex to obtain good response time.

Evidently, these two difficulties cannot be resolved independently. However, since most Illiac applications seem to be I/O bound, rather than "CPU" bound, design decisions about the management of the memory hierarchy seem more important for efficiency considerations.

In summary, the following sections provide a comparison of the Illiac IV programming languages both from a usual point of view (language design, implementation, and usage) and from the point of view of producing programs that make efficient use of the Illiac IV resources (management of the memory hierarchy and management of the CU-PE complex). The list of the following sections can be viewed as a list of design issues for parallel machines like Illiac IV.

Any programming language defines some abstract machine for its user. The purpose of this abstract machine is to hide (in part or totally) the target machine and provide the user with facilities close to his problem domain, in order to diminish the conceptual distance between the initial problem specifications and the resulting program. Among the languages examined, there are four distinct types of "abstract" machines presented to the user by the four languages considered:

1. APPLE presents generalized array and vector computations. This simple revision of the APL does not require any knowledge of the Illiac to produce a working program (whether efficient or not).
2. IVTRAN presents a FORTRAN machine with some one-dimensional parallel facilities. IVTRAN can be considered in two different ways. Since the IVTRAN compiler accepts standard FORTRAN and attempts to isolate DO loops that can be executed in parallel, IVTRAN can be regarded as a FORTRAN compiler that generates code for the Illiac. Unfortunately, the techniques used by the current compiler to extract parallelism are very restricted, and the code generated for a FORTRAN program is very often code running in one PE at a time. The parallelizing part of the IVTRAN compiler must be considered as a tool to improve programs. On the other hand, IVTRAN can be regarded as a FORTRAN based language with some parallel facilities. The computational model provided by the full IVTRAN language as implemented is an extension of FORTRAN where some parallel operations can be performed along *one* dimension of arrays of arbitrary size. IVTRAN requires some knowledge of the Illiac architecture for the allocations of arrays and the alignment of operands.
3. CFD is a FORTRAN-based language which requires the user to know that the Illiac is composed of 1 CU and 64 PEs, organized in linear order. There is a very clear distinction between control variables and vector aligned variables. However, CFD hides little of the Illiac from the user; the MODE must be manipulated directly, transfers between CU and PE memory must be programmed, and the limited arithmetic capabilities of the CU are reflected in CU arithmetic expressions.
4. GLYPNIR, an ALGOL-like language, requires less detailed knowledge of the Illiac IV than CFD. There are no limitations on CU arithmetic (the programmer, however, should make sure that at least one PE is enabled when such expressions are evaluated). GLYPNIR differs importantly from the other languages in that it presents the Illiac IV as a set of PEs

operating simultaneously and does not constrain the user to either a vector or an array approach.

From the point of view of the abstract machines provided by CFD, GLYPNIR, and IVTRAN, three serious criticisms can be made.

First, all three languages fail to abstract more than the CU-PE complex of Illiac. The two-level memory hierarchy is not part of their respective computational models. Management of the disk memory is left entirely to the user and only very low level facilities are provided to transfer data between I4DM and PE memory.

Second, CFD and GLYPNIR do not hide enough of the Illiac. They force the user to think directly in terms of Illiac parallelism (e.g., 64 simultaneous operations).

The last criticism applies to programming languages besides CFD, GLYPNIR, and IVTRAN. In the case of Illiac languages, none of them provide features that are at the same level of abstraction as the user problem domain. Admittedly, it is difficult to provide, in the same language, a facility like "layers of the atmosphere" to one user and a facility like "particle" to another. However, it is not being too demanding to require some mechanism that would allow each user to define the additional abstractions that fit his problem.

As an APL-like language, APPLE provides very high level capabilities for vector and array processing. Because of the generality and the highly dynamic behavior of some of its features, there are major difficulties in implementing such a language efficiently on Illiac. One should not neglect, however, the importance of such primitives to design large programs. They allow the programmer to concentrate his attention on high level optimizations instead of attempting to organize cleverly very low-level code.

The use of vectors and arrays in IVTRAN, CFD, and GLYPNIR is more primitive, often reflecting the physical limitations of the Illiac, but also allowing the programmer various degrees of control on the use of the machine resources.

3.4.1. PE Variables

The only way to obtain parallelism in IVTRAN and CFD is through the use of arrays. Both IVTRAN and CFD provide arrays of up to three dimensions. Parallelism is obtained by applying the same operation to elements of an array that lie across PEs. Data types are limited to integer and floating point. Furthermore, CFD restricts the first dimension of arrays to be less than or equal to 64.

In GLYPNIR, where the Illiac IV is explicitly presented as a set of 64 processors operating simultaneously, a PE variable defines a collective name for a set of 64 "simple variables" distributed across the PEs. Similarly, a PE vector defines a set of 64 vectors of identical size. The basic data types provided by GLYPNIR are similar to the data types of IVTRAN and CFD, but the ALPHA "type" provides an escape hatch for the representation of other quantities.

3.4.2. PE Variables Memory Allocation

CFD and GLYPNIR PE structures map directly only the physical memory. A one-dimension 64-element CFD array is equivalent to a GLYPNIR PE variable. A two-dimension CFD array is equivalent to a GLYPNIR PE vector. In CFD, the first dimension of an array lies across PEs. Any other storage structure (e.g., skewed array) that may be needed by the programmer must be implemented on top of the available structures, and each reference to such "application structures" in the text is done by indicating the CFD or GLYPNIR variables (which stand for areas of PE Memory) along with an adequate subscript denotation. The abstraction of the "application structures" is ultimately lost in the program text.

IVTRAN offers a much more powerful scheme for array storage where the programmer can choose which array dimension lies across the PEs and can specify skewing or alignment of other dimensions. There are two problems with IVTRAN array allocation that often force the programmer to restructure his arrays to obtain some efficiency. The first case consists of restructuring an array A (2,30) into B (60) so that all elements lie in one PE row. The second case consists of controlling the allocation of distinct arrays through EQUIVALENCE or DEFINE's to align them and avoid inefficient routing during computation. Thus, like GLYPNIR and CFD, IVTRAN often forces the programmer to recode his problem in a notation which no longer indicates the logical structure of the data used in the computation. A main drawback of all these languages is that they do not provide any facility to pack many data items in various fields of the same world. Only GLYPNIR enables packing with the ALPHA data type, but the field manipulations can become tedious.

3.4.3. Array Addressing

The selection of array components in CFD and GLYPNIR reflects directly the addressing structure of the I4. Each array reference selects one element in each PE. The position of the elements selected in different PEs may differ if the index expression includes some quantity local to each PE. This feature is especially important to implement nontrivial algorithms. However, other accessing methods must be programmed with additional control structures, e.g., proper MODE setting to access a single element, and explicit loop to iterate overall elements of a two or three dimensional array.

IVTRAN provides a different approach. First, any array element can be addressed separately, as in FORTRAN. All mode operations are hidden from the user. Second, the "*" notation allows references to entire cross-sections (note that an IVTRAN cross-section is a very restricted form of submatrix) for component-wise operations; depending on the allocation, cross-section operations may be parallelized. Finally array references within a DO FOR ALL loop denote "simultaneous" access to an entire row of the array. IVTRAN does not allow "local indexing" as in GLYPNIR and CFD, since the array allocation is supposed to alleviate the need for numerical applications. However, this restriction prevents simultaneous access to array elements that lie across PEs, when the index set of these elements is not regular (e.g., not a row, column, or diagonal of a skewed two-dimensional array), and forbids the implementation of nontrivial control schemes.

3.4.4. Routing

The simultaneous evaluation in many PEs of an expression involving terms that are stored in various PEs require routing of the operands. Syntactically, this routing is entirely transparent to the IVTRAN user. The main drawback of this approach is that it is very difficult to estimate the routing cost of an IVTRAN program, and thus to be able to modify storage structures to improve the performance.

In GLYPNIR, the communication of values among PEs is accomplished by specifying a routing expression in an assignment, or by using an intrinsic function with the appropriate routing expression. When a GLYPNIR assignment contains a routing specification, the expression on the right-hand side of the assignment is evaluated in the source PEs, although these PEs may not be part of the current MODE setting when the assignment statement is entered. GLYPNIR allows a different routing distance to be specified at each PE, and this provides a great amount of flexibility, but is rather inefficient.

The CFD approach to routing is much more restricted. Routing is implied in an expression like A(*) + B(*+3) where each element of B is transferred 3 PEs to the left before being added to an element of A. Computation is entirely done at the destination PEs. The routing distance must be identical for all elements being routed.

The most important limitation of routing in CFD and GLYPNIR is that only circular transfers (PEs are arranged on a ring) are available. Clever programming seems required to make the PE ensemble look like a square (ends off) of a torus. More elaborate data manipulation functions (e.g., perfect shuffle) require important programming effort.

In summary, two categories of array processing can be distinguished between IVTRAN, on the one hand, and GLYPNIR and CFD on the other. The IVTRAN approach is to provide storage schemes that are as general as possible without indicating precisely the costs of using these structures. GLYPNIR and CFD provide very low level storage structures that reveal entirely the Illiac structure, but for which the implicit computational costs are low and well defined.

Both approaches are flawed because they only provide a fixed set of storage schemes that do not always correspond to the logical structures dealt with by programs. The implementation of a matrix using a skewed storage in GLYPNIR or CFD requires a complex notation to be used every time a row, a column, or a diagonal of the matrix is accessed. Similarly, IVTRAN requires obscure notation if two consecutive elements of a vector have to be stored in the same PE. It is obvious that no language can or should provide all possible structures. At the machine level, there are few possible schemes in addition to the ones provided by the above languages. However, the use of these structures through an entire program leads to obscure notation and represents an important loss of abstraction. It would be preferable to provide a scheme allowing the programmer (1) to define the logical structures in terms of the basic storage structures of the machine, in one part of the program, and (2) to refer to the logical structures through the rest of the program. This hiding mechanism should alleviate much of the program complexity, while retaining control over its efficiency.

In the context of this discussion, scalar processing means the set of facilities offered by the various languages to perform computations other than component-wise simultaneous operations (parallel processing). A scalar expression evaluates to a single value. The elements of scalar expressions are usually elements of what are called CU variables in CFD and GLYPNIR, although this need not be.

Scalar processing in CFD is limited to the arithmetic capabilities of the Illiac CU. Complex scalar expressions in CFD must be explicitly performed in the PEs. Things are a little bit easier on the programmer in GLYPNIR where arbitrary scalar expressions can be expressed. The only problem is that at least one PE must be enabled to evaluate properly subexpressions involving floating-point arithmetic. IVTRAN is even simpler.

An important disadvantage of the IVTRAN language is the provision for numerous type conversions in expressions, which almost defeats the purpose of type and hides the complex transformations that take place during execution. Strong data type checking at compile time, as in GLYPNIR, has been shown to eliminate many programming errors without going through extensive debugging runs, and enforces an explicit notation throughout the program. For these reasons, this approach is preferable.

The main drawback of all the languages reviewed is the lack of distinction between control structures that affect the instruction stream (i.e., modifying the instruction fetch by the CU) and control structures that affect the data streams (i.e., modifying or selecting the set of PEs that should execute forthcoming instructions). This is especially true of IVTRAN where IF statements within a DO FOR ALL are interpreted differently from regular IFs. This is also true of some control structures of GLYPNIR. For instance, the GLYPNIR IF statement can be used for two different purposes. On the one hand, it can be used to signify the conditional execution of some statements depending on the single boolean value of some CU expression, as in visual "sequential" programming languages. On the other hand, a GLYPNIR IF statement can be used to signify the execution of a second sequence of statements in a complementary set of PEs. Only one branch is meant to be executed in the first case, while the two branches are executed sequentially in disjoint sets of PEs in the second case. Unfortunately, both cases are handled identically in GLYPNIR and unnecessary mode manipulations occur when the first type of IF is meant.

The distinction is made more clearly in CFD where two kinds of IDs, scalar and vector, are provided. The drawbacks of CFD are the restrictions on the selection expressions and the fact that such IFs can accommodate only a single statement. To restrict the execution of a series of CFD statements to a subset of the PEs, MODE manipulation is required. This feature is also available in GLYPNIR but, fortunately, it can be avoided most often when programming in this language. The problem with an assignment to the pseudo-variable MODE in CFD, as in MODE-SOMEPESONLY, is that it is a highly dynamic feature that modifies the meaning of the statements that follow. This kind of notation is dangerous (for example, it remains in effect when a branch is taken, which complicates debugging) and some other syntactic device (e.g., FOR < PE EXP > DO control structure in GLYPNIR, or indexing with a control vector in APPLE) should be preferred.

Another problem with GLYPNIR and CFD concerns those control structures that indicate iteration over subarrays. In many instances (consider the addition of two n × m matrices), the looping statements require the specification of indices and of index sequences unnecessarily. This kind of overspecification reduces further the amount of abstraction available in both languages. The "*" (array cross-section) construct of IVTRAN prevents the need for such overspecifications.

Although the management of the memory hierarchy seems to be a critical factor in the overall performance of an Illiac program, neither GLYPNIR nor CFD offers facilities beyond BUFFER IN, BUFFER OUT types of statements for data transfer between the disk memory (I4DM) and PE memory. They only provide access to the primitive facilities of the machine. IVTRAN provides most of FORTRAN I/O, but at a prohibitive execution cost.

There are two related aspects to the I/O problem on Illiac. The first is creating I4DM areas from TENEX files according to user-supplied map. Within an area, many distinct logical entities may be interleaved, so that all operands required by some iterated step of the program can be loaded in one single I/O request at execution time. The second aspect consists of the various transfers between I4DM and PE memory during execution. The efficiency of a program depends on the relative position of locations addressed by successive requests to the I4DM. The mapping mechanism that enables the user to distribute data over the physical disk space requires much knowledge of the program behavior and timing in order to produce a suitable map. Not only do the current languages fail to include the memory hierarchy of the Illiac in their computational models, but they also fail to provide the user with any help. Buffered I/O seems a minimum, with the compiler inserting I/O requests in the generated code as early as possible. Second, an estimation of the computation times between successive requests could enable the compiler to provide an initial map for the user (note that this may not always be possible). Furthermore, it should be easy for the user to obtain run-time statistics on program behavior in order to facilitate improvements of the initial mapping.

This lack of I/O structuring facilities is the major problem of all languages reviewed. A minor problem is the lack of list directed, possibly formatted I/O in GLYPNIR and CFD. The only type of I/O statements currently offered by these languages implements transfers between areas in PE memory and areas in disk memory. It is not possible to produce directly any readable output (program log, intermediate results for debugging purposes, or simply final results), or to input data in character form. IVTRAN provides this facility, but very inefficiently. It should be possible to restrict these features so that most of the formatting and conversions can be performed by pre- and post-processors operating on TENEX (which is what users have to do currently to obtain any readable output).

IVTRAN appears to be the most sophisticated of the various languages reviewed. Its parallelizing processor does provide some help in converting a FORTRAN program to a running Illiac program. However, the user should be warned that the capabilities of the parallelizer are limited and that usually much program manipulation is required to obtain a program which is at all efficient. Recoding in IVTRAN is strongly recommended. All these operations require a good knowledge of the inner workings of the IVTRAN compiler. IVTRAN seems to provide a reasonable debugging package.

Compared to IVTRAN, GLYPNIR offers limited debugging facilities and run time checks. On the other hand, the compile time evaluation facility and the macro facility of GLYPNIR are important program development tools that are not provided by any other Illiac language. These facilities assist the development of programs in a systematic fashion, without losing much of the initial abstraction and at no cost in run time efficiency. As for CFD, no similar facility is provided to support program development or testing.

There are a number of program development and maintenance tools that are unavailable to the Illiac user, to cite a few: test data selection, program prover utilities, and symbolic dumps, for program validation; program transformation (source to source program "optimization") and performance prediction utilities, for program enhancement. Although some of these tools are just being understood and implemented for "sequential" languages, they are widely recognized as being relevant for software production. There is no reason why their benefits could not be exercised in the production of Illiac software.

There are three important points that need to be considered seriously before any new language for parallel machines like the Illiac can be proposed.

1. Software development for the Illiac is not much different from software development for other machines: this means that any new language for Illiac should be designed to be part of a complete programming system including extensive program development tools. Serious restrictions on the language may be required to make these tools possible.
2. The management of the Illiac memory hierarchy is an important factor

of the efficiency of the Illiac applications program: this hierarchy should be manageable in the language itself; the supporting software should facilitate the optimum use of these resources.

3. There are many ways a given program can be implemented in parallel: a language should not force a user to view problems only in terms of vectors or only in terms of simultaneously executing PEs. No language of manageable size can provide all desirable facilities to all users. This means that a new language should enable program-defined extensions (abstractions) while leaving the user a good deal of control over the efficiency of the generated code.

3.5. PERFORMANCE

Tables 3, 4, and 5 contain timings for the execution of three commonly used vector operations on the CDC 7600, CDC Star 100, Illiac IV and CRAY 1.

Also Figure 17 displays performance by the Star 100, Illiac IV and CRAY 1 for the operation:

$$V(*) = A(*) * (B(*) + C(*)) \qquad (1)$$

The units used to measure performance in the tables as well as Figure 17 is MFLOPS or millions of floating point operations per second.

3.6. SEISMIC ANALYSIS APPLICATION: A THREE-DIMENSIONAL FINITE DIFFERENCE CODE FOR SEISMIC ANALYSIS ON THE ILLIAC IV PARALLEL PROCESSOR

3.6.1. Empirical Evidence

Since empirical evidence for the complex earthquake fault behavior, deep within the earth, is normally gathered from motions at surface stations, there has been increasing interest in the computerized prediction of the ground motions which would result from postulated earthquake fault models. In addition, two-dimensional models cannot adequately represent the complex three-dimensional effects surrounding a fault. Therefore, a number of researchers have developed three-dimensional seismic codes. One such code, TRES, was developed by Systems, Science and Software for their UNIVAC 1108 computer.[1] Unfortunately, this code, like most three-dimensional codes, required excessive amounts of computational time to run. For even a moderately sized problem (51 × 51 × 101 finite difference mesh, 253 time steps), 15 min of computer time were required for each time step on the full grid. To reduce run times to reasonable levels, a decision was made to implement the TRES code on one of the world's most powerful computers, the Illiac IV.

This paper describes the implementation and some of the results obtained.

TABLE 3
Vector Operation Timings
V(I)=A(I)+B(I)

Operation		\	Vector length				
V(I)=A(I)+B(I)		5	10	50	100	500	1000
FORTRAN	CDC 7600	1.3	1.5	1.6	1.6	1.6	1.6
RDALIB	CDC 7600	2.55	3.73	5.55	5.75	5.95	6.03
STAR 100	64 bit	1.7	3.3	13.2	20.8	39.1	43.9
STAR 100	32 bit	1.8	3.5	15.3	26.6	64.4	78.4
ILLIAC IV	64 bit	1.45	2.91	14.53	18.13	24.90	25.88
ILLIAC IV	32 bit	1.45	2.91	14.5	29.1	46.29	49.80
CRAY 1		10.0	14.5	22.9	23.0	23.6	23.6

TABLE 4
Vector Operation Timings
V(I)=A(I)*B(I)

Operation		\	Vector length				
V(I)=A(I)*B(I)		5	10	50	100	500	1000
FORTRAN	CDC 7600	1.4	1.6	1.7	1.8	1.8	1.8
RDALIB	CDC 7600	2.42	3.46	6.3	6.5	6.84	6.88
STAR 100	64 bit	.8	1.5	6.0	9.7	49.6	66.3
STAR 100	32 bit	1.0	1.9	9.0	16.4	49.6	66.3
ILLIAC IV	64 bit	1.45	2.90	14.5	16.25	24.9	25.9
ILLIAC IV	32 bit	1.45	2.90	14.5	29.0	46.3	49.8
CRAY 1		9.8	14.3	22.7	22.75	23.5	23.5

TABLE 5
Vector Operation Timings
V(I)=(A(I)+51)*S2

Operation		\	Vector length				
V(I)=(A(I)+51)*S2		5	10	50	100	500	1000
FORTRAN	CDC 7600	2.1	2.4	2.7	2.8	2.8	2.8
RDALIB	CDC 7600	—	—	10.0	10.6	11.1	11.1
STAR 100	64 bit	1.1	2.0	8.2	13.2	25.5	28.9
STAR 100	32 bit	1.3	2.5	11.3	20.3	56.1	71.8
ILLIAC IV	64 bit	2.40	4.80	24.0	26.55	39.7	41.0
ILLIAC IV	32 bit	1.47	2.937	14.7	29.37	60.1	67.21
CRAY 1		—	—	59.3	60.2	63.9	64.4

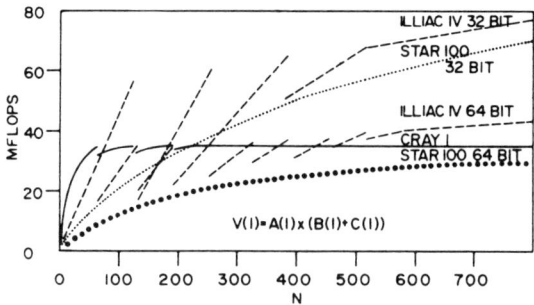

FIGURE 17. Performance of various supercomputers on the operation V=A* (B+C).

To facilitate understanding of the details of the implementation, a description of the numerical algorithms used by TRES will first be presented.

3.6.2. The Tres Computer Program

The problem predicting ground motions resulting from earthquake sources may be divided into three smaller problems. The first step is to simulate the earthquake source and collect motion data on a sphere surrounding the fault. The second step is to use this data to define an "Equivalent Elastic Source". The Equivalent Elastic Source is a collection of coefficients of a spherical expansion (in terms of Bessel functions, trigonometric functions and associated Legendre functions), where each coefficient is a function of frequency. The third step is to use these Equivalent Elastic Source coefficients to analytically predict the motion at selected sites. The TRES code is only concerned with the first of these steps. Thus, TRES simulates the earthquake faulting process, calculates the resulting wave motion in the earth surrounding the fault, and collects divergence and curl histories on a spherical surface surrounding the fault.

The Fault Model — Currently, the only fault model is a bilateral strike-slip fault using a stick-slip rupture mechanism.[2] For this type of fault, the fault plane is vertical and the motion involves symmetric horizontal slip of the sides of the fault relative to one another. The rupture is initiated at the center of the fault, the focus, and spreads radially at a specified rupture velocity until the limit of the fault plane is reached. The grid is split on the fault plane. Until rupture occurs, the two nodes must move identically. After rupture, they are free to slide relative to one another in the plane of the fault. While they are sliding, the force between them is just the kinetic friction. When the relative velocity drops to zero, the rupture is said to heal and no further relative motion between the two nodes is permitted. The maximum size of the fault plane was 4×6 in the UNIVAC 1108 version of the code.

The Plastic Zone — The fault plane is surrounded by a zone in which inelastic behavior is permitted. The size of this zone was 9 × 11 × 9 in the UNIVAC 1108 code. The material behaves as if it were perfectly elastic until the yield strength of the material is exceeded. The Huber-Von Mises-Hencky yield criterion is used to determine incipient yield. When this shear-distortion energy limit is violated, the material behaves in a perfectly plastic fashion. The usual stress-strain relationships are replaced by the Prandtl-Reuss equations. That is, the rate of plastic strain is proportional to the state of stress and the elastic strains are considered to not exceed the yield surface. Since the stress is no longer proportional to strain, six stress components are carried with each node in the plastic zone. In addition, the work done by the plastic deformation is calculated. So a seventh plastic variable is carried with each node (the current integrated total of the plastic work).

The Elastic Zone — The remainder of the grid (surrounding the plastic zone) is treated as ideally elastic. In the UNIVAC 1108 version of the code, the total grid is limited to a maximum of 51 × 51 × 101 nodes. Six variables are carried at every node in the grid. These are the three displacements at each node and the three velocities. Thus, a grand total of 13 variables are carried at each node in the plastic zone, and 19 variables for the split nodes in the fault plane.

The Boundary Conditions — The initial conditions in TRES consist of zero displacement and velocity. However, a uniform state of horizontal shear stress is permitted. On the six surfaces of the grid, either the force or the displacement must be constrained to be zero. Since the three coordinate directions on each surface may be independently prescribed, symmetry conditions may be imposed. In fact, it is customary to apply symmetry conditions on both planes normal to the fault, thus treating only a quarter of the fault. The current fault algorithm does not permit a symmetry condition across the plane of the fault (even though such a condition could be formulated for this class of fault). This is the reason for the final dimension being double the first two.

The Computational Algorithm — The basic computational cycle in TRES consists of integration for one time step. The cycle starts by numerically approximating the derivative of the displacement field to obtain the strain field. A central difference of displacement values at adjacent nodes approximates the partial derivative at the midpoint. By combining these partial derivatives and averaging, an estimate is obtained for the strains at the center of the block of material determined by eight nodes. The constitutive relationships (Hooke's Law in the elastic region) are then used to obtain the stress at the center of the block. Then the equilibrium relationships (Newton's Law) are used to determine the acceleration. The partial derivatives of the stress required in the equilibrium relationships are obtained in a fashion similar to the strains. In this case, the stresses at the centers of the eight blocks surrounding the node are used to estimate the acceleration at that central node. The velocity, V, and the displacement, U, of a node are obtained as

$$V(T+.5 \times DT) = V(T-.5 \times DT) + DT \times A(T), \text{ and}$$
$$U(T + DT) = U(T) + DT \times V(T+.5 \times DT) \qquad (2)$$

where A is the acceleration, T is the current time, and DT is the time step. This calculation is analogous to the centered difference technique used for the strain and acceleration calculations. Although this description covers the more important aspects of the calculation, it must be noted that there are other features in the code to treat damping, plasticity and to control a form of instability observed to occur in such centered difference techniques of a rectilinear grid.[3]

3.6.3. Implementation of the Illiac IV

The Illiac IV (I4TRES) code is designed to handle a substantially larger grid than the UNIVAC 1108 TRES. The maximum size of the fault plane is increased from 4×6 to 32×32. Similarly, the maximum dimensions of the plastic zone have been increased from $9 \times 11 \times 9$ to $32 \times 32 \times 64$. Finally, the maximum dimensions for the full grid have been increased from $51 \times 51 \times 101$ to $80 \times 80 \times 160$. The goal of this project was to implement on the Illiac IV a code which was computationally equivalent to TRES but executed an order of magnitude faster on a grid four times as large. Consequently, the algorithms were redesigned to maximize the number of PEs in use at any time, to minimize routing costs, and to minimize Illiac IV disc memory latency time.

User Input and Output — Since reading card input is a highly serial process and because more flexibility in input was desired, no direct input is done in the Illiac IV program. Rather an interactive preprocessor was written to aid the user in preparing the program's input data. This preprocessor, the I4TRES File Editor prepares a file in Illiac IV binary word representation which is moved to I4DM at the start of a run and is the sole source of input information for the Illiac IV code. Similarly, since the creation of formatted output in a program is highly serial and since extensive post-processing was to be done, no formatted output is prepared by the Illiac IV program. Rather binary output files in the user's machine word format are prepared and transferred to the user. Thus, the Illiac IV time is not spent on these highly serial tasks, but is reserved for the highly parallel computational tasks.

Computational Methodology — The basic computational strategy is to calculate results in the direction in parallel. Using J, K, and L as the indices in the X, Y, and Z directions of the grid, respectively, results for 60 J indices are all calculated simultaneously. A second calculation is then used for the remaining 20 indices. With this technique, the two outermost PEs are not used. The two PEs adjacent to the main data block (PE 1 and 62) are used to make it appear as if there were actually 62 variables. Values from the beginning of the block of 20 or the end of the block of 60 are moved into these PEs so the correct differences can be obtained. Boundary conditions are created by turning off the boundary PE during displacement calculation to enforce a constant zero displacement or by loading values which produce zero difference, strain, and consequently stress, to enforce a load free boundary condition.

Data Base Design — A key element in the design is the data layout within an I4DM page. Each page in the main data base contains the three displacements and three velocities for the nodes at all 80 J indices, for 2 K indices and for 1 L index, and is referred to as an elastic page. Recalling that a page is 16 rows in PE memory, the first four rows contain displacements for all J's and the first K; the next four contain the corresponding velocities. The remaining eight rows contain the displacements and velocities for all J's and the second K in the same format. Within each group of four rows, the first row carries 60 values of the X component of displacement or velocity surrounded by two zeros on each side (i.e., in PEs 0, 1, 62 and 63 when moved to PE memory). Similarly, the second and third rows each contain 60 values of the Y and Z components, respectively. The fourth row contains the values for the remaining 20 J indices for all three components.

The stress data for the plastic zone is contained in another block of pages. Each of these plastic pages contains the six stresses and the plastic work for all 32 J's, 4 K's and 1 L. The first four rows contain values associated with the first K index, the next four for the second K index, etc. The six stress components and plastic work are packed into four rows with half in PEs 0-31 and the remainder in PEs 32 to 63.

Data Management Scheme — In TRES, J's were scanned most rapidly, K's next and L's last. The net result was that every point had to be read three times per cycle. To minimize the number of I4DM accesses in I4TRES, a different order was used. In I4TRES, of course, all J's were processed simultaneously; this was the parallelization. However, only a quadrant of the K's was scanned at a time. Thus, 20 K's were scanned most rapidly, L's were next, and quadrants were scanned least rapidly. The net result of this strategy was most variables only had to be read once. Only the K's at the quadrant interfaces (e.g., K = 19, 20, 21, 22) were read twice.

In both codes data was moved to and from mass storage into and out of buffers in core. The buffers were configured to permit the minimum number of data transfers for the selected scanning order. In the UNIVAC 1108 version of TRES, the working data buffer contained all 51 J's, 5 K's, and 3 L's. This buffer was used in a circular fashion with 3 K's for computation, 1 K for input, and 1 K for output. A somewhat different system was used in I4TRES. In I4TRES, the input data buffer contains all 80 J's, 24 K's, and 3 L's (36 elastic pages). Results are calculated and placed in the output data buffer which contains all 80 J's, 20 K's, and 1 L (10 elastic pages). (The variables at all 24 K indices cannot be updated for two reasons: first, to update a node all 26 of the surrounding nodes must be present to allow differencing. And second, only full pages of data are updated.) A plastic input data buffer is used similarly. It contains all 32 J's, 24 K's, and 2 L's (12 plastic pages) and a plastic output data buffer contains all 32 J's, 20 K's, and 1 L (5 plastic pages).

The integration method is explicit. (Calculations are based only on values at the previous time step.) So new values cannot replace the old ones until all calculations requiring that value are complete. In the UNIVAC 1108 TRES, two separate files were maintained. During a cycle, data (values at T) were read from one file and results (value at T + DT) were output to the other. Their roles were then switched for the next time step. For the Illiac IV version, the two file scheme was replaced with a dynamic disc allocation scheme which requires only about half as much space. The key to this technique is keeping two copies of the interfacing nodes only. For instance, if K = 1 to 20 have just been updated, a copy of the old K = 19 and 20 page must be kept to permit differencing at K = 21. The way this is implemented on Illiac IV disc memory may be described as follows. Imagine for each L that an empty page is in position zero and the data follow in positions 1 to 40. After updating, the results for K = 1 to 20 (10 elastic pages) are written in positions 0 to 9. Thus, position 9 contains the new values for K = 19 and 20 and position 10 still contains the values from the previous time step. After updating the complete grid, every value is one position in front of where it was in the previous step. The complete data base is treated as circular so that when a page goes off the front of the data base, it is wrapped around and added to the vacancy left at the back.

3.6.4. Results Obtained with the I4TRES Program

The I4TRES code was compared to the UNIVAC 1108 TRES code in three test cases. The first test case was plane wave propagation and consisted of nine subcases, one for each coordinate direction of propagation and each coordinate direction of motion. The second test case involves a smoothly varying load on a small, approximately circular area on the surface. This test case permits comparison with two dimensional (axially symmetric) simulations. The third test case is an actual earthquake simulation. In each case, the numerical results obtained with I4TRES on the Illiac IV were identical to those obtained with TRES on the UNIVAC 1108 in all of the five digits printed. The solution times on the Illiac IV are about $1\frac{1}{2}$ min per time step, whereas on the UNIVAC 1108 they were 15 min. Moreover, the number of nodes has been increased from approximately a quarter million to over one million nodes (an overall speed up to approximately 60). These solution times are for an Illiac IV code before optimization. Optimizing Illiac IV disc memory organization to minimize latency and overlapping computation and input/output operations would be expected to make further substantial reductions in run time.

3.7. LANDSAT

The data from the Landsat Multi-Spectral Scanner satellite consist of frames containing about 7.5 million picture elements (pixels). Each pixel has four component channels, one for each of the four bands of spectral data collected. Each of these components is stored as an 8-bit integer and represents an intensity in a particular frequency band. These frequency bands are green, red, far-red, and near-infrared. For data analysis, two procedures, clustering and classification, are often done on large data sets for which the computation time on conventional computers is excessive. The computations required for clustering and classification are quite parallel, and thus well suited for Illiac IV use; processing turns out to be only a short job on the Illiac IV.

3.7.1. Landsat Data Analysis

The analysis of Landsat Multi-Spectral Scanner data on the Illiac IV generally consists of looking at small areas of data, identifying characteristics of pixels representing particular ground covers, and then applying these characteristics to large areas to obtain the distribution of ground covers in these larger areas. Often, the first step in analysis is to perform a cluster analysis on the smaller areas of data. This divides the data into classes and assigns each pixel to some class. Also, a statistics file is generated giving the means and variance-covariance matrix (for each of four channels) for each of these classes. Using available information for small areas of land known as "ground truth areas", a correspondence of those classes with known ground covers is obtained. The statistics file is then used in the classification process to assign the pixels of the full data set to the various classes, and hence to the ground cover categories.

Due to memory size limitations on the Illiac IV, a maximum of 64 categories may be used for classification in order that the tables be core-contained. While this is sufficient for most applications, it still means that care must be taken in the selection of categories corresponding to the various ground covers. Once a statistics file is obtained which seems to yield classification with sufficient accuracy, the entire large area, possibly an entire frame, is classified using the Illiac IV.

3.7.2. Illiac IV Implementation

Two processes are involved in Landsat data analysis on the Illiac IV. These are clustering and classification and are discussed in separate subsections.

Clustering

The clustering technique used is taken from LARSYS as developed at the Laboratory for Applications of Remote Sensing (LARS) at Purdue University and applied to the Illiac IV by the Center for Advanced Computation at the University of Illinois. The clustering algorithms can be divided into the following four steps:

Step 1 — Initialization: The initial mode centers for the categories are used to determine the sample mean and variance of the area to be clustered.

Step 2 — Category assignment: The square of the Euclidean distance is determined from each pixel to each category mean, and the pixel is assigned to the closest category.

Step 3 — Category migration: If Step 2 did not change the assignment of any of the pixels (and the first time through Step 2 always changes the assignment of all pixels), Step 4 is executed. Otherwise, the old category mean values are replaced by the mean of all the pixels currently assigned to that category. The algorithm then returns to Step 2.

Step 4 — Variance-covariance calculations: To complete the statistical description of the categories, the variance-covariance matrix for each category is calculated.

The data is presented to the Illiac IV as a two-row header containing various information about the data file followed by the pixels. The pixels are stored in 32-bits and thus two per PE. The first step is to convert each channel value into a floating-point value so that all further computation may be done in 32-bit floating point.

The summing required for calculation of mean value in Step 3 proceeds first in parallel down the PEs for all rows. Next, the sums in the two parts of each PE are added and finally the entire sum is computed by routing and adding (as is well known, only six routing steps are needed). The category assignment (Step 2) is, of course, easily done in parallel for 128 pixels at a time since in this phase the pixels are handled quite independently. Thus, the parallelism of the Illiac IV is used quite effectively in the inner loop processes. Note that the ability of the CU to broadcast information is useful in this process since the CU can broadcast the mean of each cluster in turn and the PEs can calculate the distance from that cluster to its assigned pixel. Finally, when the process converges, the output cluster and statistics files are created. The output cluster file has 16 bits for each pixel. The creation of this file requires several routing steps in order to combine two rows of categories into one.

Since clustering is an iterative process, the current implementation of the ILLIAC IV is core-contained. This means that the maximum number of pixels which may be processed is 40,704. However, since it is sometimes useful to be able to cluster more pixels, CAC experimented with "weighted clustering." This is based on the observation that within an area of Landsat data, pixel values tend to be repeated many times.

Thus, if each pixel is stored once along with the number of occurrences (or weight) of that pixel, and the clustering formulas are modified appropriately to take this weight into account, the same resultant statistics are obtained so that more pixels may be clustered.

Classification

Classification is the process of assigning a category to each pixel based on a statistics file obtained by cluster analysis. Currently, two classification algorithms are implemented on the Illiac IV: the statistical maximum likelihood classification as adapted from LARSYS and the simple minimum distance Euclidean classifier (equivalent to Step 2 of the above cluster analysis). For both methods, the classification of any one pixel is entirely independent of that of any other pixel. Therefore, the procedure is parallel and may proceed very rapidly on the Illiac IV. Also, the classification procedure is not iterative, so there is no reason for the data to be core-contained. The system can handle as much data as will fit on the Illiac IV disk.

An enhancement of classification is "masked classification" in which different statistics files are used on different areas of the data. Which statistics file to use is determined by a Mask file fitting the area to be classified. The mask file is generated by digitizing boundaries between different types of terrain. Such a procedure should be of use in areas where the terrain varies widely, as in certain areas of California where the transition from agricultural valleys to mountains is abrupt. The process of determining, for each PE, the mask field to which a pixel contained in the PE belongs is not particularly well-suited to Illiac IV processing, but this is a relatively minor loss of parallelism in the complete context of processing Landsat data. This process is implemented by using the enable/disable capability of the Illiac IV. Only those PEs with pixels requiring statistics File 1 are enabled, then only those requiring File 2, and so forth.

3.8. DIGITAL PROCESSING OF SYNTHETIC APERTURE RADAR DATA ON ILLIAC IV

Under the sponsorship of the NASA SEASAT Program, and in collaboration with the Jet Propulsion Laboratory and Lockheed Missiles and Space Company, NASA-Ames investigated digital processing methods for Synthetic Aperture Radar (SAR) data. Several computationally intensive algorithms were studied that are appropriate for implementation on the Illiac IV. Two algorithms have been experimentally executed on the Illiac IV.

3.8.1. Synthetic Aperture Radar Concepts

Consider a radar on board an aircraft with the antenna oriented broadside to the aircraft. The distance between an object on the ground and the line of flight is called the slant range. The direction parallel to the line of flight is called the azimuth direction. The radar echoes received from pulses transmitted at regular intervals can be processed to obtain a radar image of a strip on the ground that is parallel to the ground track of the aircraft. The image correlates the radar brightness of illuminated objects with their slant range and azimuth positions.

The resolution in the slant range direction is a function of the time-bandwidth product of the transmitted pulses. High resolution in the slant range direction can be achieved by high-powered pulses of short duration or by pulses of longer duration, and moderate power that have a linear FM modulation applied to the carrier frequency. Radars of the latter type are known as "chirped" radars. The return signal of a chirped radar has to be processed by a matched filter or some other equivalent process in order to compress the signal and obtain a correlation of power with slant range equivalent in resolution to an unmodulated pulse of much shorter duration and higher power. All of the radars considered here are of the chirped variety.

By displaying the echo return of a pulse as a line image, where the distance along the line is correlated with the slant range, a radar image can be made by placing the line images of successive echoes alongside one another to form a strip. The azimuth resolution for an image formed in this way is generally not very good since it is limited by the width of the radar beam's main lobe. For most airborne radar systems with a slant range of 10 km, the azimuth resolution cannot be expected to be much better than 200 m. By recording both the phase and the amplitude histories of the return echoes of an airborne radar, it is possible by processing this information to obtain an image with a much higher resolution in the azimuth direction. A SAR is an airborne radar where such information from the return signals are processed to obtain a high resolution radar image. A stationary object on the ground that is being imaged by a SAR is illuminated in succession by a large number of individual pulses. Each pulse reflected back by the object is received by the antenna at different positions along the line of flight. These signals can be processed to simulate the signal that would have been produced if all the signals reflected from the object had been reflected simultaneously and had been received by a long linear array of antennas placed along the line of flight. The long array of antennas is equivalent to a single antenna of the same length. It is from this synthetic antenna that the SAR gets its name.

Digital Processing of SAR Data

SAR video data are SAR signal data after demodulation with the carrier. On the Illiac IV, this data is in 8-bit words. The sequence of words produced by sampling the return signal of a single pulse is called a range vector. N of these vectors are processed at a time to obtain an image of a piece of the strip illuminated by the radar. If each range vector has M elements, then the digital SAR data can be viewed as an M × N array. The row vectors of such an array are called azimuth vectors. The position of each element in the array is correlated with a specific slant range and azimuth position.

One method for determining the amount of signal received from the position correlated with the array location (i,j) is (1) to compute up to a scale factor the expected value of the elements in a subarray containing (i,j) that would have been produced by a point target located at that position, and then (2) to multiply the elements of the subarray by the values of the corresponding elements for the point target and then summing the products. The probability that a radar point target is located at the position correlated with (i,j) is proportional to the absolute magnitude of the resulting number. The entire array could be processed in this way to obtain the image array. This method is, in most cases, not a practical way to process SAR data due to the large number of computations required.

A faster procedure for processing SAR data that is equivalent to the above method is based on the Fourier transform. The components of the transformed vector are multiplied by the corresponding components of the range filter vector. The inverse Fourier transform is then applied to the resulting vector to obtain a new range vector. Each azimuth vector in the new array has associated with its row index a specific slant range, as before, but now the data in the row are correlated with the same slant range as well.

The remainder of the process correlates the azimuth data with azimuth position. Each azimuth vector is processed in the same way as the row vectors. The matched azimuth filter vector depends on slant range. Normally the coefficients of the azimuth filter vector change slowly with respect to change in the slant range, so that the azimuth filter vector is recomputed only every eight azimuth vectors.

A digital radar map can be obtained from the array of correlated azimuth vectors by replacing each element by its amplitude squared. The array may be further processed to correct for known geometric distortions and to normalize the power by computing the average power for elements of a row, and then dividing each of the elements in the row by the average.

During 1977, NASA-Ames, in cooperation with the Jet Propulsion Laboratory and Lockheed Missiles and Space Company, developed an experimental Illiac IV SAR data processing program that was run successfully a number of times. The SAR processing program was developed to gain experience with digital techniques for processing SAR data that might be applicable to the NASA SEASAT Program. Another reason for developing the program was to experiment with various algorithms for azimuth correlation of SAR data. The input to the program is a 512 × 6144 array of SAR video data. Each column of the array is a range vector of length 512. Each data element in the array is an 8-bit word. There are three main parts to the program; each part will be described in detail.

Range Correlation

Step 1 — Initialization: The 512 × 6144 array of 8-bit words is read onto Illiac IV disk memory (I4DM). The input file occupies 384 pages of I4DM. There is also a control file that contains radar parameters and program processing parameters.

Step 2 — Input blocks of vectors: Twelve pages of input data into PEM. The twelve pages fill 192 rows of memory. Each row contains data for one range vector since in the 8-bit format, eight elements of each range vector can be stored in one 64-bit word. The vectors are expanded to 64-bit representation. Each range vector now occupies eight rows, and so the 192 vectors fill a total of 1536 rows, which is three fourths of PEM.

Step 3 — Correlation: The matched range filter vector is computed and stored (first time only). For each range vector replace the vector by its Fourier transform using an FFT subroutine, then filter the transform vector by multiplying each component by the corresponding component to the filter vector, finally use the FFT subroutine to get the inverse vector. This last vector is the new range correlated vector.

Step 4 — Output blocks of vectors: The 192 range correlated vectors are written to linear output file on I4DM. Each 192 vectors take up 95 pages of I4DM.

Step 5 — Loop: Steps 2 through 4 are repeated until all 6144 range vectors have been processed (loop 32 times).

The output from the range correlation process can be regarded as an 8×96 array of blocks, each block being a 64×64 subarray. In terms of the 8×96 array of blocks, the next part of the program reads into PEM one column of blocks and transposes the elements in each block. The blocks of transposed data are then written over the old data in I4DM.

The Illiac IV has an extensive repertoire of logical and data manipulation instructions which are useful in reformatting data. In particular, the shift operator used in subfield extraction in Step 2 has been implemented in the hardware by a barrel shift network which is a fast and efficient method of shifting data within a word — the instruction takes only one cycle to shift up to 64 bits.

Transposition of 64×64 Subarrays

Step 1 — Input: Thirty-two pages are read from output file of the preceding part. (64 vectors by 512 rows.)

Step 2 — Transpose: The 512 rows are sorted into eight 64×64 arrays. Each array consists of 64 successive rows, each row being a 64 element segment from a different range vector where each column of elements in the 64×64 array belongs to the same azimuth vector. Each 64×64 array is then transposed.

Step 3 — Output: The 512 rows (32 pages) are written to the same area on I4DM from which the input was read.

Step 4 — Loop: Steps 1 through 3 are repeated until all of the output from the correlation phase has been processed (repeat 96 times).

Transposition on the Illiac IV is essentially a matter of selecting diagonals of a matrix. Diagonals of a matrix are fetched by having each PE index into a different memory location: for example, if PE i modifies the memory address broadcast by the CU by adding i, then the main diagonal of a matrix will be fetched; adding i + 1 will fetch the first subdiagonal, routing the fetched information two PEs and storing in location indexed with i − 1 will put this subdiagonal in the first superdiagonal position. Note that not all PEs are required to move a given subdiagonal: for example, moving the subdiagonal with 50 elements leaves 14 PEs unused. However, these 14 PEs can be used to move the superdiagonal with 14 elements. The end-around nature of the routing network in the Illiac IV means that these 14 elements will not be lost when the 50 elements are aligned, but instead are themselves properly aligned in their destination PEs.

Azimuth Correlation
Step 1 — Input: Sixteen azimuth vectors are read from I4DM into PEM. This is accomplished by reading in one page out of every 32 pages of the output file of the preceding subprocess, or in terms of blocks, reading in a quarter of the rows from the row of blocks. Since each page consists of 16 azimuth vector segments with each segment belonging to a different azimuth vector, the segments have to be separated and placed in nonadjacent rows in order to end up with the 16 vectors being stored in PEM in the proper positions. This process results in the 16 azimuth vectors being in PEM as if they had been read in a linear sequential order.

Step 2 — Correlation: The azimuth filter vector is computed and used to filter the first eight vectors. A new filter vector is computed and used to filter the final eight vectors. The squared amplitude for each component of the 16 azimuth vectors is scaled and converted to an 8-bit representation to get the final azimuth vectors (each vector consisting of 6144 8-bit words).

Step 3 — Output: The 16 azimuth vectors to final output file on I4DM.

Step 4 — Loop: Steps 1 through 3 are repeated until all 512 azimuth vectors have been processed (repeat 32 times).

The final output array consists of 512×6144 8-bit words. Each word represents the power reflected from a specific location.

The only portion of Step 2 which is not completely parallel is the computation of the scaling factor for each vector; this factor is an average of the elements of the vector, and hence requires summing across all PEs. This is actually a negligible loss of parallelism. In fact, the most noticeable inefficiency in all three processes is the latency in disk access.

3.9. FAST FOURIER TRANSFORM

The Fourier transform, the coefficients of a series expansion for a function in terms of sines and cosines, has quite properly found broad applicability throughout the scientific computation community. A wide range of applications have employed the Illiac IV-implemented FFT algorithm. In the realm of general digital image processing, two-dimensional Fourier transforms are used for enhancement, compression, texture classification, quality assessment, cross-correlation, and a host of other operations. One simple example is contrast improvement. Here, the Fourier transform of an input image is used to ascertain the spatial frequency content of that image. If the zero frequency component of that transform is set to zero, the inverse of this "filtered" transform will be a new rendering of the image with the background haze removed.

The Illiac IV program to perform a two-dimensional FFT is called TWDFFT. TWDFFT operates on arrays of complex numbers. The complex numbers are stored one to each 64-bit Illiac IV word with the components coded as 32-bit floating-point numbers. The real and imaginary parts of each complex number are stored in one 64-bit word. Matrices are assumed to be stored in contiguous rows of PEM. When rows of matrices are longer than 64 elements, the elements of the rows are stored in contiguous rows of PEM.

TWDFFT requires four scalar arguments and a row-aligned vector array. Two integers, M and N, are required to specify the dimensions of the array to be transformed. M specifies the first dimension or extent of the array in the "across PE" direction. N specifies the second dimension or number of blocks of M/64 rows. Both M and N must be powers of two, not less than 64, and their product cannot exceed 65,536. Two other parameters indicate whether a forward or inverse transform is to be performed and whether an error occurred; the error conditions are incorrect parameters specification or array extent beyond the end of PEM. Approximately 86 rows of PEM are required for storage of the code and tables needed for the TWDFFT subroutine package.

The FFT is always computed down PEs; this avoids any routing during the transform process. Before processing in the dimension "across PEs", the data are transposed. Essentially, transposition requires a linear number of routes whereas the FFT across PEs requires N log N routes, i.e., more than linear (N) routes, hence transposition is generally faster (except in extreme cases, such as 64×2 transforms, for example). Aside from the issue of data rearrangement, the FFT is a completely parallel operation for which the Illiac IV, as well as most other parallel computers, is ideally suited. In fact, the FFT's innermost loop can be tightly coded so that the Illiac IV runs at peak efficiency; all bookkeeping overhead can be masked by the floating-point operations by taking advantage of the CU's capability of performing overhead computation while the PEs are performing arithmetic operations.

3.10. LINEAR PROGRAMMING IMAGE ENHANCEMENT

The applications described to this point have all been executed on the Illiac IV computer. On the other hand, none of them absolutely required a machine of the speed and capacity of the Illiac IV. We describe in this section an image processing application that is completely impractical except on machines in the Illiac IV class. The linear programming software necessary to perform this processing was not implemented on the Illiac IV, but a detailed design has been completed. Furthermore, this application is characterized as compute-intensive, which made it particularly attractive for Illiac IV efficiency.

3.10.1. Statement of the Problem

Consider a point light (source object) in space focused by a lens onto an image plane. Unfortunately, even with a perfect lens, the image of that object is not a point but rather a blur of some finite size. For a perfect lens the theory of optical diffraction tells us how to calculate the size of the blur and the intensity distribution within the blur as a function of the wavelength of the light, the lens diameter, and the lens focal length. For a real lens the intensity distribution within the blur approximates that of a perfect lens. This intensity distribution, when normalized, is termed the point spread function (PSF) and is used as one way to characterize the quality of a lens. If we were to measure this intensity distribution at a 5 × 5 array of points in the image plane using an arbitrary scale, with high numbers indicating brighter spots, we might obtain the following values:

$$
\text{Array A} \quad
\begin{array}{ccccc}
0 & 1 & 1 & 1 & 0 \\
1 & 1 & 2 & 1 & 1 \\
1 & 2 & 4 & 2 & 1 \\
1 & 1 & 2 & 1 & 1 \\
0 & 1 & 1 & 1 & 0
\end{array}
\quad \text{(array sum = 28)}
$$

Similarly, if we were to measure the intensity distribution in a plane (termed the object plane) containing the point light source and parallel to the image plane, again in a 5 × 5 array and using the same scale we would obtain:

$$
\text{Array B} \quad
\begin{array}{ccccc}
0 & 0 & 0 & 0 & 0 \\
0 & 0 & 0 & 0 & 0 \\
0 & 0 & 28 & 0 & 0 \\
0 & 0 & 0 & 0 & 0 \\
0 & 0 & 0 & 0 & 0
\end{array}
$$

Hence the object of Array B (a point light source) when imaged is spread into the distribution shown in Array A.

Now consider two adjacent point light sources in the object plane, with the second twice as bright as the first. Measurement in a 5 × 6 array in the object plane would produce:

$$
\text{Array C} \quad
\begin{array}{cccccc}
0 & 0 & 0 & 0 & 0 & 0 \\
0 & 0 & 0 & 0 & 0 & 0 \\
0 & 0 & 28 & 56 & 0 & 0 \\
0 & 0 & 0 & 0 & 0 & 0 \\
0 & 0 & 0 & 0 & 0 & 0
\end{array}
$$

The image plane intensity distribution of the left light source is

$$
\text{Array D} \quad
\begin{array}{cccccc}
0 & 1 & 1 & 1 & 0 & 0 \\
1 & 1 & 2 & 1 & 1 & 0 \\
1 & 2 & 4 & 2 & 1 & 0 \\
1 & 1 & 2 & 1 & 1 & 0 \\
0 & 1 & 1 & 1 & 0 & 0
\end{array}
$$

while the right light source produces:

$$\text{Array E} \quad \begin{matrix} 0 & 0 & 2 & 2 & 2 & 0 \\ 0 & 2 & 2 & 4 & 2 & 2 \\ 0 & 2 & 4 & 8 & 4 & 2 \\ 0 & 2 & 2 & 3 & 2 & 2 \\ 0 & 0 & 2 & 2 & 0 & 2 \end{matrix}$$

The combined effect in the image plane of both light sources is simply the sum of the two Arrays D and E:

$$\text{Array F} \quad \begin{matrix} 0 & 1 & 3 & 3 & 2 & 0 \\ 1 & 3 & 4 & 5 & 3 & 2 \\ 1 & 4 & 8 & 10 & 5 & 2 \\ 1 & 3 & 4 & 5 & 3 & 2 \\ 0 & 1 & 3 & 3 & 2 & 0 \end{matrix}$$

This illustrates the principle of superposition wherein the combined effect in the image plane of an array of light sources in the object plane is just the linear aggregate of the effects of each source separately. Hence, by induction, one may intuitively grasp what the image of an arbitrary array would look like — a defocused version of the object plane.

F can also be obtained from Arrays A and C by what is termed a convolution. Commonly, however, we have knowledge of Array F and Array A and we seek Array C. This is termed a deconvolution and is an important tool in image enhancement. In the ideal example above, deconvolution is not too difficult. In real data cases, however, Array F and Array A are known only approximately since noise and measurement errors are included. In the above example, if we lump all the noise and errors of Array F into a 5 × 6 array termed N, then we can express the problem mathematically as A # C + N = F. Given A and F, without knowing N, find C where # represents a convolution. In the mathematical literature, A is also called a kernel. We ignore here the errors associated with A.

A large body of literature has evolved about how to solve this problem. Obviously, without additional information there is the trivial solution C = 0, N = F. This is avoided by imposing additional conditions such as (F/N) > 2 (signal-to-noise ratio). One school of thought believes that the best approach involves Fourier transforms and matched filters.

Before one may answer the question of the best approach or optimum filter, a measure of goodness must be specified, i.e., if one uses two different methods and obtains two different answers, C1 and C2, one must be able to decide which is preferred. Unfortunately, there is no unanimous answer from the image analysis community. Dozens of image quality measures are in regular use and none correspond perfectly to photointerpreter performance for all images, all interpreters, and all analytic tasks.

3.10.2. Linear Programming Approach

An approach and some preliminary experimental results are available that seem to indicate some advantages of linear programming (LP), at some cost, compared with competing techniques:

- Quantitative measure of success provided
- Sensitivity analysis available
- Convenient change of Figure of Merit
- *A priori* knowledge systematically incorporated
- Simultaneous processing of multiple images of same object
- Positional variations in PSF accommodated

In summary, the convolution expression gives rise to a series of constraint equations consisting of a series of coefficients times the elements of the object array plus an element of the noise array set equal to an element of the image array. A set of object array elements is found to optimize the objective function. One example of an objective function to be minimized is the sum of the noise array elements.

One of the drawbacks of this approach is that a lot of computing time and storage are required. A 512×512 image (about the size of a TV screen) gives rise to an LP model with 262,000 rows and 262,000 columns (approximate). For comparison, a 4000 row 10,000 column 1% dense LP problem on a 360/75 requires 6 to 8 h to solve. Clearly the technique is practical only for relatively small images.

A second drawback is that many of the Figures of Merit commonly used and desirable to optimize are nonlinear, e.g., entropy where $H = -$ (sum of) $P(i) \log_2 P(i)$. The point is that H is nonlinear; the $P(i)$ are not important here. To accommodate these terms, nonlinear programming would be needed. It appears, however, that much progress can be made within the linearity restriction.

3.10.3. Practical Issues

The Illiac IV offers the computational power to make this algorithm practical. A 256×256 digital image gives rise to an LP model of about 65,000 rows and 65,000 columns. But for a small kernel, say 5×5, this matrix is only 0.038% dense; that is, only 1.6M nonzero values appear. This is substantially less than the capacity of the Illiac IV disk memory.

Dr. Charles Pfefferkorn and Dr. John Tomlin have designed an Illiac IV implementation of the Product Form of the Revised Simplex Linear Programming algorithm. Although this design was not implemented on the Illiac IV, the design shows that a 32,000 row model is comfortably addressable and larger models can be solved if the density is low.

Solving linear programming models is an iterative process. The Pfefferkorn/Tomlin design estimates that the Illiac IV would require about 1 s per iteration.

The rule of thumb is that an n equation model will require 3 n iterations to solve. This suggests that to solve a 64,000 equation model would require 54 h of Illiac IV time to solve. Fortunately, the rule of thumb does not apply in this case. The 3n iterations estimate pertains to cases where there is no prior knowledge about the solution. Here we know that the image is a reasonably good guess at what the object looks like. This prior information may well reduce the run time by a factor of 50, so that an optimal enhancement could be obtained on the Illiac IV in about 1 h.

The basic parallelism of the Illiac IV (the 64 processing elements acting in lockstep) is usually fully utilized by image processing applications; the major exceptions are processes which require reduction operators (operators that take information from a vector and produce a scalar, such as summing the elements of a vector) and the processes that perform many data-dependent tests in parallel, such as a texture algorithm. Reduction operators account for very minor amounts of the computation in image processing applications.

The ability of each PE to index its access to its own private memory is useful in a number of algorithms requiring transposition of data or table look-up. For the Illiac IV applications, however, these processes are only minor or incidental subtasks and hence independent indexing has been of relatively minor importance.

The major architectural aspect of the Illiac IV system with respect to image processing has been the relatively small local scratch-pad memory and the long latency time to access the main storage device (a high bandwidth disk system). This arises from the frequent characteristic of image processing: few operations per data element per pass over the image. Even when the algorithm is compute intensive, as in the classification algorithm, the limited storage serves to limit the range of alternatives for the algorithm (number of categories in the classification, for example), and hence the potential power of the process. In addition, there are some applications that are too large for the disk memory (eight million 64-words), and hence cannot be handled efficiently on the Illiac IV system (since the support system is not configured for frequent, interactive use by an Illiac IV program; it is designed instead for staging jobs onto and off of the Illiac IV processor).

The major precision used for the image processing applications on the Illiac IV has been 32-bit floating point, given a choice of 8-bit integers and 64-bit floating point. The one-dimensional nearest-neighbor routing network, with end-around connection, has been sufficient for Illiac IV image processing applications; in the case of transposition, the end-around feature has been beneficial. The main reason for the adequacy of the one-dimensional connection in a two-dimensional image processing is that 64 processors is a small number when dealing with images measuring 512 pixels on a side, for example. An array of thousands of processors (microprocessors) however, must accommodate more flexible connection patterns for routing data.

3.11. COMMENTS ON SOME CASE STUDIES

The following are some selected observations on several programs which have been written for the Illiac. The intention is that they may shed some light on the nature of scientific computing that may be amenable to parallel computation. These case studies are important both for the problem formulation strategies, program design decisions and coding techniques.

3.11.1. Sparse Matrix Multiply

The following three paragraphs are a somewhat edited quotation from a report on three-dimensional stress wave simulation for the Illiac, authored by Gerald Frazier and Christian Peterson (DNA 331F report by Systems, Science and Software), pages 48 and 49.

> The time stepping process for this problem consists of the calculation U=V+A*W for each time step. The first term V is a vector and its calculation involves vector operations which require no interaction among the Illiac PEs. As a result, it is easily computed in parallel. Similar operations are involved in the calculation of the vector W. The significant calculation is the multiplication of the vector W by the large sparse matrix A. This multiplication accounts for almost all of the computation time that is required to complete one numerical time step. A sophisticated but simple mechanism has been developed to perform the sparse matrix multiply in parallel. The non-zero terms of A lie in 3×3 submatrices of A, no more than 27 such submatrices in any row of A. These are arranged on disk so that when read into memory each arrives in the PE which contains the three elements of W which enter into the computation of the product of the submatrix of A and W. Furthermore, as successive terms of A are read from disk the matrix row numbers increase monotonically (but not necessarily sequentially). This is done so that the sparse matrix multiply can be completed in the order of ascending row number.
>
> The first submatrix to arrive in each PE from the disk is multiplied by the appropriate three components of the vector W and the results are accumulated in a buffer along with the row number identifier. This operation allows some PEs to work ahead on other row numbers. Since several rows may be processed simultaneously, a look-ahead buffer is maintained in each PE which contains both the elements and their row numbers. Since rows will continuously be completed as new ones are started, the buffer need only be large enough to contain the maximum number to be worked on at one time in any given PE. On the average, all of the multiplies for about 2.4 rows of the sparse matrix multiply are completed at a time.
>
> During the matrix multiply, a test is made to see if all contributions from the sparse matrix multiply are ready to be summed for the node numbered n. If all of the row numbers from the submatrix multiply are greater than n, then all contributions for n are calculated (all PEs are now working on contributions to higher node numbers). The contributions for n are then summed and added to the other terms to obtain the advanced nodal displacement U(n). This displacement vector is stored in PEk, where k=n mod 64. If the contributions from row n+1 are completed, then node n+1 is also advanced in time, otherwise the next submatrix multiply in line for each PE is performed. The parallel submatrix multiplies, row sums, and disk reads continue until all of the A matrix has been processed and all nodes have been advanced in time. The entire operation is repeated for each time step.

Some points are worth mentioning here. First of all is the surprise that the matrix-vector product is not programmed as vector operations but rather as separate processes (the Illiac is being used not as a vector processor, but as multiple processors, each working largely in its own "context"). The difference in this case is essentially 99 vector component-wise multiplies (of vectors of length 3N, where N is the number of mesh nodes) plus aligning and summing the 99 result vectors, versus 27N matrix-vector products (involving 3×3 matrices) plus aligning and summing the 27N vectors (of length 3). The vector formulation costs about 18% more storage — the added padding of zeros is necessary for alignment purposes — plus the concomitant increase in arithmetic — the multiplications by the padding zeros; the use of zero here is exactly analogous to its use for positional notation in number systems. On the other hand, the vector formulation eliminates the control structure which tests to see when all information for updating each node has been assembled and can be combined. It also eliminates the buffer management for these intermediate results. The real subtlety of the problem lies in the aligning and summing involved in the two approaches, plus the possible necessity (based on small core memory) to partition long vectors, but we will leave the matter here.

The non-vector approach does lend itself to matrices which arise from arbitrarily connected grids. But the automatic grid generation used by this project generates grids which are unions of regions homeomorphic to a cubical lattice, hence the structure of the matrix A will have large blocks along its diagonal where the above vector approach will hold, and its off-diagonal blocks, most of which are identically zero, will have an analogous vectorizable structure.

3.11.2. A Model for Disaster

The Tensor code (Final Report of the Tensor/Illiac IV Project, ARPA Order 1839 (UCRL-51467) by Tad Kishi, 1973) is based on a grid which moves with the material; the solution at a grid point involves information from nine neighboring points nearest to it. Here whatever regularity exists in the grid at the beginning of the simulation is rapidly destroyed over the iterations, so a vector formulation of the sparse matrix is clearly inappropriate. The next question is, can an Illiac-type architecture, viewed as each processor working in its separate context but doing roughly the same thing, provide a suitable environment for such calculations? Or is this a formulation best suited for some other type of computer?

Unfortunately, the project gives no answer, since it was a complete failure. In fact, the charitable thing would be to forget this fiasco entirely, but since a computer is what it appears to its users to be, it is important to consider this project, if only as a study in cognitive psychology.

"Bound by the primary requirement to reconfigure an existing production code, the development of effective parallel processing methods for the Illiac computer system has been an exceedingly difficult one. It could not have been accomplished by a simple translation of the existing FORTRAN code to a comparable language for the Illiac. The FORTRAN listing of the Tensor code is a poor substitute for documentation. It is next to impossible to understand the Tensor code or to derive effective algorithms for parallel processing from a code that was programmed in assembly language for a conventional computer and then brute force converted to FORTRAN. The task has only been accomplished by reformulating and reexamining the basic finite difference equations. Unfortunately, neither a consistent nor complete set of equations of the existing code was available and had to be redeprived (sic) by members on the ARPA Tensor project." (One can only wonder what the sequential code has actually been computing all this time.) (p. 3)

To seal the project's fate, it was decided to code in an assembler language. The reasons given were that the higher level languages were undergoing development and hence (1) did not generate reasonable object code (which is irrelevant; bad code can be selectively tuned) and (2) their programming support was minimal at best. The result of this decision was predictable. "Once a course of action was decided upon, it was literally embedded in 'cement'. Programming in assembly language left little or no flexibility in our code development." (pp. 3-4) Thus the conclusions drawn by this project were largely due to the propagation of poor early design decisions. A stunning example of this occurred when the program was restructured, proving "that skewing of data, which we originally believed to be essential for efficient boundary calculations, was immaterial. To reconsider the skewing of data at this point in our code development was next to impossible. This is the price one pays when a code of this complexity is programmed in assembly language." (p. 14) There was an even greater price: the code never ran. "Two simulation runs have been attempted in this configuration. The code has crashed in loop 1 in the k=0 boundary routine. The results have been evaluated, but there are no plans to continue debugging." (p. 15)

What were the perceived problems of programming this formulation on the Illiac? There were essentially three. First, "The inherent geometric structure of the 64-PE Illiac computer system imposes an artificial boundary (modulo 64) on the grid system and must be contended with throughout the program for an array not commensurate with this base." (p. 6) Second, "considerations of the boundary calculations ... required skewing as a fundamental requirement of the problem logistics for efficient PE usage. However, a given storage assignment for one phase of the calculation may not be suited for another part of the calculation." (p. 7) And third, "the calculational procedures of the slip lines for the Illiac array processors require extensive movement of data across the PEs in order to meet the nearest neighbor requirements for the nine-point difference scheme. This is the result of the change in the nearest neighbor relationship with time. Thus the values necessary for interpolation may be in some arbitrary assignment across the processing elements." (p. 63)

The first perceived problem is illusory; it is solved by logically programming in a system of N processing elements and then simulating N processors using 64 or fewer processors (this is what a higher level language should be able to do). As seen above, the second problem actually turned out to be a red herring, and probably a costly one at that. The third problem, which is the heart of the matter of whether this formulation can be effectively used on an Illiac-type computer, arises from assuming a fixed data structure; but if the grid moves with the physics of the process, it seems reasonable to entertain the notion that its representation moves with the computation of the algorithm; this probably won't solve the problem, but it might mitigate its presumed seriousness. Another possible approach would be to use a grid structure fine enough so that slip lines and any other physically interesting phenomenon could be derived from calculations performed on the fixed grid — this would be an example of using raw computational power in place of the potentially staggering overhead of bookkeeping and routing of information needed for a more sophisticated formulation. This solution may not be aesthetically pleasing, but it might be the best cost-effective method (or even the only technologically feasible method for very large models). Since the purpose of computing is insight, the only question is whether this insight should be derived directly from the mechanics of the algorithm or be inferred from the results of the calculations.

Notice that all three problems have a common thread: the vagaries of the programming language, in revealing all of the machine characteristics, have given the greedy programmer more than enough rope to hang himself in trying to pull the last bit of speed out of the machine. This is a very serious problem, since it detracts from the real issues. "Skewing and the pseudo 64-PE boundary are new experiences and add to the difficulties in visualizing parallel processes in the Illiac." (p. 7)

3.11.3. Monte Carlo Methods on the Illiac

The real problem with the slip-line is the interaction among dynamically varying groups of nodes, and the attendant bookkeeping necessary to locate specific nodes or assemble the necessary information. Monte Carlo methods which are formulated so that interaction among constituent elements are implicit can effectively minimize this overhead problem, but at the expense of substituting an apparent "randomness" in the control-flow. That this substitution can be successful on the Illiac must certainly be one of the ironies of contemporary computing, since "conventional wisdom" had held that the single-instruction stream was the constraining factor to the efficient utilization of the Illiac, which does not obviously lend itself to branch-driven programs. (Conventional wisdom also ignored completely the impact of the memory structure on effective data utilization, which probably will be the constraining factor once more experience with the Illiac is reported.)

A successful Monte Carlo code for the Illiac is reported in SAM-IV: a three-dimensional Monte Carlo radiation penetration code for the Illiac IV by E. S. Troubetzkoy, M. H. Kalos and H. Steinberg of Mathematical Applications Group, Inc., DNA 3303F, 1973. Of particular interest are the mechanics used to implement a disorderly control flow (one which takes many different branches when executed successively of different data by a sequential computer).

"The major difficulty with attempting to implement a Monte Carlo code ... on the Illiac lies in the intrinsic disorderly nature of Monte Carlo logic.... The order and nature of the physical events have little, if any, correlation from (particle to particle). The naive approach of following 64 histories simultaneously is therefore not feasible as the parallelism breaks down almost immediately. Our approach is to initiate many histories in each PE, and hold all of them in abeyance until any calculation is required" — that is, until enough PEs have particles upon which the same calculation can be performed. (p. 10)

The basic idea here is reminiscent of the control mechanism in a production system, or Markov algorithm, where, at least conceptually, processes are activated in an associative manner whenever certain specified conditions in the data base arise. In the Monte Carlo program, certain computations are performed whenever a certain amount of parallelism is possible.

3.11.4. Conclusions

A general statement of the philosophy underlying the successful programming strategy would be: divide the problem formulation into as many independent steps as possible — steps which would have to be executed repeatedly on varying data by a sequential computer — and then at each point of the parallel computation, choose to execute that step which will utilize the greatest amount of parallelism. The ultimate success of any code seems to lie in the ability to minimize the overhead of bookkeeping, either implicitly (as for example, when the computation required for a particular node is known to be completed when all PEs are working on computations involving higher numbered nodes) or explicitly (as where the formulation is in theory without any dynamically varying interrelationships among distinct components; that is, the aggregate effects of interest can be viewed as data reduction which can be done without regard to order and in a cumulative fashion, and hence lends itself well to homogeneous parallel processing).

One of the unifying characteristics of these three projects is their unwillingness to view the Illiac as a vector computer. This may be because of the small random access memory or because of the short nature vector length. Or it could be a (perhaps deserved) infatuation with a sequential program. However, if one generalizes the notion of a vector operation from component-wise scalar operations to more complex operations on structured components, then these programs may be interpreted as attempts to simulate generalized vector computations.

3.12. THE EFFECTS OF THE ILLIAC IV SYSTEM ON COMPUTING TECHNOLOGY

This section, based on an internal memo at the Institute for Advanced Computation by G. Feierbach and D. Stevenson in August 1976, outlines some of the contributions of the Illiac project to computer science and technology. Sixteen distinct advances in four categories are described.

3.12.1. Component and Manufacturing Technology
3.12.1.1. Major Impetus to ECL Development

The I4 system was the first large scale use of ECL integrated circuits. The circuits developed for the I4 system were subsequently improved by TI and used for their ASC computer. (Of the 33 IC types used in the I4 main frame, 14 ASC parts can be directly substituted. It is questionable whether TI would have built the ASC computer had the development of the IC family not been underwritten by the Illiac requirements.)

3.12.1.2. Test Bed for Design Automation

The circuit cards in the Illiac main-frame were designed using a design automation system. This was the earliest successful large scale use of design automation outside of IBM. The Illiac contract provided both the financial resources and the level of difficulty to mature this process significantly. This is now a widespread practice in the computer industry.

3.12.1.3. New Contribution to Logic Circuitry

The barrel switch is a major circuit innovation in the Illiac that enables full word length shifts in one machine clock. This is used for floating-point normalization and alignment and for shifting in general. Current supercomputer designs incorporate the barrel switch in one form or another. It has become popular enough that Fairchild has created an Isoplanar TI ECL part (F100158) which is essentially an 8-bit slice of a barrel switch.

3.12.1.4. First Significant Use of Semiconductor Memory

The 256×1 bipolar RAMs in the I4 PE memories are the first use of bipolar semiconductor memories in a large scale computer main memory. Since thin film memories were also considered (even prototyped) but rejected in favor of semiconductor memories, this was probably a significant development for the Illiac IV became the father for the first commercially available semiconductor memories offered by Fairchild.

A minor additional note: by using an interlocking mesh for power and ground distribution on the memory PC boards and judicious placing of ground strips between signals requiring isolation, it was possible to arrange the memory on a two sided PC board for a significant cost savings. Up to that time it was felt mandatory to have separate power and ground planes in addition to the circuit layers.

3.12.1.5. Definitive Contribution to Interconnection Technology

The system was the first to make use of extremely dense belted cables which are soldered to paddle card PC boards using infrared light and then covered with epoxy. These cable assemblies are a major constituent of the system but have been responsible for very few failures. The current state of the art in cabling (excepting fiber optic technology) cannot do better today.

3.12.1.6. A Major Milestone in Multilayer PC Cards

The I4 control unit PC cards are 16" × 20" and have as many as 12 layers. Not only was this the first successful utilization of large multilayer laminated boards, but it is still a state of the art achievement.

3.12.2. Machine Architecture

3.12.2.1. Definitive Demonstration of Array Approach to Computation

An operational Illiac validates the design concept of array processors; it has become the standard against which to measure proposals for increasing computer speed through architectural innovations involving replication of components.

3.12.2.2. Synchronous Control to Focus Research on Efficiency of Computation

The Illiac demonstrates the sufficiency of a single instruction stream to control the multiple data streams encountered in scientific computing. In a single stroke, this approach (via the route instruction) solves the problem of synchronizing processor communication. This has permitted research to focus on the efficient utilization of the array using the single instruction stream, in contrast to the case of asynchronous, independent processors, where research has focused largely on synchronization issues and only recently turned to the efficiency of algorithms in such an environment.

3.12.2.3. First Large Scale Computer to be Microprogrammed

The I4 control unit contains a ROM driven microprocessor which converts single instructions into a sequence of enable signals for the PEs. At the time the I4 was designed, the only significant machines to be microprogrammed were the lower model numbers of the IBM 360 series. The prevailing opinion in the computer community at the time was that microprogramming was slow and that fast main frames could only be designed using hardwired logic. Today, major supercomputers (including the Star-100 and ASC) contain micro-coded control logic.

3.12.2.4. Synchronization of Independent Disk Drives

All the I4 disk memories are synchronized to within 2 degrees of a revolution. This was formerly thought to be impossible on theoretical grounds: a continuous feedback mechanism requires instantaneous acceleration and this was felt impractical to obtain without very complex detection and control circuitry and elaborate sensors and control effectors. The method actually used in a startlingly simple use of an oscillator as a virtual disk. Some manufacturers have shown interest in utilizing this innovation since it makes possible very high bandwidth synchronous transfers from multiple drives.

3.12.2.5. Exhaustive Simulation as a Realistic Diagnostic Tool

PESO is a PE simulator that runs on the Illiac (even when some PEs are down). The ratio of the computer power of the entire I4 system to the simplicity of a single PE makes it possible to simulate completely the complex operations of a single PE in a few milliseconds of I4 time. This has opened the door to a novel diagnostic technique not possible on other machine architecture: exhaustive simulation of all possible single gate faults. Over 4000 cases can be tested at the same time so that within about 5 min a list of possible fault locations that match the failure symptom is in the hands of a technician. About one third of the PE faults show up in this manner, making it a powerful tool in system maintenance which would otherwise be unavailable.

3.12.2.6. Test Bed for Future Machines

The Illiac IV has taught some important lessons which will have significant impact on future parallel processors. In particular, the processor interconnection scheme has been found to be wanting. It is both inflexible and difficult to program.

Research in this area has focused on the optimum interconnection scheme and on the most efficient way to use a given interconnection pattern. All this has been predicated on the assumptions that the connection network must be fixed (hardwired) and that each processor can be connected to only a few other processors (because of fan-out limitations or cost considerations). These assumptions are no longer valid since there are other alternatives than interconnection schemes based on cabling, and the next generation of array computers should re-focus the attention that the Illiac has inadvertently misdirected.

Further, the Illiac IV is a fixed configuration with no self-repair capability. Current research into self-repairing processors (multi-processors such as C.MMP and array processors such as PEPE) are inadequate as a base for massive computing power required by scientific computation because those prototypes in practice admit only extremely narrow bandwidth paths of information flow among processors. Further systems will have modular configurations for improved problem matching and will be able to switch ailing PEs out and good PEs into the configuration all under software control.

3.12.3. System Architecture

To quote from Bouknight et al.,[4] "It should be remembered that the Illiac IV project was initially directed toward experimenting with the feasibility of building a massive hardware configuration." In a word, the result is yes, it is feasible.

The I4 system is a massive implementation of the concept of a functionally distributed operating system. It can be viewed as the culmination of a progression which started with early computers originally designed to execute efficiently different types of computing tasks, joined together to execute different steps of the same computing job (e.g., the front end user interface preparing the job for a large number-cruncher). Historically, these were incorporated into the main-frame design of more recent large computers. The I4 system approaches the problem by dedicating functionally separate mini-computers and memory module buffers to the independent functions of system support. For example, file transfers to prepare jobs for execution are handled by a separate mechanism from the one in charge of the movement of program data from backing store to I4 processor memory during program execution. The advantages of this approach are fault tolerance (jobs which require only part of the system can run whenever this part is available, whether or not the whole system is working) and technology independence (as technology advances are made, system components can be enhanced on a module basis). An additional benefit is that when modifications are to be made to the system to add unanticipated capabilities, at most only the relevant modules which are to interface with the new capability need be modified (or replaced); this is in contrast to the more usual situation where the maximum system capability is determined by the initial main-frame design.

3.12.4. Applications
3.12.4.1. New Horizons in Solvable Problems
The size and speed of the Illiac makes feasible the solution of many computational problems which were computationally intractible when the machine was originally designed. The essential reason for this is the large memory and the high bandwidth between this memory and the processing power (more conventional super-computers which have access to large backing store disks suffer from a narrow bandwidth between this store and the processing unit, resulting in very large problems being essentially I/O bound — this is especially true of the CDC Star-100). The situation is exacerbated by the general rule of thumb that for many scientific problems, larger data bases (for a finer resolution of the physical phenomenon) both take longer to pass through the data base and, more importantly, have to pass through the data base more often (because the iteration process converges more slowly or because smaller time steps have to be taken).

3.12.4.2. Spurring the Development of New Algorithms
The concept of an array computer had provided a model for developing parallel algorithms, but the announcement that a powerful computer was to be based on this concept unleashed a spate of activity in the area. At the present time, more research has been based on the array model of computation than on any other, save for the classic von Neumann (or sequential random access) computer and the Turing machine.

3.12.4.3. Rethinking Problems for Parallel Processors Pays Dividends on Other Processors
A machine architecture which is a radical departure from conventional sequential computers (as the Illiac is) encourages users to re-formulate, or re-code, their problems to make use of the additional capabilities. Before the Illiac

(and the Star) were available, some of the re-formatted codes were debugged in a CDC-7600, whereupon it was found that they ran faster in their new parallel-formulated versions than in the original sequential version. The reason for this unexpected phenomenon is that parallel formulation leads to short compact code sequences and regular memory accessing, and these two characteristics describe code which the 7600 is particularly efficient at executing. As a result of this experience, the design philosophy and algorithms originally designed for the Illiac are being adopted as codes for the 7600.

REFERENCES

1. **Bache, T. C., et al.,** A deterministic methodology for discriminating between earthquakes and underground nuclear explosions, *Final Report to Advanced Research Projects Agency under Contract No. F44620-74-C-0063,* July 1976.
2. **Cherry, J. T., et al.,** A deterministic approach to the prediction of free field ground motion and response spectra from stick-slip earthquakes, *Earthquake Engineering and Structural Dynamics,* Vol. 4, 1976, 315.
3. **Maenchen, G. and Sack, S.,** The tenson code, *Methods in Computational Physics,* Vol. 3, Academic Press, 1964.
4. **Bouknight, W. J., et al.,** *Proc. IEEE,* April 1976.
5. **Feustal, E. A., et al.,** Future trends in computer hardware, *Proc. AIAA Computational Fluid Dynamics Conf.,* July 1973.

4 THE MPP

4.1. THE MPP DESIGN

The Massively Parallel Processor[1,2] (MPP), an advanced computer architecture termed single-instruction stream multiple-data stream (SIMD), showed promise of delivering enormous computational power and at lower cost than other existing architectures of its day. Its computational element, the array unit, consists of a 128 × 128 array of small 1-bit processors, each containing 1,024 bits of local memory, and having nearest neighbor connectivity. A secondary storage unit, the staging memory, holds 32 Mbytes of data and connects to the array memory via an 80 Mbyte/s data path. An array control unit broadcasts control signals to all processors in the array unit. The MPP is a back-end processor for a VAX-11/780 host, which supports its program development and data needs (see Figure 18).

The MPP was built for the Goddard Space Flight Center by Goodyear Aerospace Corporation and delivered in May 1983. At that time, the construction of a digital processor using the very high degree of parallelism embodied in the MPP had not been previously attempted.

4.1.1. MPP Software

Since its delivery to Goddard, an extensive language system and a unique operating system have been implemented for the MPP. Dozens of teams of scientific investigators who are developing, testing, and running parallel algorithms rely on this system software repertoire daily. The initial high-level language implemented in 1983 was Parallel Pascal. This language was designed to be independent of computer architecture, thus allowing portability of applications programs between diverse parallel computers having Parallel Pascal compilers. Experience gained in the development and use of this approach revealed that the MPP's 128-by-128 square grid architecture could not easily be hidden from the programmer by using current compiler writing

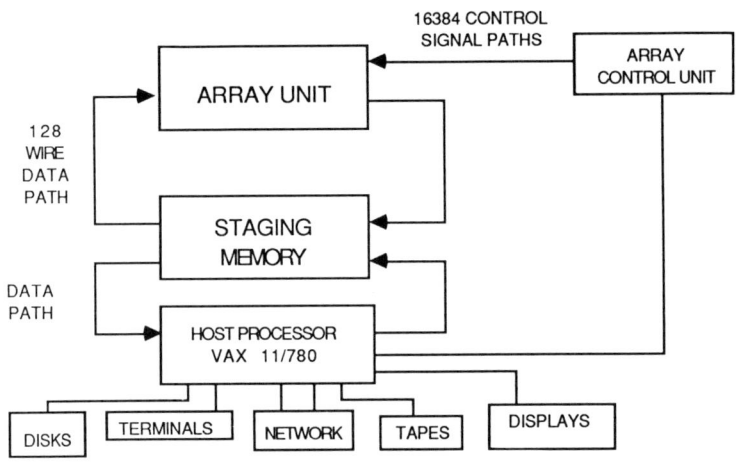

FIGURE 18. MPP system.

technology. A modified language, MPP Pascal, was then implemented that is architecturally dependent, possessing important semantic features that allow the programmer to make very efficient use of the hardware's capabilities. The MPP Pascal compiler[3] is capable of producing highly optimized code and is flexible enough to allow easy modification. The MPP is also programmable in assembly language.

The MPP operating system provides interactive debugging aids in addition to support for running applications code. The software that performs these tasks is shared by all MPP users, greatly reducing the demand on the host's main memory. The debugging aids include performance monitoring, error reporting or MPP hardware detected faults, breakpointing, single-step and status display. A first-come-first-served queue is the central arbiter controlling user access to the MPP.

All MPP applications programs must be prepared as two parts. One part runs in the MPP control unit; the other part runs on the host. They are linked together through a message passing system. A master/slave control relationship exists between the MPP and the host. The host resident program is the highest level of control. This program interacts with the user and starts MPP programs. MPP programs, in turn, use the host as an I/O server, directly accessing the host's disk and image analysis terminals through an extensive set of I/O service routines.

A device driver that communicates directly with the MPP hardware runs at the lowest level in the host operating system. This driver is the hub of the entire system, controlling the execution of programs in the MPP, as well as the flow of data throughout the system. The bulk of the operating system interacts directly with this device driver to accomplish tasks in support of a running application such as initializing the hardware, loading programs, starting and stopping programs, reading and writing data, and delivering messages between running programs in the host and the MPP.

A number of libraries of computational subroutines are supported. One type holds more than 270 microcoded subroutines that define the actual instruciton set of the MPP. These include the basic arithmetic and transcendental functions, as well as multi-precise arithmetic and special user-written instructions. A second type of library holds MPP Pascal callable subroutines including fast Fourier transforms, a random number generator, a sort computation package, linear algebra routines, and utility programs. Another set of libraries holds I/O subroutines that control the movement of data within the system.

For many applications, having only 1024 bits of memory available to each of the 16,384 MPP processors has been a serious constraint. The memory chip technology available in 1980 when the system was designed imposed this limitation. As an alternative to an expensive hardware upgrade, a Bit-Plane I/O software package was developed that treats the staging memory as individual bit planes.[4] A system was implemented that provides each processor with 16K bits of virtual array memory. The penalty is an increase in memory access time from 0.1 to 25 µs per bit plane. However, many applications benefit from this virtual memory, as they effectively overlap computation with data transfer. In addition to Bit-Plane I/O, another system, SMM I/O, gives the user access to the powerful data reformatting capabilities of the staging memory.

Two MPP simulation environments have been developed and distributed to remote user sites. One is the MPP Simulator, which supports the development, testing, and refinement of MPP Pascal, or assembly language applications programs on any VAX operating under VMS. It allows a user the convenience of a local dedicated MPP that doesn't need be shared with other users. In addition, its use at remote sites off-loads program preparation work from the MPP/VAX system at Goddard. Code that runs on the Simulator will run on the MPP after adjusting any references to the size of the array unit (usually simulated as a 16×16 array to speed execution).

A second simulation tool, the Parallel Pascal Translator, takes Parallel Pascal source code as input and produces equivalent serial Pascal source code as output. The serial Pascal can be compiled and executed using a standard Pascal compiler system. The Translator allows the development, testing, and refinement of applications programs on most computers that have a Pascal compiler.

4.1.2. MPP Hardware
4.1.2.1. Array Unit

The Array Unit (the 128×128 processing element (PE) array) supplies the MPP's computational power. Each PE has a local 1,024-bit random access memory and is connected to its four nearest neighbors—north, south, east, and west. Opposite array edges can be connected together to form either a plane, a horizontal cylinder, a vertical cylinder, or a torus. Arithmetic and logic in

TABLE 6
Speed of Typical MPP Operations

Operation	Execution speed, MOPS*
Addition of Arrays	
8-bit integers (9-bit sum)	6553
12-bit integers (13-bit sum)	4428
32-bit floating-point numbers	430
Multiplication of arrays	
8-bit integers (16-bit product)	1861
12-bit integers (24-bit product)	910
32-bit floating point numbers	216
Multiplication of array by scalar	
8-bit integers (16-bit product)	2340
12-bit integers (24-bit product)	1260
32-bit floating-point numbers	373

* Million operations per second.

each PE are performed in bit-serial manner. All operands are located in the 1024-bit local memory. The cycle time is 100 ns. Table 6 shows the raw computing speeds for selected arithmetic operations. The data-bus states of all 16,384 PEs are combined in a tree of inclusive-OR logic elements whose single wire output is used in the Array Control Unit for operations such as finding the maximum or minimum value in parallel of an array.

MPP Processing Element—A single PE is shown functionally in Figure 19. The P-register, together with its input logic, performs all Boolean logic functions on two variables and can also receive data from the P register in any one of its four nearest neighbors. The A, B, and C registers, the shift register, and associated logic form an arithmetic unit. The G register controls masking of arithmetic, logic, and routing operations. (Unmasked operations are performed in all PEs. Masked operations are only performed in those PEs where the G register is set.) The S register is used to shift data to and from the staging memory without disturbing PE operations. A custom integrated circuit (IC) holds eight PEs, exclusive of the 1,024 bits of random access memory which is on a separate IC chip. This chip, containing a 2-row by 4-column array of PEs, uses high speed complementary MOSFET (HCMOS) technology. A 1-bit wide bidirectional data-bus connects the memory and the internal components of the PE.

4.1.2.2. Array Control Unit

The Array Control Unit broadcasts control signals and memory addresses to all PEs in the array unit and receives array unit status bits. It is designed to perform bookkeeping operations (address calculation, loop control, branching,

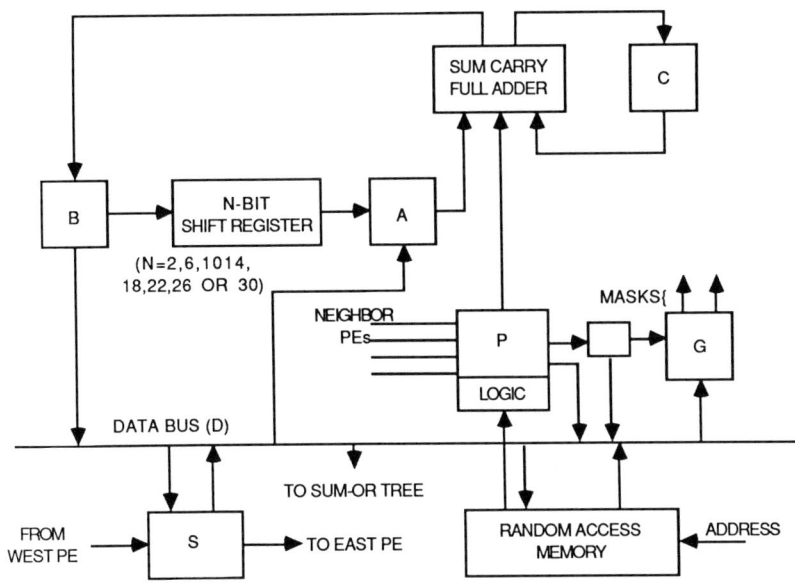

FIGURE 19. Functional units of one MPP PE.

subroutine-calling, etc.), and control the array unit simultaneously. It contains three parts: (1) the main control unit, (2) the PE control unit, and (3) the I/O control unit.

The Main Control Unit executes the application program stored in its program memory. It performs the scalar arithmetic operations required, calls the PE control unit for all array logic and array arithmetic operations, and calls the I/O control unit for all I/O operations. Both sets of calls are queued to await execution while the main control unit moves on to generate other calls.

The PE control unit generates all array unit instructions except those pertaining to the S register (data I/O). It executes microcoded routines stored in its program memory to perform all array operations required by applications programs.

The I/O control unit controls the shifting of I/O data through the array unit S registers, as well as the transfer of I/O data between the S registers and the array unit memory. It executes I/O channel control programs stored in the main control unit's program memory. The speed of typical MPP operations is shown in Table 6.

4.1.2.3. Staging Memory

The MPP system includes a staging memory for buffering array unit data. This memory provides both the "corner turning" functions, which converts conventional byte- or word-oriented data into the bit-plane form needed by the array unit, and the "multi-dimensional access" function, which allows large multi-dimensional arrays of data located in the staging memory to be read out or written in along arbitrary orderings of array dimensions. The current capacity of the staging memory is 32 Mbytes and is upgradable to 64 Mbytes.

Data moves between the array unit and the staging memory via 128 parallel lines. The upper limit on the transfer rate is 1.28 billion bits/s. Goddard's MPP currently supports 0.64 billion bits/s. Data movement in both directions can be overlapped with processing.

4.1.2.4. Host Processor

A DEC VAX-11/780 computer manages data flow between MPP units, loads programs into the Control Unit, executes system test and diagnostic routines, and provides program development facilities. The MPP is interfaced to the VAX through a 5 Mbytes/s DR-780 channel. Remote access to the VAX is provided through ARPANET, SPAN, TELENET, SITNET, and dial-in.

4.2. PARALLEL FORTH LANGUAGE

4.2.1. Introduction

The MPP, as viewed by the FORTH user,[5,6] consists of essentially three main components: the main control unit (MCU) (the scalar processor and controller of the array), the array unit (ARU) (for parallel processing of data), and the staging memory (STG) (primarily used for I/O and as a large external bit-plane memory).

If every processor had to perform every instruction given to it, it would be of little use as a general purpose computer. Conditional processing alleviates this problem. Conditional processing (such as the execution of an 'IF...ELSE...THEN' statement) on the array divides the processor into two groups of processors — those processors for which the condition is true and those for which the condition is false. Since processors can be individually told not to execute the current instruction, the processors for which the condition is true will only execute those instructions between the IF and the ELSE and those processors for which the condition is false will only execute those instructions between the ELSE and the THEN. Thus, through prudent use of conditional statements, the processors can be programmed to perform a range of different functions within the same general time span.

Parallel FORTH is implemented as simply and as straightforwardly as possible. A Uni-FORTH system[7] is implemented on the MCU. Parallel extensions have been added to the kernel under a new vocabulary called PARALLEL. Context switching has been simplified so that the FORTH word '{' switches to the parallel vocabulary and '}' switches back to the vocabulary that was in use before the switch to the parallel vocabulary. This allows the user to redefine serial words as analogous parallel words under a parallel context, thus making it easier for the FORTH user to remember the new parallel words. For example, '+' normally means to add two numbers that are on the data stack, but in the parallel context, '{ + }' means add two 128 × 128 arrays of numbers on the array stack, which is in the ARU memory.

Two new stacks have been added to parallel FORTH that are not in serial versions of FORTH. These two stacks are the array stack (A) and the mask stack (M). The mask stack is not normally used or seen by the FORTH programmer. It is used to facilitate nested conditional statements, such as 'IF...ELSE ...THEN' or 'BEGIN...UNTIL'. The array stack is extensively used by the FORTH programmer, since it is the parallel equivalent of the MCU's data stack. Most operations that can be performed on elements of the data stack have corresponding operations that can be performed in parallel on the array stack, such as +, *, DUP, DROP, and ROT. There are a few other operations that are peculiar to the array stack.

The following sections will discuss in more depth the parallel operations that have been implemented to extend FORTH into the realm of parallelism.

4.2.2. Vocabulary and Data Definition

In MPP Parallel Forth there is a vocabulary called PARALLEL. All new parallel words are in this vocabulary except PE control unit (PECU) primitive words and mask stack operations. As pointed out in the introduction, '{' and '}' are used to enter and exit the parallel vocabulary. The following is a definition that will manipulate the MCU data stack:

```
: MULTADD * + ;
```

While the next definition manipulates the ARU array stack:

```
: MULTADD { * + } ;
```

Parallel variables can be allocated in either the array or the staging memory. If a user wants to allocate a 128 × 128 array of 7-bit values named AR1 in the staging memory, the following is used:

```
7 STG VARIABLE AR1
```

If a user wants to allocate a 128 × 128 array of 11-bit values in the array memory named AR2, the following is used:

```
11 ARU VARIABLE AR2
```

The definition of parallel constants is similar to defining variables, except the user puts an array on top of the array stack and then executes the statement:

```
13 ARU CONSTANT CON1
```

to create a 128 × 128 array of 13-bit constants. Likewise, vectors and arrays of 128 × 128 arrays may be defined with VECTOR and MATRIX, respectively. A vector of 20 8-bit 128 × 128 arrays can be defined with the following statement:

```
20 8 ARU VECTOR VEC1
```

The word ALLOT will allocate variable space in either the stager or the array if it is preceded by the word STG or ARU, respectively. ALLOT is used by all the above-mentioned definition words.

4.2.3. Parallel I/O

Parallel files can be stored on the host in either matrix or image format. Each format allows for 8, 16, and 32-bit values. A matrix format file contains multiple arrays of 128 × 128 values. An image format file contains multiple images of 512 × 512 values. The following command,

```
CHANA IMAGE8OPEN WHAT.DAT
```

opens the file 'what.dat' as an image file of images with 8-bit values. LOAD is then used to read the matrix or image into a previously defined staging memory array. An image from an image file or 8-bit values should only be loaded into a VECTOR or MATRIX that has at least 8 × 16 or 128 bits allocated to it. The command to read a matrix into a stager array is:

```
V1 3 LOAD
```

This loads the third matrix of the current file into variable V1. To store an image into a file, the word STORE is used (i.e., V1 3 STORE).

4.2.4. Memory Operations

Memory operations are used to move data between the three MPP memories: the MCU, the ARU, and the staging memory. The word '@' fetches arrays from array variables in the stager and the ARU memory and puts them on the array stack. The word '!' stores an array from the array stack into an array variable in the stager or ARU memory. The word 'SCALAR' takes a value from the data stack in the MCU memory, broadcasts it to all PEs, and produces an array on top of the array stack that has the same value for all elements of the array. Also, when the context is the parallel vocabulary, any literals are compiled into constants that will be sent to the top of the array stack as a scalar value during execution. Operations such as GMAX, GMIN, and GOR can reduce an array of values into a scalar value that can be put onto the data stack.

4.2.5. Array Stack Operations

Most array manipulation occurs on the array stack. The array stack consists of a stack of descriptors in MCU memory and the actual bit plane stack in the ARU memory. The array stack is manipulated by operations very similar to those used on the data stack. These operations consist of words such as DUP, DROP, SWAP, OVER, ROT, PICK, and ROLL. In addition to the standard stack operations there are also operations that are peculiar to the array stack. They consist of the following words: -NDROP, NDROP, A@, >A, ZERO, EXTRACT, SLIDE, EXG, CROSS, and TOPOLOGY.

NDROP drops the top n elements of the array stack. -NDROP skips the first n1 elements of the stack and drops the next n2 elements of the array stack. 'A@' copies the descriptor off of the MCU array stack onto the data stack. A parallel array descriptor consists of two values: the address of the least significant bit plane of the array (LSB) and the number of bit planes in the array (LEN). '>A' creates an array of n bit planes on the array stack, where n is taken from the top of the data stack. ZERO is the same as '>A' except the bit planes are initialized to zero. EXTRACT extracts a field of bits from the second element of the array stack and inserts it into a field of the same size in the top element of the stack. SLIDE slides the top element of the array stack across the array of PEs. EXG exchanges data in the top elements of the array stack among PEs of the ARU. CROSS exchanges data from the top elements of the array stack with the second elements of the array in different PEs of the ARU. The TOPOLOGY operation changes the topology of the ARU.

4.2.6. Arithmetic, Logic, and Comparison Operations

All the operations in this section deal primarily with the elements on the top of the array stack. Basically they are analogous to the corresponding operations that operate on the top of the data stack. The difference is that operations on the array stack perform 16,384 operations at the same time instead of one at a time and values on the array stack can have variable numbers of bits instead of a fixed number such as 8, 16, or 32.

Normally operations on the data stack are either single or double precision. On the array stack, however, operations are classified as either fixed or variable precision. A fixed precision operation requires that both operands have the same length and that their result is the same length. A variable precision operation may operate on operands whose lengths are different. The result of such operations has a length that is dependent on both the specific operation and the length of the operands. All basic operations discussed here have a fixed precision operation. Only a few operations have both a fixed and a variable precision form of operation. These operations are +, −, *, /, MOD, and /MOD. Their variable precision forms are ~+, ~−, ~*, ~/, ~MOD, and ~/MOD.

The result of a ~+ or a ~− operation has a length equal to one plus the maximum of the two operands. The result of a ~* operation has a length equal to the sum of the length of the two operands. The length of result of a ~/ operation is the difference between the length of the dividend operand and the length of the divisor operand. Note that the length of the dividend must be larger than that of the divisor. The result of the ~MOD operation has a length equal to the length of the divisor operand. Since the result of the ~/MOD operation is the result of the ~/ operation followed by the MOD operation, the lengths of the results are the same as described for ~/ and ~MOD.

The fixed precision only operations are MAX, MIN, ABS, NEGATE, 1+, 1–, 2/, 2*, AND, OR, XOR, and NOT. Three special operations find the aggregate result and place it on the data stack. These global operations are global maximum (GMAX), global minimum (GMIN), and global or (GOR).

Comparison operations differ slightly from the other operations in this seciton in that they result in a value of length 1. These operations are <, =, >, 0<, 0=, and 0>.

4.2.7. Control Operations

Control operations cause certain portions of code to be executed on some data and not on others. Parallel control is quite different from serial control. In serial control, condition evaluation determines whether or not a certain piece of code will be executed. In parallel control, both the code corresponding to the true condition and the false condition may have to be executed. Some of the processors must be turned off during the execution of the code for the true condition, then turned on for the execution of the code for the false condition. This is accomplished with a mask bit. It is set to one in processors whose data satisfy the condition, and to zero in those whose data do not. Thus only those processors that satisfy the condition execute the code for the true condition. The mask bit is then complemented and only those processors that did not satisfy the condition will execute the code for the false condition. As with execution of serial conditions, parallel conditions can be nested. Therefore, there is a mask stack. Mask stack primitive operations are used to implement the operations in this section.

The basic conditional structure is the IF...ELSE...THEN statement. The IF word duplicates the top element on the mask stack, takes the least significant bit of the top element of the array stack, and ands it to the top element on the mask stack. The ELSE word complements the top element of the mask stack. And the THEN word drops the top element of the mask stack.

The parallel conditional loop structure is also somewhat unusual. It continues to execute as long as there is a processor that has not met the condition to terminate the loop. The two types of loops are the BEGIN...UNTIL and the BEGIN...WHILE...REPEAT. The BEGIN word duplicates the top element of the mask stack. The REPEAT word marks the end of the loop. The WHILE word ands the least significant bit of the top element of the array stack to the top element of the mask stack and terminates the loop if no processor has the top element of the mask stack equal to one. The UNTIL word is the same as WHILE except the least significant bit of the top element of the array stack is complemented before it is anded to the top element of the mask stack.

Note that only certain operations are maskable. Therefore, one should be aware that operations may execute when the processor was masked out because the operation was not maskable. Generally, only operations that do not change the number of elements on the array stack or the order of the elements on the array stack are maskable. Thus, most stack manipulation operations and two operand operations are not maskable. See the MPP Parallel FORTH Word Reference for more specific details.

4.2.8. PECU and Mask Stack Primitives

The PECU and mask stack primitives are not meant to be used by the general FORTH programmer. They are used by the primary parallel FORTH words to initiate actions to be performed in the ARU. If it is necessary to use them, they are described in the MPP Parallel FORTH word reference.

4.2.9. MPP Parallel Forth Word Reference
4.2.9.1. Context Changing Words

'PARALLEL' is the vocabulary that contains all the words that act on data in the MPP array unit.

{

This changes the context of word searches to the 'PARALLEL' vocabulary.

}

This returns the context to that specified prior to the change to the 'PARALLEL' vocabulary.

4.2.9.2. Arithmetic Words

+ (A: a1(n) a2(n) — A: a3 (n))

This adds a1 and a2, which are the same size and produces a3, which is that size.

- (A: a1(n) a2(n) — A: a3(n))

This subtracts a2 from a1, which are the same size and produces a3, which is that size.

* (A: a1(n) a2(n) — A: a3(n))

This multiplies a1 by a2, which are the same size and produces a3, which is that size.

/ (A: a1(n) a2(n) — A: a3(n))

This divides a1 by a2, which are the same size and produces a3, which is that size.

MOD (A: a1(n) a2(n) − A: a3(n))

This divides a1 by a2, which are the same size and produces the remainder a3, which is that size.

/MOD (A: a1(n) a2(n) − A: a3(n) a4(n))

This divides a1 by a2, which are the same size and produces a3, the remainder, and a4, the quotient, which are of that size.

MAX (A: a1(n) a2(n) − A: a3(n)) {maskable}

This finds the maximum of a1 and a2, which are the same size and produces a3, which is that size.

GMAX (A: a1(n) − S: m) {maskable}

This finds the global maximum of the a1 and places it on the data stack as a scalar value.

MIN (A: a1(n) a2(n) − A: a3(n)) {maskable}

This finds the minimum of a1 and a2, which are the same size and produces a3, which is that size.

GMIN (A: a1(n) − S: m) {maskable}

This finds the global minimum of the a1 and places it on the data stack as a scalar value.

ABS (A: a(n) − A: a(n)) {maskable}

This finds the absolute value of 'a' and replaces "a' on the stack.

NEGATE (A: a(n) − A: a(n)) {maskable}

This finds the 2's complement of the value of 'a' and replaces 'a' on the stack.

1+ (A: a(n) − A: a(n)) {maskable}

This increments the value of 'a' and places it back on the stack.

1- (A: a(n) — A: a(n)) {maskable}

This decrements the value of 'a' and places it back on the stack.

2/ (A: a(n) — A: a(n)) {maskable}

This shifts a(n) to the right.

2* (A: a(n) — A: a(n)) {maskable}

This shifts a(n) to the left.

~+ (A: a1(n) a2(m) — A: a3(max(n,m)+1))

This adds a1 to a2 and produces a result, a3, which has a size that is the maximum of the sizes of a1 and a2, plus 1.

~- (A: a1(n) a2(m) — A: a3(max(n,m)+1))

This subtracts a1 from a2 and produces a result, a3, which has a size that is the maximum of the sizes of a1 and a2, plus 1.

~* (A: a1(n) a2(m) — A: a3(n+m))

This multiplies a1 by a2 and produces a result, a3, which has a size that is the sum of the sizes of a1 and a2.

~/ (A: a1(n) a2(m) — A: a3(n-m))

This divides a1 by a2 and produces a result, a3, which has a size that is the difference of the sizes of a1 and a2.

~MOD (A: a1(n) a2(m) — A: a3(m))

This divides a1 by a2 and produces the remainder, a3, which has a size of a2.

~/MOD (A: a1(n) a2(m) — A: a3(n-m) a4(m))

This divides a1 by a2 and produces a remainder, a3, which has a size the same as a2 and a quotient, a4, which has a size that is the difference of the sizes of a1 and a2.

4.2.9.3. Logical Words

```
AND  (A: a1(n) a2(n)  —  A: a3(n)  )
```

    This ands a1 and a2, which are the same size and produces a3, which is that size.

```
OR  (A: a1(n) a2(n)  —  A: a3(n)  )
```

    This ors a1 and a2, which are the same size and produces a3, which is that size.

```
GOR  (A: a1(n)  —  S: m  )  {maskable}
```

    This finds the global 'or' of the a1 and places it on the data stack as a scalar value.

```
XOR  (A: a1(n) a2(n)  —  A: a3(n)  )
```

    This xors a1 and a2, which are the same size and produces a3, which is that size.

```
NOT  (A: a(n)  —  A: a(n)  )  {maskable}
```

    This finds the complement value of 'a' and places it back on the stack.

4.2.9.4. Comparison Words

```
<  (A: a1(n) a2(n)  —  A: a3(1)  )
```

    This determines if a1 is less than a2, and produces a bit plane that is 1 where it is true and 0 where it is false.

```
=  (A: a1(n) a2(n)  —  A: a3(1)  )
```

    This determines if a1 is equal to a2, and produces a bit plane that is 1 where it is true and 0 where it is false.

```
>  (A: a1(n) a2(n)  —  A: a3(1)  )
```

    This determines if a1 is greater than a2, and produces a bit plane that is 1 where it is true and 0 where it is false.

```
0<  (A: a1(n)  —  A: a2(1)  )
```

This determines if a1 is less than 0, and produces a bit plane that is 1 where it is true and 0 where it is false.

`0= (A: a1(n) — A: a2(1) )`

This determines if a1 is equal to 0, and produces a bit plane that is 1 where it is true and 0 where it is false.

`0> (A: a1(n) — A: a2(1) )`

This determines if a1 is greater than 0, and produces a bit plane that is 1 where it is true and 0 where it is false.

4.2.9.5. Stack Operation Words

`DUP (A: a(n) — a(n) a(n) )`

Duplicates the top element on the array stack.

`DROP (A: a(n) — )`

Drops the top element on the array stack.

`NDROP (S: n A: a(m). . .a(p) — A: a(m) a(q) )`

Drops the top n elements of the array stack.

`-NDROP (S: n1 n2 A: a1(p1). . .a2(p2). . .a3(p3) — A: a1(p1) a2(p2). . .a3(p3) )`

Skips the first n1 elements of the array stack and drops the next n2 elements of the array stack.

`SWAP (A: a1(n) a2(n) — A: a2(n) a1(n) )`

Swaps the top two elements on the array stack.

`OVER (A: a1(n) a2(n) — A: a1(n) a2(n) a1(n) )`

Copies the second element on the array stack to the top of the stack.

`ROT (A: a1(n) a2(n) a3(n) — A: a2(n) a3(n) a1(n) )`

Moves a1 to the top of the array stack.

PICK (S: m A: a1(n). . .a2(n) — A: a1(n). . .a2(n) a1(n))

Copies the mth element of the stack to the top of the stack. (1 PICK is the same as OVER.)

ROLL (S: m A: a1(n). . .a2(n) a3(n) — A:. . .a2(n) a3(n) a1(n))

Moves the mth element of the stack to the top of the stack. (3 ROLL is the same as ROT.)

DEPTH (— S: n)

Returns the number of elements on the array stack.

A@ (A: a(n1) — S: n2 n1)

Copies the first descriptor on the array stack onto the data stack.

>A (S: n — A: a(n))

Creates an element on top of the array stack that has n bit planes.

ZERO (S: n — A: a(n))

Creates an element of size n that has a value of zero onto the top of the array stack.

EXTRACT (S: m1 m2 n A: a1(n1) a2(n2) — A: a1(n1) a2(n2))

Extracts a field of n bits of a1(n1) starting at m1 and places it in a2(n2) starting at m2.

SLIDE (S: n d A: a(p) — A: a(p))

Slides 'a' n PEs in the direction designated by d. East if d = 0, west if d = 1, south if d = 2, and north if d = 3.

EXG (S: m1 m2 n d A: a(n) — A: a(n))

Exchanges elements of 'a' n PEs apart in the direction designated by d. East/west if d = 0, south/north if d = 2. The addresses of mask bit planes are m1 and m2. The mask m1 determines which PEs accept data during the east or south portion of the move and m2 determines which PEs accept data after the west or north portion of the move.

```
CROSS (S: m1 m2 n d A: a1(n) a2(n)  -  A: a1(n)
   a2(n) )
```

Exchanges elements of a1 with a2 n PEs apart in the direction designated by d. East/west if d=0, south/north if d=2. Elements of a1 move to the east or south and elements of a2 move to the west or north. The addresses of the mask bit planes are m1 and m2. The mask m1 determines which PEs accept data during the east or south portion of the move and m2 determines which PEs accept data after the west or north portion of the move.

```
TOPOLOGY (S: n  -  )
```

The number n designates the topology of the array when an EXG, SLIDE, or CROSS is performed.

Topology	North/South Connection	East/West Connection
0	None	None
1	None	Cylinder
2	Cylinder	None
3	Cylinder	Cylinder
4	Open-spiral	None
5	Open-spiral	Cylinder
6	Closed-spiral	None
7	Closed-spiral	Cylinder

4.2.9.6. Memory Operation Words

```
@ (S: m  -  A: a(n) )  {maskable}
```

Moves an array variable described by a descriptor at address m onto the array stack.

```
! (S: n A: a(n)  -  )  {maskable}
```

Moves an array from the array stack into an array variable described by a descriptor at address m.

```
SCALAR (S: <scalar value>  -  A: a(n) )  {maskable}
```

Broadcasts a scalar value into array 'a' of all PEs.

LITERAL

Compiles a constant into a word that will be placed onto the array stack during execution, or will immediately place it on the stack during interpretation.

LIT

This is the execution time routine used to place a constant, compiled into the code, onto the array stack.

DESC (S: n — S: n2 n1)

Fetches the descriptor at address n and places it on the data stack. The address of the least significant bit plane (LSB) of the variable is n2 and n1 is the size of the variable.

4.2.9.7. Control Words

IF (A: a(n) M: m1 — M: m1 M: m2)

Creates a new layer on the mask stack that is the result of anding the least significant bit plane of 'a' and m1.

ELSE (M: m — M: m)

Complements the value of the top element of the mask stack.

THEN (M: m —)

Drops the top element of the mask stack.

BEGIN (M: m1 — M: m1 m1)

Duplicates the top element of the mask stack.

WHILE (A: a(n) M: m1 — M: m1)

or

(A: a(n) M: m1 —)

Ands the least significant element of 'a' and m1. If no element of m1 is equal to 1, the loop is terminated.

REPEAT

    Marks the end of a BEGIN...WHILE...REPEAT loop.

UNTIL (A: a(n) M: m1 — M: m1 M: m2)

or

    (A: a(n) M: m1 —)

    Ands the complement of the least significant element of 'a' and m1. If no element of m1 is equal to 1, the loop is terminated.

4.2.9.8. I/O Words

MATRIX8
MATRIX16
MATRIX32

    File types for files that contain 128 × 128 arrays of 8-, 16-, or 32-bit values.

IMAGE8
IMAGE16
IMAGE32

    File types for files that contain 512 × 512 images of 8-, 16-, or 32-bit values.

LOAD (S: n2 n1 —)

    Loads an array n1 or image n1 from the currently opened file into the designated bit plane described by the descriptor at address n2.

STORE (S: n2 n1 —)

    Stores the designated bit planes described by the descriptor at address n2 into an array n1 or image n1 of the currently opened file.

4.2.9.9. Defining Words

VARIABLE (S: n f —)

    Allocates an n bit plane variable array in either the stager or the array.

CONSTANT (S: n f A: a(n) —)

Allocates an n bit plane constant array in either the stager or the array and loads it with the top element of the array stack.

```
VECTOR (S: m n f — )
```

Allocates a vector of m n bit values in either the stager or the array.

```
MATRIX (S: m1 m2 n f — )
```

Allocates an m1 × m2 matrix of n bit values in either the stager or the array.

4.2.9.10. Compiler Words

```
ALLOT (n f — )
```

Allocates n bit planes in either the array (ARU) or the stager (STG).

```
ARU
```

Indicates that the desired variable will be allocated in the array.

```
STG
```

Indicates that the desired variable will be allocated in the stager.

4.2.9.11. PECU Primitive Words

```
PECU (S: <PECU address> — )
```

The word 'PECU' takes an address off the MCU data stack and places it in register 'SPE', which starts the PECU at that address.

```
S>C (S: <64 bit scalar> <LSB of scaler> - - - )
```

The word 'S>C' loads the LSB of a scalar into PE0. The 64-bit scalar value will be loaded into the common register from the data stack.

```
C>S ( — S: <64 bit scalar> <LSB of scalar> )
```

The word 'C>S' stores the 64-bit return register A value on the data stack followed by the value 64.

```
A>PE2  ( A: <descriptor> — )
A>PE4  ( A: <descriptor> — )
A>PE6  ( A: <descriptor> — )
```

Takes two descriptors from the array stack and places them into registers PE2-PE3, PE4-PE5, or PE6-PE7, respectively. Each descriptor consists of a 1-bit LSB and a 16-bit size.

```
S>PE2  ( S: n1 n2 -)
S>PE4  ( S: n1 n2 -)
S>PE6  ( S: n1 n2 -)
```

Takes two words from the data stack and places them into registers PE2-PE3, PE4-PE5, or PE6-PE7, respectively.

4.2.9.12. Mask Stack Operations

```
A>M  ( A: a(n) M: m1 - M: m1 )
```

Ands the least significant bit plane of a(n) to m1. The mask stack pointer is maintained in PE1.

```
M>A  ( M: m1 - M: m1 A: a (1) )
```

Copies the top bit plane of the mask stack onto the top of the array stack.

```
MDROP (M: m1 - )
```

Drops a mask from the mask stack.

```
MDUP (M: m1 - M: m1 m1 )
```

Duplicates top element on mask stack.

4.3. PARALLEL PASCAL DESIGN

4.3.1. Motivation

Parallel Pascal is a high-level matrix language designed for the user of a parallel matrix processor. The decision to design a new language reflects a degree of dissatisfaction with facilities available in existing languages. For some reason (or combination of reasons) no single existing language was judged to be suitable for parallel matrix processors.

To judge the suitability of a language, it is necessary to consider the functions it is to serve. There are three goals of a programming language: it is a design tool, a vehicle for human communication, and a vehicle for instructing a computer. The language which is chosen for a particular application should be one which satisfies all of these criteria.

Programming can be considered to be the act of mapping a problem into machine code. This mapping occurs at two levels. The original problem is translated by a human into a program in some language, and then this program is translated by a compiler (or assembler) into machine code. Each translation involves the loss of information — the program contains less information than the problem and the machine code contains less information than the program. Unfortunately, these relationships are often dual in nature — a language which facilitates programming by humans (and communication among them) will often be more difficult to compile into machine code.

In recent years, a great deal of emphasis has been placed upon the use of a set of techniques collectively referred to as "structured programming". These techniques encourage careful, regular, modular designs, thereby facilitating the construction of programs which are highly reliable and maintainable.

Taking the above factors into consideration, a "good" language is one which facilitates communication among humans and between humans and machines, one which permits expression of a problem without undue loss of information, one which can be compiled into reasonable efficient machine code, and one which encourages structured programming techniques.

Having determined what factors are necessary for a "good" language, the development of Parallel Pascal can now be considered.

4.3.2. Parallel Pascal Specification

Since none of the available languages were entirely suitable for implementation on a parallel matrix processor, the design of a new language was undertaken. The design goals of this new language (which eventually became Parallel Pascal) for a parallel matrix processor were as follows:

- The language should be efficiently implementable. A principal reason for using a parallel processor is to obtain the maximum possible execution speed; a language whose implementation is costly significantly diminishes the advantage of parallel processors relative to more conventional (and familiar) sequential processors.
- The language must permit the direct specification of parallelism. This relates strongly to the previous objective — the direct specification of parallelism produces more efficient programs than the extraction of inherent parallelism by a compiler.
- The language must be easy to learn and use. Such a language facilitates communication among humans and between programmers and computers. A language which is difficult will be avoided by its users whenever possible.
- The language should not require the user to have an intimate understanding of the hardware upon which it is implemented.

Because Pascal is (by design) efficiently implementable and easy to learn and use, it was chosen as the basis for the new parallel matrix language. The resulting language was therefore named "Parallel Pascal". The following criteria were used in the specification of Parallel Pascal:

- Parallel Pascal is an extension to standard Pascal. As such, it should be fully upward-compatible; that is, any Pascal program should also be a valid Parallel Pascal program.
- Parallel Pascal extensions to Pascal should be consistent with the design philosophy of Pascal. The design should be orthogonal and the new features should not detract from the careful program construction permitted by Pascal.

When deciding upon extensions to a language, it is necessary to consider the applications for which the language will be used. Someone once said that a general-purpose system (or processor, or language) is one which does many things but which does none of them well. To avoid the trap of implementing everything that anyone would possibly want, new features were considered in light of the desired applications area — image processing and dense matrix numerical algorithms (e.g., partial differential equations).

The following sections describe the Parallel Pascal extensions to standard Pascal. The development of each extension will be discussed.

In order to satisfy the design objective that parallelism be directly expressible, a suitable data structure must be chosen. Since this specification should be as compatible as possible with standard Pascal, it is instructive to first consider the data structuring provided by Pascal. Indeed, Pascal's flexible data type facility is one of its most significant features.

The most basic Pascal data types are the predefined primitive types "integer", "real", "char", and "Boolean", and the *scalar types*. A user-defined scalar type associates with the type name a set of distinct identifiers. This permits the programmer to use mnemonic names rather than arbitrary integer constants, which in turn improves program readability and facilitates compile-time error checking.

The range of values which may be assigned to a scalar may be restricted by defining a *subrange type*. A subrange type definition comprises a base type (either a user-defined scalar type or a primitive type other than "real") and a range of legal values. Hence, if type "x" is defined by

type x = 1 . . 5;

then a variable of type "x" may legally take on only the values 1, 2, 3, 4, or 5. Like simple scalar types, subrange types aid in program documentation and compile-time error checking. Also, subrange types provide information to the compiler about the amount of storage required for a variable of that type; in the above example only 3 bits of storage need be allocated to specify any legal value in type "x". Providing there is hardware support, the compiler may choose to adjust the space allocated to a subrange type depending upon the available hardware representations.

A *power set* may be defined for a scalar or subrange type. A power set in Pascal is conceptually the same as a set in mathematics; it is a collection of elements, the composition of which changes at runtime. The base type (the subrange or scalar type over which it is defined) determines the items which may belong to the set.

There are two data structuring facilities in Pascal, the *array* and the *record*. An array is a homogenous ordered set of items. The elements of an array may be of any type: scalar, subrange, set, array, or record type. Associated with each array component is an "index"; the range of this index may be specified by a scalar or subrange type. A record is a non-homogenous collection of items. The components of a record may be of any type, and may occur in any order. There is also a provision for the overlapping use of storage by allocating elements which are mutually exclusive into the same storage area; this is achieved through the use of a *variant record*.

Pascal also provides *pointer types*. These are defined by the compiler and initialized at run-time by the user-controlled dynamic storage allocation routine ("new"). They contain addresses and may be copied and compared (for equality), thus permitting the construction of data structures such as linked lists, trees, etc.

The target architecture for Parallel Pascal — a parallel matrix processor — consists of a set of identical execution units which perform the same operation at the same time. The hardware thus appears as an ordered collection of homogenous processors. This organization maps naturally into the array structure provided by Pascal; hence, the array was chosen as the vehicle for the expression of parallelism in Parallel Pascal.

Often a parallel matrix processor will be closely coupled to a more conventional processor. For example, the Massively Parallel Processor contains a main control unit which is a conventional 16-bit minicomputer. In addition, the MPP is attached to a host machine, a VAX-11/780. In such an environment, it may be more efficient to perform scalar operations on one processor and matrix operations on another. This in turn is reflected in the assignment of storage to variables used in the program — those variables which are used in a scalar fashion may be physically located in a different memory than those used in an array fashion. Parallel Pascal provides for this situation by permitting an array to be declared *parallel:*

xxx: *parallel array* [1 . . 5] *of* integer;

The *parallel* keyword is a means by which the programmer can advise the compiler that the array ("xxx" in this case) will be heavily used in a parallel fashion. Some compiler implementations may choose to ignore this (e.g., if there is only one type of memory). The concept that the user is advising the compiler about the implementation is similar to the *register* keyword in the language C.

It is important to note at this point that, aside from the possible difference in physical storage, arrays declared as *parallel* are syntactically and semantically equivalent to "ordinary" arrays in Parallel Pascal.

Having chosen the array as the vehicle for expressing parallelism, it is necessary to specify the manner in which that parallelism is to be expressed. The logical starting place is the building block of any computational language — the assignment statement. It must be possible to specify the evaluation of array quantities in a simple, direct form.

Standard Pascal provides an array assignment statement; if "a" and "b" are the same type then the statement

$$a := b;$$

specifies that each element of "b" is to be assigned to the corresponding element of "a". A natural extension of this concept is to allow arrays of the same type of participate in arithmetic operations, for example, given

    a,b: *array* [1 . . 5] *of* integer;
    i: integer;

the statement

$$a := a + b;$$

would achieve the same result as

    for i := 1 *to* 5 *do*
      a[i] := a[i] + b[i];

While expressions involving identical arrays are useful, they are limited in the range of problems to which they can be applied. Several deficiencies are evident. First, it is necessary to be able to select a portion of an array (for instance, a row or a column) rather than the entire array; hence, necessary to allow arrays of different types (but identical shapes) to be combined in an arithmetic expression.

A number of schemes have been proposed for array indexing, as described above in the discussion of other parallel languages. The array indexing facilities which are provided must be powerful enough to solve useful problems, while remaining simple enough to efficiently implement. The choice of indexing mechanisms should therefore begin with the simplest and proceed toward the more complex.

In standard Pascal, each array index may be specified by a scalar constant or expression. This is the simplest form (and least parallel) of indexing permitted in Parallel Pascal. When an array is indexed by a scalar its rank is (conceptually) reduced by one. When a one-dimensional array is indexed by a scalar the (logical) type of the result is a pointer to a scalar.

It was stated above that standard Pascal permits arrays participating in an assignment statement to be unsubscripted. This is actually a special case of a more general feature in standard Pascal — it is possible to elide (omit) the rightmost indices in an array assignment, provided the resulting expressions are of the same type. For example, given the definition

var a,b: *array* [1..5,1..10] *of* integer;

both of the following are legal assignments in standard Pascal:

a := b;
a[i] := b[1];

The first statement assigns to each element of "a" the value of the corresponding element of "b". The second statement performs this action only on the first row of "a" and "b". It has the same effect as:

for i ;= 1 *to* 10 *do*
    a[1,i] := b[1,i];

Parallel Pascal extends this to permit the omission of any index; hence, in Parallel Pascal the statement

a[,1] := b[,1];

assigns to the first column of "a" the values contained in the first column of "b". The use of a scalar index effectively reduces the rank (number of dimensions) of an array by one; hence "a[,1]" is considered to be a vector.

The ability to select a row or column, as opposed to an entire array, is useful; however, it is often desirable to further restrict the number of elements which participate in an operation. A common requirement is the selection of a subset of a row, column, or both. Standard Pascal provides no symbolism to directly express this concept; hence, it is necessary to introduce a new construct to the language.

The simplest subset of a set of array indices is a consecutive range. For example, given an array with five elements, one may wish to access elements 2, 3, and 4. Standard Pascal permits the use of a *subrange* in type definitions to specify a range of values which a variable may possess. Parallel Pascal entends this concept by defining a *subrange constant*. The Pascal *const* statement may be used to define an identifier as a subrange constant:

const rangeconst = low..high;

A subrange constant may be added to a scalar expression and used as an array index. The most desirable syntax for this would be

arr[scalarexpression + low..high]
(or)
arr[scalarexpression + rangeconst]

where "rangeconst" is an identifier defined as a subrange constant. Unfortunately, due to the recursive-descent implementation of most Pascal compilers, this syntax introduces complications when a compiler parses the program. In deference to the implementation the symbol "@" is used to represent the addition of a subrange constant:

arr[scalarindex @ low..high]
(or)
arr[scalarexpression @ rangeconst]

Given the following definitions:

const
    rr = 1..5;
    cc = 2..4;

var
    a,b: *array* [1..10,1..10] *of* integer;
    i,j: integer;

the following two code sequences achieve the same result:

(* with subrange indexing: *)
a[0@rr,0@cc] := b[1@rr, 3@cc];

(* without subrange indexing: *)
for i := 1 *to* 5 *do*
  for j := 2 *to* 4 *do*
    aj[i,j] := b[1+i, 3+j];

Subrange indexing does not alter the rank of an array. Thus, while "a[1,]" is a 10-element vector, "a[0@1..1,]" is a 1×10 matrix.

Other languages provide additional array indexing facilities, such as indexing by a logical set or a vector. These indexing notations are powerful, but on a processor with a limited interconnection network (e.g., a mesh network) their implementation can be very expensive. For this reason, set and vector indexing were excluded from the specification of Parallel Pascal.

The ability to elide indices and use subrange constants for array indexing brings with it an associated problem: what combinations of array expressions are legal? In standard Pascal, the operands of an arithmetic expression must be *type compatible*. A subrange is type compatible with its base type, and integers are type compatible with reals. (An integer may be converted to a real number with no loss of information. Thus, if an integer expression is used where a real expression is required, e.g., on the right-hand side of an assignment statement or as an argument to a function or procedure, Pascal automatically converts the integer expression into a real expression. Since it is not true that any real may be converted to an integer with no loss of information, Pascal prohibits the opposite case — using a real expression where an integer expression is required.)

Parallel Pascal preserves the Pascal concept of type compatibility. Because of the array indexing, scalar type compatibility alone is insufficient to determine the conformability of array expressions. It is necessary to also consider the rank (number of dimensions), size, and indices of each array expression. The specification was designed to meet the following goals (note that the term "array" below may refer to an entire array or a subset created according to the indexing facilities described above):

- Arrays of the same type should be compatible.
- A scalar of the same base type as an array should be conformable to that array. This implies that the scalar is effectively replicated into an array of identical type.
- Arrays which differ in rank (number of dimensions) or size are not compatible. Recall that indexing with a scalar "compresses" a dimension out of the array.
- Arrays which have the same index ranges, but whose element types are different are compatible if the element types are compatible.
- Arrays which are the same size and shape should either be compatible, or it should be possible to make them compatible with little effort.

The first requirement above preserves the standard Pascal array assignment statement:

a := b;

where "a" and "b" are of identical types. The second requirement allows the use of scalars with array expressions; e.g., given that "a" and "b" are of the same type, and "c" is the same type as the elements of "a" and "b", then the following statement:

a := b + c;

adds "c" to each element of "b" and stores the result in the corresponding element of "a".

Arrays are required to be the same shape and size to prevent situations such as:

var
    a: *array* [1..5] *of* integer;
    b: *array* [1..6] *of* integer;
b := a;

Allowing arrays which are the same type except for the elements, which are compatible, allows common constructions such as

var
    a: *array* [1..5] *of* integer;
    b: *array* [1..5] *of* real;
b := a;

The biggest difficulty which arises from the generalized indexing mechanisms is the compatibility of arrays whose sizes and shapes are identical, but whose index ranges are not:

var
    a: *array* [1..5] *of* integer;
    b: *array* [2..6] *of* integer;
a := b;

This problem becomes more severe when the control-flow facilities of Parallel Pascal (which are discussed later in this chapter) are used. In order to prevent ambiguity in these cases, two arrays with nonidentical index ranges are compatible only if the elements of at least one are specified explicitly by subrange indexing. Hence, given the "a" and "b" defined above, the following assignments are all legal:

a := b[0@2..6];
a[0@1..5] := b;
a[0@1..5] := b[0@2..6];

The type compatibility rules described above extend in the intuitive way to multiple dimensions. If "c" and "d" are defined by

var
    c: *array* [1..3,1..5] *of* integer;f
    d: *array* [1..8,1..7] *of* integer;

then the following assignments are all legal:

c := d[2@2..4,0@2..6];
c[0@1..3,] := d[0@6..8,0@2..6];
c[1,] := a;

(Note that in the last example the ranks of "c[1,]" and "a" are the same because the scalar indexing reduced the rank of "c" by one.

Pascal provides a number of *standard functions* and *standard procedures*. These perform various services, including type conversion (e.g., *trunc, ord*) arithmetic functions (e.g., *sin, sqrt*), and input/output procedures.

Many of the standard functions perform simple transformations, for instance:

var x, y: real;
x := sqrt(y);

It is quite natural to think of these functions as extensions to the set of operators provided by Pascal (e.g., "+", "−"). Since Parallel Pascal allows the operators to act upon arrays as aggregates, it is only natural to extend this feature to the standard functions. Thus,

var
    x, y: *array* [1..16] *of* real;
    i: integer;
x := sqrt(y);

is effectively the same as

for i := 1 *to* 16 *do*
    x[i] := sqrt(y[i]);

These standard functions are, in a sense, "generic": they may be used with arrays of any shape. The value returned by the function has the same index ranges as its argument. Since these functions operate independently upon each array element, they are called *elemental functions*.

While the elemental functions are useful, the effective use of Parallel Pascal requires the use of functions which alter the structure of arrays in a more complex fashion. These functions are referred to as *transformational functions*.

The first type of array restructuring which Parallel Pascal provides is the reordering of array elements. Certain image processing algorithms (e.g., convolution) require the capability to move data within an array. This movement may take two forms: a "shift", in which data is shifted off the edge of the array and zeros are brought in from the other end, or a "rotate", in which data is moved within an array such that data shifted off one end will reappear at the other. Parallel Pascal provides both these functions:

shift(array, i_1, i_2, i_3, ...)
rotate(array, i_1, i_2, i_3, ...)

where *array* is the array name and i_n is the magnitude of the shift along dimension n. (Row major order is used.)

In addition to shifting (rotating) data, it is sometimes necessary to transpose two dimensions of an array. This is performed by the *trans* function:

trans(array, dim_1, dim_2)

This effectively swaps two index ranges. For instance, given the definition:

var
    x: *array* [0..7, 3..4, 6..10] *of* integer;
    y: *array* [6..10, 3..4, 0..7] *of* integer;
    i,j,k: integer;

then the statement

    y := trans(x, 1, 3);

is equivalent to

for i := 0 *to* 7 *do*
    for j := 3 *to* 4 *do*
        for k := 6 *to* 10 *do*
            y[k,j,i] := x[i,j,k];

The second major type of array manipulation is the alteration of the number of dimensions of an array. Some array operations may require that a array with N dimensions be combined with a array with N + 1 dimensions. (One example is the computation of the matrix product $\overline{Ax}$ where $\overline{A}$ is an n × m matrix and $\overline{x}$ is an m-element vector. In this case, it would be desirable to multiply all rows of $\overline{A}$ by $\overline{x}$ simultaneously, or, equivalently, to perform an element-by-element multiplication of $\overline{A}$ and an n × m matrix $\overline{B}$, each of whose rows are the vector $\overline{x}$.) The *expand* function can be used to expand an array along a new dimension.

    expand(array, dim, newidx)

Array is either a scalar or an array. Let N be the number of dimensions of *array* (zero if *array* is a scalar). *Dim* must be an integer constant in the range 1 to N + 1. *Newidx* is a subrange or the name of a subrange type (note: it is *not* a subrange constant). The array is replicated along a new dimension of type *newidx* which is inserted before dimension *dim*. For example, given the definition:

> var
>   x: *array* [0..7,8..15] *of* integer;

the result of

> expand(x,2,5..7)

is a matrix with dimensions [0..7,5..7,8..15] in which the value of [i,j,k] is the same for all 5<j<7.

The last type of array operation which is frequently required is the *reduction* of an array — applying an operator over a set of dimensions. For instance, one might wish to accumulate the sum along all of the rows of an array. Reduction functions have the general form:

> func(array, dim_1, dim_2, dim_3,...)

where *array* is the array to be reduced and each dim_i is a constant expression specifying a dimension along which the reduction is to be performed. (These are required to be constants so that the compiler may determine the shape of the result.) The following reduction functions are provided:

sum	arithmetic sum
prod	arithmetic product
all	Boolean AND
any	Boolean OR
min	arithmetic minimum
max	arithmetic maximum

Standard Pascal provides several mechanisms for controlling the flow of execution. The most basic (and often overlooked) mechanism is sequencing — assignment statements are executed one at a time, in the order they appear. At a higher level, the flow of control may be altered by one of the following mechanisms:

Procedures—A Pascal program consists of a set of procedures and functions (called "subroutines" here for convenience). A *subroutine call* (a *procedure call* or *function call*) diverts the flow of control to one of these subroutines. When execution of the subroutine is complete, control returns to the statement following the subroutine call.

Repetition—A statement (or a group of statements) may be executed several consecutive times by using the Pascal *while, repeat-until,* or *for* constructs. For the *while* construct a Boolean expression is evaluated before each iteration of the controlled statement(s); as long as this controlling expression is *true* the repetition continues. The *repeat-until* construct performs the test after each iteration rather than before it; when the termination condition is satisfied (the Boolean expression is *true)* the iteration stops. The *for* statement uses an index variable; this variable is assigned an initial value and successively incremented or decremented until it reaches a final value. The controlled statement (or statements) is executed for each value of the index.

Conditional—A statement (or group of statements) may be conditionally executed by placing it in the body of an *if* statement. If the controlling expression evaluates to *true* the statements are executed; otherwise, they are skipped. Optionally, an *else* keyword may be specified, followed by a second statement (or block of statements); this statement is executed if the controlling expression is *false*.

Selection—One of a set of statements (where each "statement" may actually be a block of statements) may be executed according to the value of a controlling expression. This is the *case* statement in Pascal.

"Goto"—The flow of control may be directed to any defined statement label by use of the *goto* statement in Pascal. The ability to transfer control to any label within the program has been criticized as an impediment to good program design. It was included in Pascal because of the lack of a general agreement as to what should replace it and because it is occasionally useful for breaking out of deeply nested code structures.

All of the standard Pascal control flow constructs are present in Parallel Pascal. In order to effectively deal with arrays as aggregate entities, it is necessary to extend these constructs to deal with array operations. This extension must be carefully considered to avoid adding unnecessary complexity to the semantics of the language.

The most basic form of program construction — sequencing — is essentially the same for an SIMD-class processor (such as a parallel matrix processor) as it is for an SISD-class (conventional scalar) processor. (This concept changes in an MIMD-class processor, since in that environment many isntruction streams may be simultaneously processed.) Similarly, the concept of a procedure or funciton call, and the meaning of a *goto* are unchanged. This suggests that the extensions to Pascal will be based upon its control statements: *if, case, while, repeat-until,* and *for*.

The *if* statement causes the execution of one (and possibly two) statement(s) according to the value of a controlling expression. The execution is "all-or-nothing" — either the controlled statement is executed or it is not. This is well suited to a scalar machine, but it presents problems in Parallel Pascal. It is sometimes necessary to conditionally perform some actions using only a subset of an array. Parallel Pascal provides the *where* statement to address this need.

The *where* statement has two forms:

> *where* arrayexpression *do*
>   statement
>
> *where* arrayexpression *do*
>   statement
> *otherwise*
>   statement

where "arrayexpression" is a Boolean-array-valued expression and "statement" is a Parallel Pascal statement. Some restrictions apply to the controlled "statement":

- A *goto* out of the *where* or between the two controlled statements in a *where* is forbidden. (These restrictions are imposed to facilitate the implementation of *where* statements with a conditional stack; uncontrolled use of the *goto* complicates such an implementation.)
- Array variables which appear on the left-hand side of an assignment statement must be type-compatible with the controlling array expression.

The execution of a *where* is defined as follows. First, the controlling expression is evaluated to obtain a Boolean array. Next, the first controlled statement (referred to later as the *where clause*) is evaluated. Array assignments are masked according to the Boolean array computed above. Finally, if there is a second controlled statement (an *otherwise clause*), it is evaluated. Array assignments within the "otherwise clause" are masked by the inverse of the Boolean array computed in the first step.

Where statements may be nested, provided that all of the controlling array expressions are type compatible. The effect of a *where* statement is local to the procedure or function in which it appears; that is, it does not affect the execution of any procedures or functions called from within a "where clause" or "otherwise clause".

The *where* statement provides Parallel Pascal with *conditional assignment* (or *masked assignment*). That is, all array expressions within both the "where clause" and "otherwise clause" are fully evaluated, but the results are only assigned to a subset of the array appearing on the left-hand side of an assignment statement. This allows the specification of many common problems; for instance: "Given two arrays A and B (of the same type), determine the maximum of A and B element-by-element and store the result in A." This is achieved by the statement:

where a < b *do*
    a := b;

An alternative to conditional assignment is *conditional evaluation*. A conditional evaluation scheme would cause the evaluation of all array expressions to be masked (element by element) by the controlling expression. This could be used to catch exceptional conditions; for instance, divide by zero:

where a <> 0 *do*
    a := 1/a;

While conditional evaluation provides some additional capabilities that conditional assignment does not, it introduces semantic difficulties. One problem which conditional evaluation raises is the treatment of function (or procedure) calls from within the *where* statement. If an array expression is passed to a function, what values are passed for those elements for which the controlling expression is false? Similar problems arise with the use of standard functions which alter the shape of arrays — at what point is the masking applied (for at that point the expression must be type compatible with the controlling expression)? The presence of these problems with conditional evaluation and the relative semantic simplicity of conditional assignment led to the latter's choice for the *where* construct.

The design of the *where* statement as a parallel extension of the *if* statement led to the consideration of a parallel extension of the Pascal *case* statement. The *case* statement selects one statement (or block of statements) from several depending upon the value of a controlling expression. It differs from the *if* statement in that the controlling expression is multi-valued rather than Boolean; hence, a very large number of alternatives may be selected. It was felt that a parallel version of the *case* statement would be used infrequently; in the interest of keeping the size of the language to a minimum it was therefore omitted from Parallel Pascal. If necessary, the effect of a parallel *case* statement can be achieved through the use of a series of *where* statements (in the same fashion as a standard Pascal *case* statement can be implemented by a series of *if* statements).

The only remaining control constructs to be considered are the loop structures *while, repeat-until,* and *for*. The loop is one of the biggest sources of error for programmers; therefore, adding complexity to the looping mechanisms seemed unwise. It is unclear how a *for* loop should be extended. Further, a combination of a standard Pascal *while* or *repeat-until* loop statement (perhaps using a reduction function such as "any" or "all" to use conditionals based upon entire arrays) and a *where* statement can express all of the operations that any new loop construct of moderate complexity could express.

Pascal provides a fairly minimal set of input and output procedures. Each file consists of a uniform sequence of objects of a fixed type. The file is accessed by means of a "buffer variable". Syntactically, file buffer variables are used in the same fashion as pointers. To perform output, the data is placed into the file buffer and the "put" procedure is called. To perform input, the file variable is read, after which the "get" procedure is called to advance to the next item in the file. The "eof" function may be used to determine whether a file is positioned at the end. The "reset" procedure repositions a file at the beginning and makes it available for reading, while the "rewrite" procedure repositions a file at the beginning after truncating it, and makes it available for writing.

In addition to the "get" and "put" procedures, the "read" and "write" procedures may be used. Given the file "f" and variable "x" (both having the same type) the following equivalences hold:

$$\text{read}(f,x) \equiv x := f\uparrow; \text{get}(f)$$
$$\text{write}(f,x) \equiv f\uparrow := x; \text{put}(f)$$

Files whose elements are of type "char" (i.e., those of type "text") are treated specially. The procedures "read" and "write" may be used to transfer numeric data to or from a text file — the appropriate conversion is performed. In addition, the procedures "readln" and "writeln", and the function "eoln" are provided for intelligent handling of line-formatted input.

The use of text files for mass information input and output on a parallel processor was considered highly unlikely. Therefore, it was decided that Parallel Pascal needed no additional provisions for dealing with text files beyond those provided by standard Pascal.

On the other hand, it was apparent that "binary format" input and output would be heavily used. In particular, the limited main memory of a matrix processor implies that a great deal of data movement will be performed during the execution of a program. This subject falls in a "gray area" between the specification of the language and its implementation, for the manner in which the main memory of the parallel processor is managed directly affects the type of input and output required. For these reasons, it was decided to retain standard Pascal input and output without extensions for the definition of Parallel Pascal. The facilities which are required for memory management can best be determined after a period of use. Additional standard functions (which can be added to the language without significant trauma) could be added at a later time if a definite need arose. Another less desirable possibility would be the inclusion of some standard procedures on a site-dependent basis. This is in fact likely in other areas of Parallel Pascal, e.g., the implementaiton of interconnection functions which are more complicated than the simple mesh network.

4.4. ISING SPIN EXCHANGE SIMULATION

4.4.1. Introduction

Ising spin simulations occur in many areas of physics and have attracted the attention of researchers since the earliest days of electronic computation.[8] They provide a very good example of how, from the point of view of algorithm design, two apparently disparate problem types can be attacked by almost the same techniques. Some classes of Ising spin calculations can require speeds far in excess of anything currently available.

The basic model is easily described. The spin variable $\sigma(i)$ is specified at the nodes of a uniform grid in two (or three) dimensions. At each grid site the spin can take on only the values +1 and −1. Spins are related to one another via the energy expression given by the Hamiltonian

$$H = -\sum_{i,j} \sigma(i)\sigma(j) \qquad (3)$$

where $\{i,j\}$ ranges over all nearest neighbor pairs of sites. In an $\ell \times \ell$ square lattice, at site i we have the local energy expression.

$$E(i) = -\sigma(i) \times (\sigma(i+1) + \sigma(i-1)$$
$$+ \sigma(i+\ell) + \sigma(i-\ell)) \qquad (4)$$

In most cases periodic boundary conditions are imposed, so that i + 1 and i + ℓ are to be determined mod ℓ.

One wishes to compute various averages with respect to the probability P(C) for a configuration of spins, C, to occur. The "classical" algorithm for using Monte Carlo methods to sample configuration space is due to Metropolis et al.[8] (called the M(RT)² algorithm for short). It consists of a series of moves through configuration space, making use of the fact that for each C, P(C) α exp(-JE(C)/kT). Here E(C) is the energy associated with configuration C, J is the coupling constant for the problem under study, T is the temperature and k is Boltzmann's constant. A site i is chosen at random and the change in energy $\Delta E(i)$ which would result in reversing the spin at that site is determined. Since only the site i and its four nearest neighbors are involved, it is easy to see that the change in energy is

$$\Delta E(i) = (E'(i) - E(i)) + (E'(i+1) - E(i+1))$$
$$+ (E'(i-1) - E(i-1)) + (E'(i+\ell) -$$
$$E(i+\ell)) + (E'(i-\ell) - E(i-\ell))$$
$$= -4E(i) \qquad (5)$$

If $\Delta E(i)$ 0, then the move is "accepted" and the sign of $\sigma(i)$ is reversed. In case $\Delta E(i) > 0$, the move is accepted with probability exp($-J\Delta E/kT$).

It can be shown that the M(RT)² algorithm defines a Markov process which samples the "correct" (Boltzmann) distribution of configuration of spins. However, for a large system several hundreds of thousands, or even millions of updates of each site must be performed in order to approach a single equilibrium configuration, and often averages over many such configurations are required. Most of the work is in generating the random numbers, since as many as 10^{12} moves of sites may be required in a single simulation. We will delay discussion of this critical aspect for the moment, and concentrate on how to modify M(RT)² in order to get an algorithm well suited to the MPP. Later we explain why this algorithm is also well suited to vector machines, and we then describe one method for vectorizing the generation and testing of random numbers.

4.4.2. Algorithms

One is tempted to think of associating processors with spin sites in a one-one manner. However, this does not make very good use of the machine, since in the M(RT)² algorithm, only one site is examined for each move. Instead of mapping sites to processors in the obvious way, we instead notice that the expressions for the energy associated with a site are similar in form to the central difference approximation used to solve the Poisson equation,

$$-\nabla^2 u = \rho$$

In the Poisson case, a grid site and its four nearest neighbors are related through the finite difference expression

$$-u(i-\ell) - u(i-1) + 4u(i)$$
$$-u(i+1) - u(i+\ell) =$$
$$(\Delta x)(\Delta y)\rho(i) \qquad (6)$$

A common strategy used in implementing iterative methods for solving the Poisson equation is to use the "red-black ordering" depicted below:

```
R B R B R B
B R B R B R
R B R B R B
B R B R B R
```

Since no pair of red sites are nearest neighbors, all red sites can be updated "simultaneously". These values can then be used to update the black sites, and so forth, alternating on each iteration between red and black sites.

The same idea can be applied to modify M(RT)². Spin flips can be attempted at all red sites or at all black sites. This is a different Markov process than M(RT)² but the same distribution is sampled. Of course it is the dynamical aspects which are of interest now, rather than merely the converged solution. This use of colors is part of the so-called multi-spin algorithm.[9] In order to implement multi-spin on the MPP, all sites of a single color can be associated with a single MPP plane of 128 × 128 processors.

The problem to which we have applied multi-spin is slightly more complicated than that of attempting to flip single sites (Glauber dynamics). Instead we want to try to exchange the spins of a pair of neighboring sites. These spin exchanges of Kawasaki dynamics calculations can be used to study phenomena such as growth of magnetic domains.[10]

Because spin exchanges are to be attempted, eight different sites are involved in each move. For example, to interchange sites k_1 and k_2 we have all of the neighbor bonds to consider. The change in energy is the sum of changes over all eight sites and is given by the expression.

$$\Delta E = -2(\sigma(k_2) - \sigma(k_1))$$
$$*\{\sigma(i_1) + \sigma(i_2) + \sigma(i_3)$$
$$-\sigma(j_1) - \sigma(j_2) - \sigma(j_3)\} \quad (7)$$

where i's indicate neighbors of k_1, and j's are k_2 neighbors.

In order to apply the multi-spin idea enough "colors" must be assigned so that no two sites of the same "color" are involved in the same move. This can be accomplished by partitioning sites into sixteen groups as is shown below.

15	16	13	14	15	16
11	12	9	10	11	12
7	8	5	6	7	8
3	4	1	2	3	4
		13	14	15	16

Each group is now associated with a single 128×128 MPP bit plane, so that a 512×512 simulation can be handled in a straight-forward way.

Notice now that the part of the expression for ΔE which is enclosed in brackets can take on only seven distinct values, namely $\{-6, -4, -2, 0, 2, 4, 6\}$. The MPP array consists of single-bit processors and Boolean operations can be done simultaneously in a single operation on all 2^{14} processors. The computation of ΔE can be reduced to Boolean operations by first associating the spin values $\sigma = \{+1, -1\}$ with Boolean values $= \{1, 0\}$ (which, in effect, makes a transcription to a lattice gas model), and next noticing that the value of the bracket part of ΔE corresponds to the sum of the 1 bits of the Boolean expression. For example,

Spin version: $\{1\ 1\ -1\ 1\ 1\ 1\}$ Value = 4
Boolean version: $\{1\ 1\ 0\ 1\ 1\ 1\}$ Sum of bits = 5

The complete correspondence is as follows:

Spin sum	Bit sum	Representation
6	6	110
4	5	101
2	4	100
0	3	011
−2	2	010
−4	1	001
−6	0	000

The bit representation for the sum of bits can itself be determined by defining summation of Boolean expressions by Boolean operations. Assume that BITS(I), I = 1,2,...,6 is an array of bit planes each consisting of all 128 × 128 sites of a given color. The bit representation of the sum consists of three more bit planes B1, B2, and B3, initially all zero. Addition is performed bit by bit with the proper rules for carries. The correct representation for the final values is obtained by executing the following loop.

```
DO 10        I=1,6
   B1 = B1 .OR. (BITS(I)
           .AND. B3 .AND. B2)
   B2 = (.NOT. (BITS(I)
           .AND. B3) .XOR. (.NOT. B2))
   B3 = (.NOT. BITS (I))
           .XOR. (.NOT. B3)
10 CONTINUE
```

The sign of the sum of neighboring spins is determined by the value of the high order bit B1. This must be combined with the value of $-2\,(\sigma(k_2)-\sigma(k_1))$ to determine when a random number needs to be compared with $\exp(-J\Delta E/kT)$. Of course, the exponential also takes only finitely many values, so that it is easy to parallelize the comparison step. The generation of random numbers can be done simultaneously if a method which allows more than 2^{14} seeds is used. We have adapted a program due to Marsaglia and Kahaner,[11] but other methods are possible. In any case, 2^{14} pairs of sites are handled in a single step. The total number of different types of nearest neighbor pairs (< 1,2 >, < 6,7 >, < 9,5 >, etc.) is 32, and one should not cycle through these types in a fixed pattern because this marching introduces false dynamics into the simulation.

4.4.3. Vector Machines

Essentially the same algorithm can be used on a vector machine such as the CYBER 205.[12] The representations for ΔE can be computed very efficiently using bit vector operations. However, vector instructions are not really the same as parallel instructions, and so the 2^{14} random numbers which are required for each step take a lot of time to generate on a vector machine. The way to alleviate this is to reduce even the comparison with random numbers to vector operations on bit arrays. Assume, for convenience, that the value of $-2(\sigma(k_2) - \sigma(k_1))$ is -4, and let $\alpha = \exp(-8J/kT)$. As in Reference 15, we create bit arrays D1 D0 of length 2^{14} where D1D0 = 11 with probability α^3; D1D0 = 10 with probability $\alpha^2 - \alpha^3$, D1D0 = 01 with probability $\alpha - \alpha^2$, and D1D0 = 00 with probability $1 - \alpha$. This is easily done by generating 2^{14} random numbers and noting where they fall in the intervals I3 = $[0,\alpha^3)$, I2 = $[\alpha^3,\alpha^2)$, I1 = $[\alpha^2,\alpha)$, and I0 = $[\alpha,1)$. The test can now be performed by computing the bits of (B1 B2 B3 + D1 D0) $\oplus$ 001. Here the operation $\oplus$ means addition modified so that the high-order bit is not "turned off" by a carry operation (e.g., 111 $\oplus$ 001 = 100). After this operation has been performed, exchanges can be made at all sites with high-order bit equal to 1.

A little reflection reveals that the above method performs exchanges with nearly the correct frequency, because the bit vectors D1D0 approximate the correct probability distribution. After D1D0 has been used for a step, the vectors must be permuted in some random fashion, and to ensure that the false periodicities are not introduced, the D1D0 vector should occasionally be reloaded by generating 2^{14} new random numbers. This, of course, is much more efficient than generating a full set of random numbers for each step.

4.4.4. Preliminary Results

The first program to implement the algorithm on the MPP was written in Parallel Pascal. However, significant portions of it were later recoded in PEARL. The result is a code which performs better than 200 million tests of spin sites per second. Since the goal is to study the long time growth of domains, very long simulation times and averages over many different configurations are required. The present code makes this a practical possibility.

As has been mentioned, in the case of spin exchanges great care must be taken to avoid introducing any false periodicities into the results. This need for care is especially acute in this study because there is considerable controversy about the long time behavior of such models.[13] Preliminary runs on the MPP indicate that this algorithm is correct.

4.5. STEREO ANALYSIS

4.5.1. Introduction

During October 1984, the Space Shuttle Challenger was flown with a Shuttle Imaging Radar instrument (SIR-B). One of the experiments during this mission was to obtain overlapping images of an area on the ground viewed from several different incidence angles. Any two of these images form "pseudo-stereo-pairs" which through a suitable geometric model can be used to compute surface elevations. This section reports results of an effort at the Goddard Space Flight Center to develop an automated algorithm for computing elevations from SIR-B image pairs using the Massively Parallel Processor.

4.5.2. Background

Historically, the derivation of elevations from stereo pairs has followed two general approaches: the contour and the profile. With the contour approach, the stereo pairs are adjusted in a viewer such that only objects at a certain height will overlap perfectly. The interpreter then traces the path of perfect overlap. In the profile approach, the spacing between the stereo pairs is adjusted until a given object overlaps perfectly in the two images. The height of the object is then obtained as a function of that spacing. Objects along a given line are usually matched with this technique and thus the term "profile". The second approach has been adopted for implementation on the MPP.

4.5.3. Difficulties in Stereo Matching

The major difficulties in detecting points or areas where perfect overlap occurs (i.e., matching of corresponding pixels in the two images) are:

1. Different brightness levels in the two images
2. Local distortions of the image
3. Low contrast areas and noise

The first difficulty is often obviated by the use of normalized correlation functions for matching grey level or edge images. The second is inherent in stereo analysis because of the different viewing angles and makes automated matching more difficult than, for instance, in the case of matching control-point chips in Landsat images. It occurs most severely in regions of rapidly changing terrain and creates a horizontally stretched or compressed area surrounding corresponding pixels in one image relative to the other. Because of the large off-nadir viewing angle and difference in viewing angle required to achieve reasonable accuracy, this problem is particularly acute in radar images. Thus, the basic clue used to determine the elevation also makes the determination of that elevation more difficult. Any techniques for correcting local distortions must take into account the fact that the distortion function can have a broad band of spatial frequencies. For example, the distortion function for a mountain range would have low frequencies, but added to these would be high frequencies caused by rock formations making up the surface. When a human observer fuses two images seen through a viewer, the low frequency information is used to obtain an initial fusion in which the eyes are brought into alignment (a technique used for automatic focusing of some cameras) and then high frequency information brings out a detailed perception of depth. The progression from low to high frequency suggests that a hierarchical approach for detecting matching pixels would be appropriate. With this approach, an initial match is performed on low frequency information in an image and then increasingly higher frequencies are incorporated to obtain the final matching of corresponding pixels.

Even with no local distortion, errors can occur due to noise, spatial periodicities and low contrast in the image. One way of reducing errors in general is to provide redundancy by computing matches at nearly every pixel in the image. Then continuity constraints on the ground surface can be used to correct local discontinuities in the elevation. Matching at nearly every pixel is a formidable task on standard serial computers. However, the architecture of the MPP is well suited to the local neighborhood operations required for matching pixels. The resulting speed allows iterations of the matching algorithm computed at every pixel in 512 × 512 images to be completed within seconds. The following sections discuss the matching techniques developed for the MPP and results obtained using the MPP algorithm.

4.5.4. Matching Technique

The matching algorithm developed for the MPP is an example of what has been termed the Hierarchical Warp Stereo (HWS) technique. Initial work on stereo analysis using a hierarchical approach was done by Marr and Poggio[14] (1979). The Marr-Poggio algorithm performs low pass filtering and edge detection on the two stereo images and then matches the edges. Filters of several bandwidths are used from 1/100 to 1/4 the highest frequency. An edge in one image is said to match an edge in the other if (1) that edge apears within a given search area, (2) the slopes of the two edges match, and (3) the direction of the change in brightness of the two filtered images is the same across the edge. The relative location of edges in the most highly (narrowest bandwidth) filtered image determines the relative positions of large objects in a scene. (For instance, mountains or mountain ranges.) These displacements are then used to define search areas for corresponding edges in the second most highly filtered image. The procedure is repeated until corresponding edges are matched in the least filtered image.

Most recent work has been reported by Quam[15] (1984) who processes multiple resolution versions of both images (by sub-sampling by powers of 2 in both directions). Starting at the lowest resolution, matches are calculated using a normalized correlation measure applied to neighborhoods or windows surrounding reference and test pixels. The disparities between corresponding pixels (i.e., difference in their location in the two images) are then used as a one-dimensional distortion function to warp one image (the test image) so that its matched pixels will be in the same location as in the other (the reference image). The warped image is then resampled at the next higher resolution. This cycle is repeated until the highest resolution images are matched. The warping operation at each iteration reduces the local distortion so that at the next iteration with the next higher resolution there is a higher probability of obtaining a good match between pixels. At the end of the process, the sum of the distortion functions from all iterations forms the disparity function used to compute elevations. Quam's algorithm also eliminates potentially bad matches at each iteration by interpolating across pixels with low values of maximum "match scores" obtained for the reference neighborhoods over the corresponding search areas in the test image.

4.5.5. Matching Algorithm on the MPP

The algorithm developed for the MPP is similar to the Quam algorithm with the following exceptions.

1. Instead of applying equal sized windows to versions of the input images at increasing resolution, decreasing sized windows are applied to each iteration to the original input images. This eliminates the need for subsampling.
2. Iterations repeat until the neighborhoods are of a size within which there is no useful information for correlation.
3. At each iteration, the net amount of warping (i.e., an updated disparity funciton) is computed. This net disparity is always applied as the warping function to the original test image which eliminates loss of information at each warp iteration.
4. At each iteration, areas where bad matches occur are detected and interpolated over. Then the disparity function is smoothed before being used for warping.

The matching algorithm consists of the following steps:

1. Preprocessing of the test image,
2. Determination of matches,
3. Removal of "bad match" areas in the disparity function,
4. Smoothing the resulting disparity function,
5. Warping the test image.

Steps 2 through 5 are repeated for each iteration. The following subsections discuss each of these steps in detail.

4.5.6. Preprocessing of the Test Image

Because of the viewing geometry, the resolutions of the two SIR-B images are different in the stereopsis direction. Thus, a linear scale change is applied to the test image so that its resolution is the same as that of the reference image. This operation is implemented using a linear warping function in the stereopsis direction. Second, the test images can be translated to reduce the absolute value of the maximum disparity between the two images. Translation is effected by making the warping function constant over the image. When this is done, the size of the initial search area can be reduced. The amount of initial warping can be determined mathematically based on the different synthetic aperture radar incidence angles or it can be determined interactively by displaying the reference and test images. In either case, the initial warping function is incorporated into the net disparity function when determining elevations.

4.5.7. Determination of Matches

For each reference image pixel, a match is performed between a neighborhood surrounding that pixel (the "template") and neighborhoods within a search area in the test image. The location of the center pixel within each neighborhood in the test image relative to the reference pixel is the disparity value associated with that neighborhood.

The measure used for matching neighborhoods is the normalized mean and variance correlation given by:

$$\text{Match Score } (k) = \frac{\sum_i (X_i - \overline{X}) * (Y_{i-k} - \overline{Y}_k)}{\text{SQRT}(\sum_i (X_i - \overline{X})^2) * \text{SQRT}(\sum_i (Y_{i-k} - \overline{Y}_k)^2)} \quad (8)$$

where X_i and Y_{i-k} are grey levels of the i*th* pixels within the template neighborhood and the k*th* in the search area respectively. The values $\overline{X}$ and $\overline{Y}_k$ are the mean values computed over the template and k*th* search area neighborhoods.

For each pixel in the reference image, the match score for all neighborhoods within the search area is computed. The pixel at the center of the neighborhood with the highest correlation value or match score is selected as the matching pixel. The resulting disparity function is, therefore, made up of integer values.

4.5.8. Removal of "Bad Match" Areas in the Disparity Function

In order for stereo analysis to produce correct topographic results, there must be a one to one correspondence between pixels in the test image and those in the reference image (at least down to the resolution required to produce the desired elevation accuracy). For synthetic aperture radar images, this means that (1) both images must be taken from the same side of the spacecraft (or aircraft), and (2) both incidence angles must be such that there is no "layover" due to large surface slopes, and that there are no "shadows". If the image having the larger incidence angle is used as the reference image and both images meet the above two requirements, it can be proven that the disparity function will always have a gradient (or slope in the stereopsis direction) between 0 and 1. If "ground range" images are used (where both images have the same pixel resolution), the slope of the disparity function will always be between 1. This result is necessary if there is to be a one to one mapping between the test image and its warped version when the disparity is used as the warping function. Since the slope of the disparity function must be between 1, a simple test for a bad match is to observe adjacent values of the disparity function and determine if there is a jump of more than 1 pixel.

The human visual system appears to have the capability of interpolating surfaces over areas where bad matches occur. This process is emulated in the

MPP algorithm by interpolating the disparity function across all areas where a bad match has been detected. The detection of bad matches and the interpolation are accomplished with the following operations:

1. Detection of "bad match pixels" (pixels where discontinuities occur)

Sudden jumps in the disparity function are detected by examining a 3×3 neighborhood surrounding each pixel. If the disparity value at the pixel differs from that of any of its neighbors by more than one, the pixel is identified as having a bad match.

2. Expansion or growth of each bad match pixel to form a neighborhood.

If the maximum difference between the disparity value at a pixel and those of its adjacent neighbors is "d", then one must interpolate the disparity function over a neighborhood surrounding that pixel whose diameter is at least "d" to satisfy the constraint that the slopes of the disparity function be less than +1. The expansion of each "bad match pixel" to form a neighborhood is done for this purpose. This is accomplished on the MPP by alternately expanding each pixel into 4- and 8-element neighborhoods. The resulting neighborhood is octagonally shaped. A diameter of 2N is achieved with N iterations. As each pixel is expanded, the resulting overlapping neighborhoods form bad match regions.

3. Interpolate of the disparity function over resulting bad match neighborhoods

Interpolation is performed using heat flow equations. The architecture of the MPP is well suited to the iterative solution of the boundary value partial differential equations typical of heat flow problems. To perform the interpolation, two-dimensional heat flow partial differential equations are applied to solve for the steady-state "temperature" or disparity in the bad match regions assuming that the bordering pixels surrounding the bad match area are held at a constant "temperature" or disparity. The equations used for obtaining the interpolated disparity at pixels [i,j] at iteration t+1 from the values at iteration t are:

$$D(i,j,t+1) = D(i,j,t) + \frac{d(D(i,j,t))}{dt}$$

where

$$\frac{d(D(i,j,t))}{dt} = \frac{\partial^2(D(i,j,t))}{\partial i^2} = \frac{\partial^2(D(i,j,t))}{\partial j^2} \tag{9}$$

The second partial differential equation reduces to

$$\frac{d(D(i,j,t))}{dt} = D(i,j+1,t) + (D(i,j-1,t) + D(i+1,j,t) + D(i-1,j,t) - 4D(i,j,t) \quad (10)$$

when it is assumed that the two-dimensional grid increments are unity. The number of iterations required to reach "steady state" is dependent on the size of the bad match regions.

4.5.9. Smoothing the Resulting Disparity Function

After interpolation, a smoothing operation is applied over the whole disparity function in order to obtain a smooth warping function with fractional pixel values rather than integer values. Smoothing is performed on the MPP by averaging over a neighborhood proportional in size to the neighborhood used to obtain the disparity function.

4.5.10. Warping the Test Image

The smoothed disparity function is used as a one-dimensional distortion function to "geometrically correct" the test image in the stereopsis direction. The brightness values in the warped image are obtained by applying a linear interpolation function to the test image data in the resampling process.

4.5.11. Interactive Operations on the MPP

The matching algorithm on the MPP has been implemented to be run in an interactive mode where parameters such as neighborhood sizes, search area, and discontinuity thresholds can be input before starting each step of the matching algorithm. At the end of a given step, the results (such as the disparity function, or the warped version of the test image) can be immediately displayed or saved on disk. This interaction provides the ability to experiment with various parameter values and quickly observe the results. In addition, the control or shell program is designed so that operations can be easily modified or added.

The operations presently implemented in the matching algorithm which can be run interactively are shown below along with the input parameters which can be selected:

1. INITIAL WARP (left edge movement, right edge movement)
2. MATCH (neighborhood size, search area size)
3. "BAD MATCH DETECTOR" (discontinuity threshold for bad match, neighborhood diameter for expansion)
4. INTERPOLATE (number of iterations)
5. SMOOTH (neighborhood size)
6. WARP

The left edge and right edge movement in the INITIAL WARP operation define the linear warp function in the stereopsis direction. If they are equal, only a translation is applied to the test image.

4.5.12. Interactive Turnaround Time

The following table shows turnaround times for the six stereo matching operations. Where times are dependent on parameters, some example parameters and the corresponding times are shown. The times are measured from the time a key is pressed on the terminal to start the task to the time the next prompt is displayed indicating that the task is finished. Times less than about a half second could not easily be measured.

The matching operation which is the most computationally expensive task has been optimized to require a time proportional to the length of the sides of the neighborhoods as opposed to the area. The smoothing operation is unoptimized and requires a time proportional to the area of the neighborhood. However, since the time required for smoothing is not prohibitively long for interactive purposes, this optimization has not been implemented.

Interactive Turnaround Time for Stereo Analysis Operations

Operation	Parameter		Time in seconds
Initial Warp			0.5
Match	11×11	Nbh. size	5
	21×21		10
	41×41		20
Detect bad match	2	Radius	<0.5
	4		0.5
	8		1
Interpolate	250	No. of iterations	2
	500		4
	1000		8
Smooth	11×11		2
	21×21		6
	41×41		21
Warp			3

4.6. SORT

4.6.1. Massively Parallel Communications

The MPP contains bit-serial processors that can perform a lot of computational operations in parallel, since it has 16384 such bit-serial processing elements. It is perhaps hard to believe that the MPP could perform very well if much data communication is necessary, since each processor can only communicate with four other processors. However there are 16,384 processors communicating at the rate of 10 million bits/s. Thus while some data may need to be communicated to a processor that is one processor away from its destination, or as many as 256 processors away, the maximum data rate of 160 Gbits/s or 20 Gbytes/s is not a fact to be ignored.

To bring this into perspective, let us look at the problem of sorting on the MPP. If we use Nassimi and Sahni's bitonic sort algorithm, the number of communication steps necessary to sort one bit plane is 14 times the width of the array (128), where a communication step takes 0.1 μs on the MPP. Thus, if there are k bits in each record of a layer of records, it will take 1792k communication steps to sort a layer of records. Therefore, if we sort 16,384 records that are 1024 bits (128 bytes) long, it will take approximately 0.2 s of communications. Note that this time does not involve any overhead due to main control or comparisons in the array. In actual testing on the MPP, it took 0.040 s to sort 96-bit planes and this value includes all overhead. This rate extrapolates roughly into sorting a 2-Mbyte file of 16,384 records 1024 bits long in 0.4 s. A VAX would have problems copying a file in less 1 min, let alone sorting the file in less than a second.

4.6.2. Sorting as a Communication Primitive

Sorting is the most versatile method of reorganizing data and can be performed at near optimal speed on the MPP. Optimal time to perform a general data reorganization is $4\sqrt{n}$ where n is the number of processors in the array. Thomson and Kung's sort takes $7\sqrt{n}$. All that is necessary to perform a general data reorganization is to determine a key for each record that will cause the records to be sorted either to the desired destination PEs or into the desired record order, and to sort the records into that order.

If all that could be done with the sort was to reorganize records of data, the sort would be of little use on the MPP, since the MPP is I/O bound to the outside world. However, if the sort could be interleaved with computations, or better yet if we could perform computations during the sort, the sort could be viewed as a data rendezvous mechanism.

Therefore the real power of it can be better utilized if we view the sort as a tool to bring together related data, rather than as a tool to merely position data. That is to say, if we can guarantee that two desired pieces of data will come together during the sort when necessary, we do not care where the records end up at the end of the sort. This is true if we desire to sum all the values of a field (the sum-field) of a layer of records that have the same key value (i.e., histogramming) or if we wish to broadcast a specific value to all records that have the same key value (i.e., table lookup).

4.6.3. Sorting Algorithms Implemented on the MPP

The sorting algorithm that was implemented on the MPP was a modified version of Nassimi and Sahni's sort. It sorts 2n records where $n = 2^{2m}$, m = 1,2,3

... is the number of processors. The following is a pseudo-code description of this sort algorithm.

Each processor contains the following variables and data structures: a row index (RINDEX), a column index (CINDEX), a shuffled row major index (SINDEX), and three data structures RECORD0, RECORD1, and RECORDT. Each data structure contains a key field, KEY. The notation 'RINDEX(i)' means the ith least significant bit of RINDEX. SINDEX is created by shuffling the bits of RINDEX and CINDEX together (i.e., SINDEX(2i) = CINDEX(i) and SINDEX (2i+1) = RINDEX(i) for i = 0, ... ,m–1). Note SINDEX(2m) = 0 for all processors.

Each processor contains 2 of the 2n records to be sorted. The routine COMPUTATION performs a comparison-exchange operation. In later sections it performs other operations as well. The following routines make up the sort.

```
routine SORT
  do i = 1 to m
  HORIZONTAL_MERGE (i,2i-1)
  if (i<m) VERTICAL_MERGE (i,2i)
  end do
  ENTIRE_PLANE_MERGE (m, 2m-1)
  COMPUTE (2m-1)
  ENTIRE_PLANE_MERGE (m, 2m)
end SORT

routine HORIZONTAL_MERGE (i,j)
  SWAP_COLUMNS (i)
  TWO_COLUMN_MERGE (i-1, j)
  SWAP_COLUMNS (i)
  COLUMN_MERGE (i-1, j)
end HORIZONTAL_MERGE

routine VERTICAL_MERGE (i,j)
  ROW_MERGE (i,j)
  COLUMN_MERGE (i,j)
end VERTICAL_MERGE

routine ENTIRE_PLANE_MERGE (i,j)
  SWAP_ROWS (i)
  COMPUTE (j)
  ROW (i-1, j)
  COLUMN (i,j)
  SWAP_ROWS (i)
end ENTIRE_PLANE_MERGE
```

```
routine ROW_MERGE (i,j)
  do k = i to 1
   SWAP_ROWS (k)
   COMPUTE (j)
   SWAP_ROWS (k)
  end do
end ROW_MERGE

routine COLUMN_MERGE (i,j)
  do k = i to 1
   SWAP_COLUMNS (k)
   COMPUTE (j)
   SWAP_COLUMNS (k)
  end do
end COLUMN_MERGE

routine TWO_COLUMN_MERGE (i,j)
  COMPUTE (j)
  do k = i to 1
   SWAP_ROWS (k)
   COMPUTE (j)
  end do
end TWO_COLUMN_MERGE

routine SWAP_COLUMNS (j)
  K = j-1
  In Processors where CINDEX(k) = 0
    move  RECORD1  to   RECORDT
        of the PE that is $2^k$
        processors to the EAST
  In Processors where CINDEX(k) = 1
    move  RECORD0  to   RECORD1
        of the PE that is $2^k$
        processors to the WEST
    move  RECORDT  to   RECORD0
        within those processors
end SWAP_COLUMNS

routine SWAP_ROWS (j)
  k = j-1
  In Processors where RINDEX(k) = 0
    move  RECORD1  to   RECORDT
        of the PE that is $2^k$
        processors to the SOUTH
```

```
    In Processors where RINDEX(k) = 1
      move RECORD0 to RECORD1
          of the PE that is 2 ᵏ
          processors to the NORTH
      move RECORDT to RECORD0
          within those processors
    end SWAP_ROWS

    routine COMPUTE (j)
      /* Comparison Exchange */
      if SINDEX(j) = 0
      if RECORD0.KEY > RECORD1.KEY
              Exchange RECORD0 and RECORD1
      end if
      else
      if RECORD1.KEY > RECORD0.KEY
              Exchange RECORD0 AND RECORD1
      end if
      end if
      end COMPUTE
```

4.6.4. Sort Aggregation

Summing during a sort (sort-summing) is a special case of sort-aggregation. This could be extended to products as well as sums or to any other function of an aggregate nature. We will use sort-summing to show how the mechanism for sort-aggregation works. A group of records are all records whose key values are considered equal. A set of records to be sort-summed may contain many groups of records. It will be shown that all records of a group will have their sum-field values summed and the result stored in the sum field of one of the records of that group.

Let us start with two sorted lists of records $A(i)$ $i = 1....n$ and $B(j)$ $j = 1....m$. Given a record in each list, $A(p)$ and $B(q)$, that have the same key value, and given that no other $A(i)$ or $B(j)$ has a key value equal to that of $A(p)$ or $B(q)$, then during the merge of these two lists of records the key values of $A(p)$ and $B(q)$ must be compared with each other to determine which will be sorted before the other. When $A(p)$ and $B(q)$ are compared, and it is determined that their key values are the same, their sum-field values are added together and the sum is placed in the sum field of the record which is considered to be the record with the greater (lesser) key field value. Zero is placed in the sum field of the other record.

Assume now that the two lists $A(i)$ and $B(j)$ contain multiple records with the same key value (i.e., of the same group) and the greatest (least) record of each group contains the sum of all the sum fields of records in that group. Following the same line of thought of the previous paragraph, the greatest (least) record of each group of each list must be compared to determine which record will be the greatest (least) record in the resulting list, at which time the sum of their sum fields is placed in the greater (lesser) record's sum field and the other record's sum field is zero filled. This process always leaves the sum of the sum fields of all the records of a group in the sum field of the greatest (least) record of that group.

To incorporate this capability into the sort algorithm we change the routine COMPUTE and add a field (TOTAL) to RECORD0 and RECORD1. Thus the routine COMPUTE is changed to the following:

```
routine COMPUTE (j)
  /* Comparison Sum */
  if SINDEX(j) = 0
  if RECORD0.KEY > RECORD1.KEY
        Exchange RECORD0 and RECORD1
  end if
  if RECORD0.KEY = RECORD1.KEY
        RECORD1.TOTAL =
        RECORD1.TOTAL  +  RECORD0.TOTAL
        RECORD0.TOTAL  =  0
  end if
  else
  if RECORD1.KEY > RECORD0.KEY
        Exchange RECORD0 and RECORD1
  end if
  if RECORD1.KEY = RECORD0.KEY
        RECORD0.TOTAL =
                RECORD1.TOTAL + RECORD0.TOTAL
            RECORD1.TOTAL = 0
        end if
  end if
end COMPUTE
```

This will cause the sum of all the TOTAL fields of each group to be placed in one of the records of each group of records.

4.6.5. Sort Distribution

Sort distribution is similar to sort aggregation in that processing is done within groups of records with the same key values. However, sort aggregation takes values from all the records of a group and places the results of the processing in one of the records of the group, while sort distribution takes the value from a specific record of the group and puts it into a field of all records of the group. We will show that the desired distribution can be performed during a sort.

Consider again the two sorted lists of records $A(i)$ $i = 1...n$ and $B(j)$ $j = 1....m$. Assume that each record has a broadcast flag which is set to unity if it already contains the desired distribution value for the corresponding key value, and a zero if this record has not obtained the correct distribution value. Suppose that there are records in each list, $A(p)$ and $B(q)$, that have equal key values, and that no other $A(i)$ or $B(j)$ have key values equal to that of $A(p)$ or $B(q)$. Assume that $A(p)$ has its broadcast flag set to one and $B(q)$ does not. During the merge of these two lists of records, the key values of $A(p)$ and $B(q)$ must be compared with each other to determine which will be sorted before the other. When it is determined that their key values are the same, $A(p)$'s broadcast field is copied to $B(q)$'s and $B(q)$'s broadcast flag is set to one. This in effect causes two copies of $A(p)$ to exist, since any other difference between $A(p)$ and $B(q)$ has no effect on the processing that we are concerned with here.

We know from the discussion of sort aggregation that if we always sum into the greater record's sum field, at the end of the sort the highest value record for each group of records of the same key value will contain the sum and, if we always sum into the lesser record's sum field, at the end of the sort the lowest value record for each group of records of the same key value will contain the sum. Thus if we broadcast the same value to both the greater and the lesser records during a comparison, then both the highest and lowest value record of each group of records with the same key value must have the distribution value.

We must show that the inbetween records of a group of records that have the same key values will also obtain the distribution value. First we must assume that the broadcast flags of the records of a group in a list must be either all ones or all zeros. Thus, for a specific key value if both lists to be merged contain records that have not been broadcast to, then the resultant merged list will contain only records that have not been broadcast to. If both lists contain records that have been broadcast to, then the resultant list contains records that have been broadcast to. The question is, if one list contains records that have been broadcast to and the other list contains records that have not been broadcast to, will the resultant list contain only records that have been broadcast to?

We know that the first and last record of a group will have been broadcast to. Let us assume however that an in-between record $(B(q))$ of the group was not broadcast to. Note that all records that have been broadcast to are effectively a duplicate of $A(p)$. Thus if $B(q)$ had ever been compared to $A(p)$, it would have been broadcast to. Yet if $B(q)$ had never been compared to $A(p)$, and $B(q)$ and $A(p)$ have the same key value, we don't know whether $B(q)$ should come before or after $A(p)$. Therefore the merge will not be complete. If it were, then $B(q)$ would have been broadcast to.

For a specific key value, the resultant list of a merge of a list that has been broadcast to and one that has not is a list that has been entirely broadcast to. Therefore the sort distribution will result in a list for which all groups that had at least one of its member's broadcast flag set to one before the sort will result in all of its members broadcast flags being set after the sort, and thus a complete broadcast has been performed.

Once again only the routine COMPUTE needs to be modified. We will add two new fields, VALUE and DFLAG, to RECORD0 and RECORD1. VALUE will contain the value to be distributed to all records of the same group. DFLAG is a flag that is set only in records that contain a valid distribution value in VALUE. Each group of records that will have a value distributed to all records of the group must have at least one record with its DFLAG set. The following COMPUTE routine will replace the comparison/exchange COMPUTE routine of SORT.

```
routine COMPUTE (j)
  /* Comparison Distribution */
  if SINDEX(j) = 0
   if RECORD0.KEY > RECORD1.KEY
        Exchange RECORD0 and RECORD1
   end if
  else
   if RECORD1.KEY > RECORD0.KEY
        Exchange RECORD0 and RECORD1
   end if
  end if
  if RECORD1.KEY = RECORD0.KEY
   if RECORD0.DFLAG = 1
        RECORD1.VALUE = RECORD0.VALUE
        RECORD1.DFLAG = 1
    else if RECORD1.DFAG = 1
        RECORD0.VALUE = RECORD1.VALUE
        RECORD0.DFLAG = 1
   endif
  end if
end COMPUTE
```

4.6.6. Merge Aggregation and Merge Distribution

As implied by the previous discussion, a sort is divided into merge steps. Thus there are corresponding merge aggregation and merge distribution steps for sort aggregation and sort distribution. If two lists are in merge aggregate order, a merge aggregation may be performed to produce one list that is in merge aggregate order. Nassimi's and Sahni's sort takes $14k \sqrt{n}$ communication steps and the corresponding merge takes $4k \sqrt{n}$ steps, where k is the number of bits per record and n is the number of processors, while Thompson's and Kung's sort takes $7k \sqrt{n}$ communication steps and the corresponding merge takes $2k \sqrt{n}$ steps, where k is the number of bits per record and n is the number of processors. Note that in either case the merge is 3.5 times faster than the corresponding sort.

140   Parallel Supercomputing in SIMD Architectures

As implemented on the MPP, the stand-alone merge merges two layers of records, RECORD0 and RECORD1, which are already in sorted order. The following routine uses the routines already described:

```
routine MERGE
  COMPUTE (2m-1)
  ENTIRE_PLANE_MERGE (m,2m)
end MERGE
```

Merge distribution leads to the concept of parallel table lookups. If the table is in sorted order and the data records are in sorted order with respect to the lookup table entry keys, the table and the data may be merge distributed together, thus passing the table parameters to the corresponding data records.

4.6.7. The Unmerge Operation

Once a table lookup is performed it is necessary to extract the table entries from between the data records. At first glance it would seem necessary to sort the table from the data. This would defeat the purpose of using a merge instead of a sort to bring the table entries and data records together, which is to save time over performing a complete sort. Thus the unmerge function was developed. One simply keeps track of the exchanges performed to enact the merge in a tracking field; the unmerge operation uses this tracking field to reverse the merge, thus separating the data from the table. The unmerge takes no more time than a merge and leaves the table and the data in the original order.

To be able to perform an unmerge, the COMPUTE routine of the MERGE operation must also be modified. Each processor will contain one additional variable (TRACK). TRACK(i) means the ith least significant bit of TRACK. Thus the COMPUTE routine of MERGE is as follows:

```
/* The variable i is initialized to zero */
/* before merge starts.   */

routine COMPUTE (j)
  /* Comparison Distribution */
  /* for anticipated unmerge */
  TRACK(i) = 0
  if SINDEX(j) = 0
   if RECORD0.KEY > RECORD1.KEY
        TRACK(i) = 1
        Exchange RECORD0 and RECORD1
   end if
  else
   if RECORD1.KEY > RECORD0.KEY
        TRACK(i) = 1
```

```
        Exchange RECORD0 and RECORD1
   end if
 end if
 if RECORD1.KEY = RECORD0.KEY
  if RECORD0.DFAG = 1
        RECORD1.VALUE = RECORD0.VALUE
        RECORD1.DFLAG = 1
   else if RECORD1.DFLAG = 1
        RECORD0.VALUE = RECORD1.VALUE
        RECORD0.DFLAG = 1
  endif
 end if
 i = i+1
end COMPUTE
```

To perform an unmerge the routine MERGE is used with the following COMPUTE routine.

```
/* The variable i is initialized to zero */
/*      before MERGE starts.             */

routine COMPUTE (j)
 /* Unmerge */
 if TRACK(i) = 1
  Exchange RECORD0 and RECORD1
 end if
 i = i+1
end COMPUTE
```

Note again that the merge and unmerge operation are of the same complexity since their only difference in their COMPUTE routine.

REFERENCES

1. **Batcher, K. E.**, Design of a massively parallel processor, IEEE Trans. Computers, C29, 836, 1980.
2. **Potter, J. L., Ed.**, *The Massively Parallel Processor*, MIT Press, August 1985.
3. **Reeves, A. P.**, Parallel pascal: an extended pascal for parallel computers, *J. Parallel and Distributed Computing*, 1, 64, 1984.
4. **Strong, J. P.**, Basic image processing algorithms on the massively parallel processor, *Multicomputers and Image Processing,* Preston, K., Jr. and Uhr, R., Eds., Academic Press, New York, 1982, 47.

5. **Brodie, L.,** *Starting FORTH,* Prentice-Hall, Englewood Cliffs, NJ, 1981.
6. **Brodie, L.,** *Thinking FORTH,* Prentice-Hall, Englewood Cliffs, NJ, 1984.
7. **Henden, A.,** *Uni-FORTH User's Guide,* Unified Software Systems, Columbus, OH, 1985.
8. **Metropolis, N., Rosenbluth, A. W., Rosenbluth, M. N., Teller, A. H., and Teller, E.,** *J. Chem. Phys.,* 21, 1087, 1954.
9. **Williams, G. O. and Kalos, M. H.,** A new multispin coding algorithm for Monte Carlo simulation of the Ising Model, *J. Stat. Phys.,* 37, 283, 1984.
10. **Huse, D. A.,** Late Stages of Spinodal Decomposition: Corrections of Lifshitz-Slyozov Scaling and Monte Carlo Simulations, preprint, 1986.
11. **Marsaglia, G. and Kahaner, D.,** Generation of Uniform and Normal Random Variables, preprint.
12. **Bhanot, G., Duke, D., and Salvador, R.,** A fast algorithm for the CYBER 205 to simulate the 3-D Ising model, *J. Stat. Phys.,* to appear.
13. **Mazenko, G. F., Valis, O. T., and Zhang, F. C.,** *Phys. Rev.,* B31, 4453, 1985.
14. **Marr, D. and Poggio, T.,** A computational theory of human stereo, *Proc. R. Soc. London,* p. 301, 1979.
15. **Quam, L.,** Hierarchical warp stereo, *Proc. Image Understanding Workshop,* p. 149, 1984.

5 DAP

The prototype DAP — "Distributed Array of Processors" — was produced in 1976 by International Computers Limited (ICL) in England. Deliveries of the first generation of DAP computers started in 1980.

The current second generation DAP systems are designed, manufactured and sold by Active Memory Technology, Inc. (AMT), headquartered in Irvine, California. AMT was founded as an independent company in October, 1986. Shareholders include founding employees, ICL and a group of venture capital firms. Research and software development activities are supported at a Reading, England facility as well as Irvine.

Among the founders are the chairman, Dr. Geoffrey Manning CBE, and the Managing Director, Rodney Hornstein. Dr. Manning was formerly the Director of the Rutherford Appleton Laboratory in Great Britain. Mr. Hornstein previously was the Managing Director of Gestetner International Ltd.

The representative DAP 510 uses a Sun or VAX host, and is programmed primarily in FORTRAN-Plus. Physically the DAP 510 uses 400 W, measures $25 \times 13 \times 15$ inches and weighs 110 pounds. In one recent placement the U.S. Army Corps of Engineers' Waterways Experiment Station in Vicksburg, Mississippi acquired a 4096 processor DAP to be mounted in a helicopter and used for real-time target recognition.

5.1. THE DAP DESIGN

In the DAP, the processors (known as "processor elements" or PEs) are arranged in a square matrix: 32×32 in the case of the DAP 500 range and 64×64 for the DAP 600. Note that the edge size of the matrix is itself a power of 2, the power being expressed in the first digit of the model identifier (i.e., 2^5 for the 500 range and 2^6 for the 600 range).

Each PE is provided with connections to its four nearest neighbors. In addition, a bus system connects all the PEs in each row and all the PEs in each column. These row and column data paths provide rapid data broadcasting or fetching facilities. The nearest neighbor connection scheme gives the high level of connectivity required for many applications (fluid flow, linear algebra, etc.).

Each processor is connected to a minimum of 32 kbits (the architecture allows up to 1 Mbit) of its own local memory. This gives a total memory configuration for the DAP 500 range (1024 processors) of between 32 Mbits and 1024 Mbits. The total memory in the DAP 600 range (4096 processors) is between 128 Mbits and 4096 Mbits.

In the DAP, the processor array is controlled by a master control unit (MCU). The MCU acts like a conventional central processing unit except that it does not itself execute all the instructions it interprets from its code memory. Many instructions are decoded by the MCU and broadcast to the entire array of processors. Such instructions are executed by all the PEs simultaneously, each operating on the data in its local memory.

It is by having a large number of processors operating simultaneously on arrays of data that the DAP attains its high performance. The PEs perform their basic operation, which usually involves fetching or storing a memory bit, within a single DAP cycle. On the DAP 510, with a cycle time of 100 ns, there are 10 million cycles per second. The 1024 processors performing logic (Boolean) operations at this rate give a total rate of 1024×10^7 or 10,240 million operations per second (i.e., 10^4 MOPs). A DAP 600 with the same cycle time would increase the final rate by a factor of 4, since the number of processors increases in this ratio. The data rate between memory and processors on the DAP 510 is 10^{10} bits/s, or 1200 Mbytes/s. This bandwidth gives the DAP a very high performance both for computing and for I/O operations.

5.1.1. The DAP Software

The DAP is programmed using conventional languages, although to take advantage of the massive parallelism, it is necessary to make extensions to current languages to support data parallel constructs, operations on arrays for example. Such extensions will be part of the standards of future languages such as FORTRAN 8X whose constructs are well suited to the DAP architecture. The current DAP version of FORTRAN is FORTRAN-Plus; it contains a number of extensions and intrinsic routines which makes it very easy to take advantage of data parallelism. An assembler language, APAL, is available for anyone who wishes to program the DAP at the bit level for special reasons, such as a requirement for unusual precision or special efficiency.

DAP programs reside in the code memory of the master control unit. The DAP may be accessed through a Sun or VAX workstation, called the DAP host, in order to take advantage of their rich interactive environment and peripheral devices. The user's FORTRAN-Plus program can be developed on the host, using a simulator and debugging tool. At run time, the DAP program

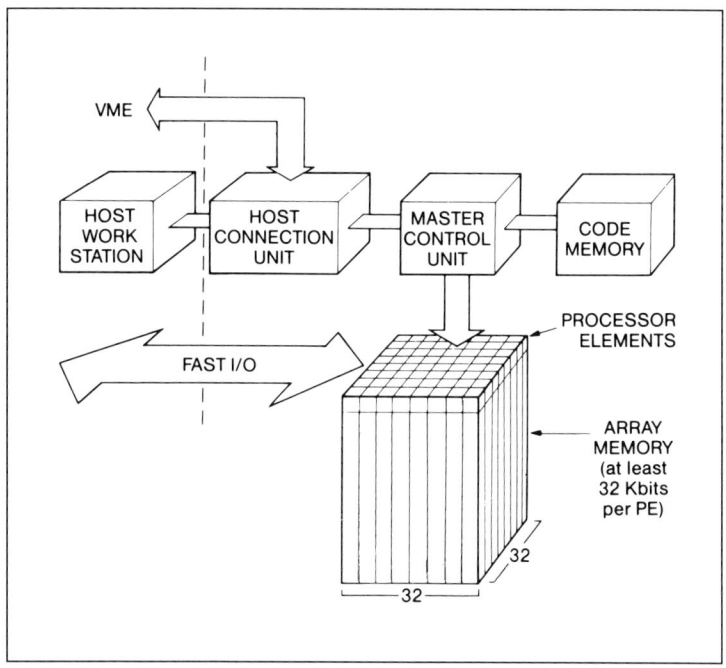

FIGURE 20. The DAP system.

is initiated and controlled from the host and can communicate with a user program on the host. It depends on the application whether it is better to run a program almost wholly on the DAP or to run only highly data parallel routines on the DAP, leaving the host program predominately in control.

Although, for many applications, it may be suitable to perform all input/output through the host and use host peripherals, there is a facility for direct I/O to the DAP internal bus using fast data channels at up to 50 Mbytes/s. This can be used to provide real-time video output of the changing data structures of a program during processing. Such I/O rates can be achieved with a negligible effect on the DAP processing speed.

A DAP system has the following main components, as outlined in Figure 20:

- Master Control Unit (MCU) and Code Memory
- Processor Element (PE) Array and Array Memory
- Host Connection Unit (HCU)
- Fast Data Channel

The Master Control Unit (MCU) is the source of instructions for the DAP. It is a 32-bit central processing unit with many conventional features such as registers, instruction counter, branch instructions, arithmetic unit, etc. The object code of a DAP program is loaded into the code memory, from which the MCU fetches instructions and interprets them. Some instructions will be executed wholly within the MCU (scalar operations using MCU registers, control instructions, etc.), others will be broadcast to the PE array to be executed by the individual PEs in parallel.

Most instructions take place in one cycle. On the DAP 500, the cycle speed is 10 MHz.

The DAP is designed for attachment to a host workstation, which is used for program development, debugging, loading, initiating, and controlling DAP programs. The host connection unit (HCU) serves as the communications gateway between the DAP and the host. The HCU incorporates a Motorola 68020 32-bit microprocessor, a SCSI port, a VME-bus interface and two RS232 serial ports. Interfacing to VAX computers is via a DR11W or DRB32 interface card.

The DAP HCU provides memory protection through two memory address boundary checkers. A 256 kbyte, or 64K word, EPROM provides code storage. And, a 1 Mbyte, or 256K word, parity-protected random access memory supports data and program code storage.

Data transfers between the HCU and the DAP are performed as memory-to-memory transfers across the VME bus. By using this architecture, any VME bus master can access the DAP memory. The HCU also supports a full VME interrupt handler.

The two RS232 communications ports offer maximum transfer rates of 38.4 kBd and 125 kBd, respectively.

A 1-kbyte FIFO buffer interfaces the SCSI controller chip to the 68020 data bus. Data transferred to and from the bus goes through the SCSI FIFO buffer which is 32 bits wide on the 68020 side and 8 bits wide on the SCSI controller chip side. Data read from and written to the registers to the controller chip bypass the FIFO buffer.

The HCU provides a calendar clock and four timers: bus time-out, system watch-dog, DRAM refresh, and system interval.

The HCU may be used for medium speed I/O transfers either to the host workstation or to directly connected devices on, for example, the VME port. However, fast data channels are also provided for data intensive applications. On the DAP 510, data may be input and output concurrently at a rate of up to 50 Mbytes/s; this would use only 4% of the DAP processor cycles.

Among the peripherals that are offered for connection to the fast data channel is a video output board to drive a high resolution color display, enabling "visualization" of the data of an application while it is actually being processed in the DAP.

The parallel processing capability of the DAP is provided by N^2 processor elements arranged in an $N \times N$ matrix (the "PE array" or simply the "array") as shown in Figure 21. Each PE is capable of performing arithmetic and logical

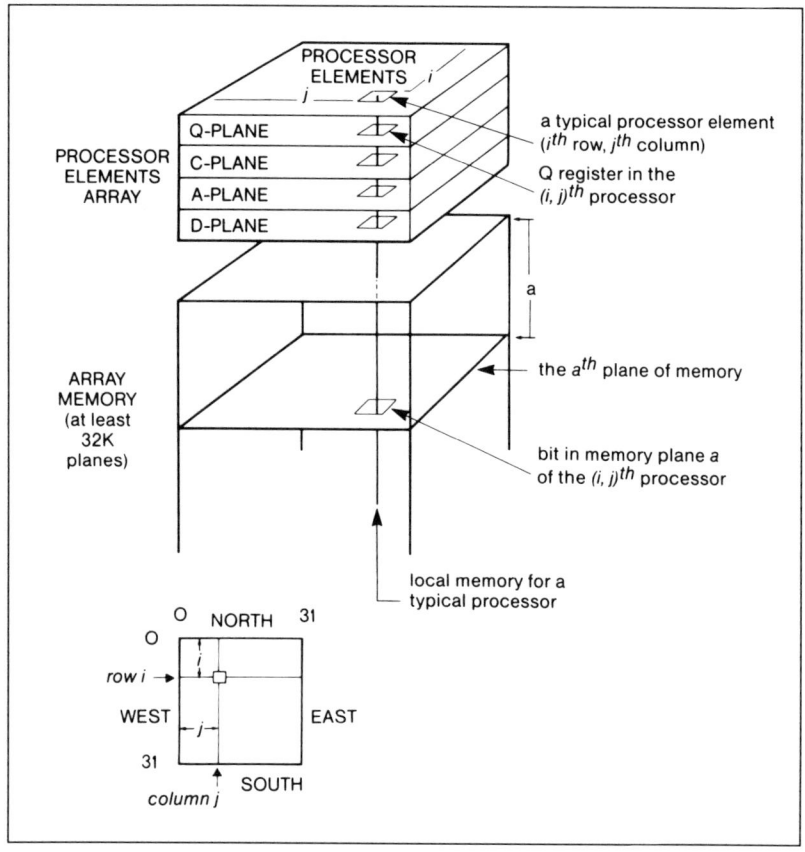

FIGURE 21. Details of the processor elements array and the array memory.

operations on operands that are single-bit values. N (the "edge size") is 32 in the DAP 500 and 64 in the DAP 600. The rows and columns of the PE matrix are numbered from 0 to N − 1. The edges of the matrix are referred to as North (row 0), South, West (column 0), and East.

Each PE is connected to the four neighboring PEs in the north, south, east, and west directions. Using these connections, data can be propagated from a register of each PE into the corresponding register of a neighboring PE in any of the four directions. Processors at an edge are connected to those at the opposite edge, thus allowing shifts to wrap around if desired. Also, the PEs in each row and in each column are connected on row and column highways to broadcast data in and out of all the processors simultaneously.

Figure 22 is a simplified diagram of one processor element. This diagram is intended only to show the main functional components and their interconnection. Each PE has three 1-bit registers, denoted A, Q, and C. They are used

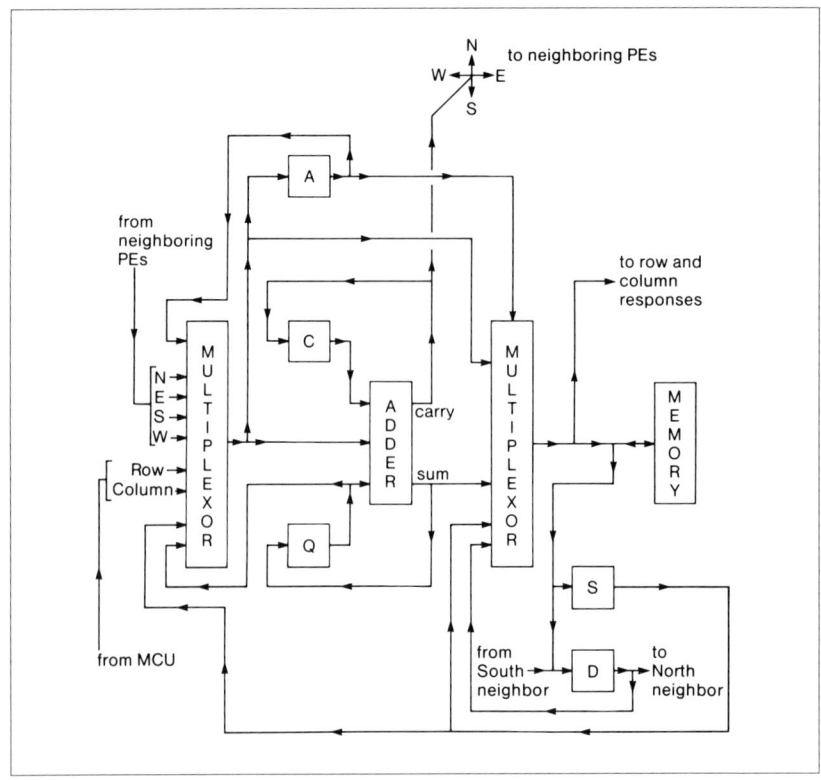

FIGURE 22. A simplified diagram of a processor element.

for a number of purposes, but it is convenient to think of the Q register as being an accumulator, the C register as a carry register, and the A register as being for "activity control" — this means it can inhibit memory write operations in certain instructions. All of the bits of a particular register over all the PEs are known as a "register plane", as shown in Figure 21.

The Q and C registers are input to an adder, but these inputs can be disabled (i.e., treated as false). In this way, the adder can take data from either, none, or both of the Q and C registers. The third adder input is selected by a multiplexer and can come from the PE memory, from the outputs of the Q or A register, from data broadcast by the MCU, or from the carry output of a neighboring PE. The A register also receives its input from this multiplexer and can either be written directly or AND-ed (masked) with the existing A register contents. For some purposes, the output of the input multiplexer can be inverted. PE outputs can be written to memory, and in some instructions this writing is conditional on the value in the A register.

The D register shown in Figure 22 does not appear explicitly in the machine-code programmer's model of the PE, but is used for input or output for the fast interface unit. Once data is loaded into the plane of D registers, it can be clocked out asynchronously without interrupting the processing.

Certain instructions both read from and write to the memory; for these instructions a register is needed somewhere in the path of the data flow. The details are implementation dependent, but an example of such a register is shown as S in Figure 22; this register does not appear in the programmer's model.

Each of the PEs has a local memory whose size depends on the version of the DAP but will be between 32K (32,768) bits and 1 Mbit. The sum total of this PE memory is referred to as the DAP array memory.

The array memory is best regarded as a three-dimensional array of bits as in Figure 21, consisting of at least 32K "store planes". Each store plane consists of those bits at the same address within each PEs memory. On the DAP 500, for example, each store plane can be regarded as 32 rows (or 32 columns) of 32 bits each. The (i,j)th bit of a store plane is located in the (i,j)th PE.

Because of the bit orientation of its hardware, the DAP is not committed to any particular representation of data. In general, the DAP regards data simply as arrays of bits, the interpretation of which (as fixed or floating point numbers, for example) is entirely dependent on software. However, the hardware supports the parallel addition on N (edge size) pairs of rows or columns held in the register planes. Each row or column is thus regarded as an N-bit integer.

Two mappings particularly well suited to the DAP structure are:

- Vector (or horizontal) mode, in which successive bits of a word are mapped onto successive bits of a single row of a store plane
- Matrix (or vertical) mode, in which successive bits of a word are mapped onto the same bit position in successive store planes

The master control unit (MCU) performs the following functions:

- Instruction fetching, decoding and address generating
- Executing certain instructions, and broadcasting other instructions to the PE matrix for simultaneous execution by all PEs
- Transmitting data between the array memory of the PE array, and the MCU registers
- Providing hardware support for DO loop instructions
- Supporting data transfer between the DAP and the host filestore or attached peripheral devices.

The object code of a DAP program resides in the code memory known as the "codestore". Instructions are all 32 bits long. The size of the codestore depends on the model of the DAP, but it is between 128 K words (512 kbytes) and 1 M word (4 Mbytes). A schematic diagram of the MCU is shown in Figure 23.

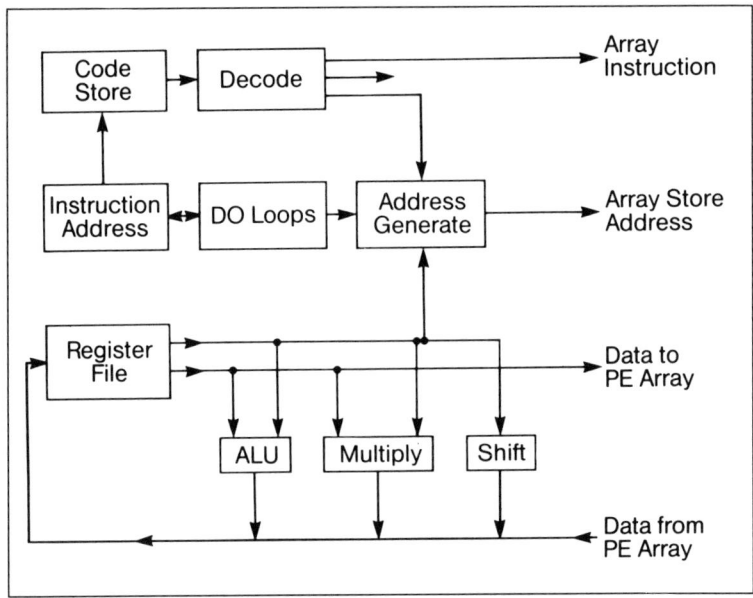

FIGURE 23. Schematic diagram of the master control unit.

The MCU has a number of 32-bit general-purpose registers that are visible to the machine language level programmer.

Registers can be loaded in various ways from the array memory, or a register's contents can be supplied as data to the memory or PEs. Logical and arithmetic operations are available to operate on the registers. Programs can test the contents of the registers and skip on certain conditions, or use them to hold link values for subroutine entry and exit. A register's contents can modify addresses or values; a register being used in this way is referred to as a modifier register. In all, there are 15 registers available to the machine-code programmer:

- M0 to M13 are general purpose registers, which can hold link values or data, and are operated upon by MCU arithmetic or logical functions. The contents of these registers can also be transferred to or from the array.
- M1 to M7 can always be used as modifiers. Register M0 is not generally available as a modifier, since value 0 in the instruction modifier field is interpreted as no modification.
- ME is the edge register, and is matched in size to the array edge dimension. In the DAP 500, since the edge size is 32 bits, ME is the same size as the other registers. The ME register is regarded as part of the array and can be used as the source or destination of data transferred to or from the array; it cannot in general take part in MCU arithmetic or logical operations or act as a modifier register.

In addition to the MCU registers, there are two 1-bit flags, the C-flag (carry) and V-flag (overflow) written to appropriately by MCU scalar arithmetic instructions on 32-bit signed or unsigned values.

There are three machine states:

0   Nonprivileged, interruptible (this is user mode)
1   Privileged, interruptible
2   Privileged, noninterruptible

The machine state affects the generation of instruction addresses and data addresses and affects the legality of certain instructions.

When in user mode, both codestore and array memory addresses are partitioned by "datum/limit" registers. Actual hardware addresses are obtained by adding the appropriate datum register, and a check is made that the limit is not exceeded. This permits protection of supervisor space from user space and also allows secure multiprogramming between user programs. User programs run in state 0 and deal only with addresses that are relative to array or code datum.

Instructions generate array memory addresses which, in the most general case, specify:

a 32-bit word
    within a row or a column
        within a store plane

Twenty (20) bits are used for the store plane address in the DAP (thus giving the architectural limit of 1 Mbit memory within each PE). This part of the effective address is known as the ADDR field.

Five (5) or 6 bits are used for the row or column address within a bit plane, depending on the model of the DAP: the DAP 500 (32 × 32) or DAP 600 (64 × 64). This is known as the INT field.

Zero or 1 bit is used to specify a word within a given row/column depending on the DAP model. This is known as the W field. (In the DAP 500, a word is the whole row or column and no separate word field is required.) Thus the full (32-bit) effective address takes the form:

```
.... ....A AAAA AAAA AAAA AAAA AAA1 1111  DAP 500
.... .AAA AAAA AAAA AAAA AAAA A111 111W  DAP 600
```

where A = ADDR field bits, 1 = INT field bits, and W = W field bits. Some instructions operate on whole store planes; in these cases the INT field of the effective address will be ignored or used for some other purpose. Some instructions operate on rows or columns rather than words, in which case the W field is not relevant to the array memory address.

In the most general case, an address is generated by a combination of four quantities:

- An address field in the instruction itself — this normally allows only 8 bits for the ADDR field
- Modification by an MCU register specified in the instruction
- Addition of a DO loop step on INT and ADDR fields
- Datum register

A hardware DO instruction is used to invoke a "DO loop", a sequence of instructions executed a specified number of times unless a premature exit is taken. These loops are used for operations such as adding successive bits of the operand. Provision of DO as a hardware feature means that there are no loop control overheads for these cases.

A feature of the DO loop is the facility for instructions operating in it to access, on successive iterations of the loop, successive bit planes, rows, columns or words of memory, or bits in an MCU register. This process is referred to as "address stepping". It is implemented by adding the DO loop iteration number to, or subtracting it from, addresses or values to give the addresses or values that are "effective" for that particular execution of the instruction. This is in addition to any register address modification. The iteration number used for address stepping is zero in the first pass of the loop and is incremented at the end of each pass of the loop.

Instructions which result in information being propagated through the nearest neighbor connections of the PEs — namely, the vector add and plane shift instructions — generate "direction, geometry and count fields". These can be specified in the instruction or picked up in a modifier register.

The "direction" is N, S, E or W. In the case of shifts, it gives the direction of the shift. In the case of vector adds, it specifies whether rows (E,W) or columns (N,S) are being added, as well as the direction of the propagation of the carry bits (i.e., which end is regarded as the most significant). The geometry specifies the behavior at the edges on shifts or carries: "cyclic geometry" implies wraparounds, "plane geometry" implies no wraparounds, with zeros being introduced at the edge. Plane and cyclic geometry can be independently specified for N/S and E/W. The count field can give the degree of shifting and the degree to which carries are propagated.

There are 16 groups of instructions. Of these, the first 14 groups contain array instructions, acted on by PEs. The operands of these refer typically to PE registers and a store plane, and the instruction will be broadcast to each PE to operate on its own bits within these planes. Some of the instructions move data between MCU registers and either 32-bit words or row/columns within the array. Note that in the latter case, fetching a row/column to an MCU register (rather than the edge register) will fetch only the least significant 32 bits; conversely, a memory write operation will expand data in the MCU register to the edge size by zeros at the top end. The edge register is, in fact, provided to assist in the writing of machine code which is compatible across different DAP edge sizes.

The other two groups of instructions (groups E and F) contain MCU scalar instructions performing logical, arithmetic and control operations. Note that status and control registers are memory-mapped, and there is no need for special instructions to perform, for example, I/O.

Medium speed I/O to the DAP can be achieved through the host or through the VME bus interface. However, fast data channels closely coupled to the DAP array are provided for data intensive applications. The DAP 500 provides transfer rates of up to 50 Mbytes/s both in and out.

On the DAP 510, the D-plane is a shift register having one bit per PE, and is thus arranged as 32 words of 32 bits each. A device coupler offering data output from the DAP may cause this register to be loaded with a specific bit plane from the array memory, and then shift the D-plane, causing successive 32-bit words to be shifted out from the D-plane. Conversely, a coupler offering data input may shift data into the D-plane a word at a time and then cause the complete D-plane to be written to a specified plane of array memory.

Loading the D-plane from the array memory, or writing it to that memory implies that the normal DAP instruction stream, which is typically using the array, is automatically suspended for one cycle while the D-plane access is performed. However, the shifting of the D-plane takes place autonomously and has no effect on the instructions stream. Thus, high speed input/output effectively proceeds in parallel with normal processing and has, at most, a few percent effect on the processing performance.

A specific system may have one or more device couplers; in the current physical implementation, there is a constraint of a maximum of four printed circuit boards for input/output, with each board normally corresponding to one coupler.

The DAP bus is composed of a 32-bit address/data bus, associated signals and an arbitration bus. It provides the mechanism whereby device couplers can receive commands or status requests from the rest of the system (usually from the MCU), or initiate loading or storing of the D-plane.

The DAP 500 includes, as an option, a VME bus backplane. The backplane is a full 32-bit address, 32-bit data implementation, and has four slots available. The maximum board size that can be used in the backplane is a 6U Eurocard. The host connection unit supplies all VME slot one functions (bus arbitration, etc.), and is capable of servicing all VME interrupt levels.

The DAP is interfaced to the VME bus as a VME memory slave with interrupt capability. The interface only supports extended VME addressing (A32) with longword accesses (D32). The interface can support transfer rates of approximately 4 Mbytes/s.

With the exception of the host connection unit, all the internal memory and control locations of the DAP are mapped into the VME bus address space. This enables, for instance, direct access to the array memory.

Access to the VME bus from user code executing on the DAP 500 is via system call to device driver software. At the kernal level, a system call of this type will cause a message to be sent to the device driver software resident in the host connection unit which actually performs the required action over the VME bus. This is necessary because the VME interface does not support the master control unit being a VME bus master.

From a programmer's point of view, a program designed to run on the DAP consists of two major parts:

- One part will run on the host machine and will be written in a standard language — generally FORTRAN or C. This will usually handle the I/O and user interaction: it will be created and compiled using the standard systems provided by the host machine, and will reside in the environment familiar to the user.
- The second part will run on the DAP. This will be written in an extended standard language which has features to handle the parallel facilities of the DAP.

The interaction between code on the host and code on the DAP is through subroutine calls which allow host code to make calls to DAP code and to send and receive data from the DAP. It will depend on the application whether it is better to program almost entirely on the DAP, or to run only the data-intensive parts on the DAP leaving the host predominantly in charge.

The programming software provided consists of several parts:

- Language System — compilers and assemblers
- Run-Time Support Systems — diagnostic tools and debugging facilities
- A Powerful Simulation System
- Program Development Systems
- Applications Support Libraries

There are two language systems for the DAP Series. These are an extended version of FORTRAN called FORTRAN-Plus and a powerful macro-assembler called APAL (Array of Processors Assembly Language). These interact with the language that the user selects to run on the host machine, typically FORTRAN or C.

FORTRAN-Plus: the version of FORTRAN designed for use on the DAP is extended with certain array processing facilities similar to those currently being considered for FORTRAN 8X. These enable the user to manipulate complete arrays as one object.

APAL: this gives the user complete control over the hardware of the DAP. In addition, it provides a macro environment for creating large, modular programs. APAL has powerful facilities for integration into FORTRAN-Plus software which enables the time-consuming parts of a program to be systematically converted from FORTRAN-Plus into APAL code if required.

A debugging facility, dapdb, gives access to the state of the user's program.

Through dapdb, the user can examine the values of variables and registers, and can view the state of the run-time stack. Break-points can be set in a user's program enabling dapdb to be entered dynamically from a running program. This is the "Program State Analysis Mode" — PSAM. Facilities exist for a user's program to trace out the values of selected variables at run time, and for a user to select the action to be taken by a failing program. Possible actions include dropping directly into PSAM.

The DAP Simulator is a fully hardware-compatible simulation system for the DAP. As such, it will run any program that the DAP will run (whether compiled or assembled) except for those that require further hardware, such as fast I/O devices. The simulator provides extra facilities such as the ability to make accurate timings of pieces of code, and to make a histogram of code usage.

Other software provided include a library maintenance utility, daplib, which enables libraries of FORTRAN-Plus object code to be built and maintained in a systematic fashion; sets of APAL macros to ease the calling of FORTRAN-Plus from APAL and vice versa; and libraries written in C and FORTRAN to allow code running on the host to call routines on the DAP.

These are libraries of subroutines designed to be called from FORTRAN-Plus. There is a general support library, which includes functions to manipulate matrices and vectors and common mathematical routines. There are also libraries for specific application areas such as image processing and signal processing.

FORTRAN-Plus is the main language provided for programming the DAP. It is a powerful implementation of recursive FORTRAN extended for the DAP. It therefore has features that are planned for the array extensions of FORTRAN 8X and allows the user to manipulate whole arrays of data as one object. The FORTRAN-Plus features map simply onto the DAP hardware, giving the user a clear abstraction of the DAP.

The most important feature of FORTRAN-Plus is the ability to manipulate complete data structures called vectors and matrices. These are extensions of FORTRAN arrays with parallel operations defined on them. (A "matrix" corresponds to a two-dimensional array, while a "vector" corresponds to a one-dimensional array.) FORTRAN-Plus includes certain syntactic extensions to provide a comprehensive environment for the manipulation of these objects. Operations on these parallel or "aggregate" objects include:

- The ability to operate on the complete object with no loop constructs (assignment, expressions)
- Selection operators that conditionally act on parts of these objects (indexing)
- Functions performing special operations, such as summing all the elements of a matrix or vector (aggregate functions)

The dimensions of the matrices and vectors used on the DAP must equal the "edge size" of the DAP (N for an N × N DAP). In FORTRAN-Plus, these dimensions are specified by "null subscripts". Therefore, on a DAP 500 the declaration

            integer X (,)

corresponds to the FORTRAN declaration

            integer X (32,32)

while on a DAP 600, it would correspond to the FORTRAN declaration

            integer X (64,64)

Future versions of FORTRAN-Plus will allow the mapping of arbitrarily-sized arrays onto the DAP.

In common with the most current implementation of FORTRAN, FORTRAN-Plus allows the user's input to be in upper or lower case; case is ignored everywhere except in character literals.

5.2. PARALLEL DATA TRANSFORMS

5.2.1. Introduction

Many computationally intensive problems have a very high degree of structured, fine-grain parallelism and benefit substantially from highly parallel execution. Such problems map efficiently onto truly parallel SIMD (single-instruction stream, multiple-data stream) processor arrays such as the AMT DAP[1] and the Goodyear MPP.[2] This is essentially because the control, synchronization and communication associated with the constituent parallel processes are implicit in the simplicity and regularity of the control and communication structure of such an architecture. To effectively use a processor array requires a mapping of the regular data structures and operations implied by the algorithms onto the regular structure of the hardware. This discussion looks at this general mapping problem with particular emphasis on an "indirect" approach known as "Parallel Data Transforms".

5.2.2. Computation on Processor Arrays

Computation on SIMD processor arrays is fundamentally different from computation on conventional serial computers. The former is characterized by a continuous cycle of activity, with each cycle comprising:

- routing of data between the different processing elements (PEs);
- arithmetic, logical or comparison operations on data local to each PE.

The local computation is easily implemented and naturally expressed using whole-array operations. For example, in FORTRAN-PLUS on the AMT DAP[3] the statement:

$$A = B + C \qquad (11)$$

where A, B, and C are arrays of the same shape will result in the addition of corresponding elements of B and C, with the results assigned to the corresponding elements of the array A. The main problem then becomes how to align the data so that such "element-by-element" local computation can be performed. This "data routing" problem is usually solved directly by providing explicit functions to route data; a simple example is the function transpose to interchange the rows and columns of a matrix.

A generalized form of many of the array processing primitives in FORTRAN-Plus has been proposed for the next FORTRAN standard.[4]

5.2.3. A Complementary Approach

Providing primitive array processing functions is a direct way to express parallel algorithms and maps very effectively on the SIMD processor arrays. It can be said to be the "mainstream" way of exploiting such systems. A complementary approach to handling the routing of data on processor arrays has been implemented on the AMT DAP and is known as Parallel Data Transforms (DPTs). A detailed introduction to DPTs is given in Flanders and Parkinson[5] and a formal description with some advanced applications in Flanders.[6]

The relation of PDTs to the direct methods of data routing described above is somewhat analogous to the relation between Fourier transform and direct methods of performing convolutions. Instead of working in the "physical space" defined by the multi-dimensional memory of a processor array, we work in a different space, the "mapping vector" space, in which the description of data mapping is greatly simplified for a significant class of mappings.

The benefits of the PDTs are as follows:

- simplification of the description of data mapping and movement, giving better insight into the mapping of an important class of problems onto parallel hardware;
- production of data routing code which executes faster than that obtained using ad-hoc techniques;
- simplification of the production of software for parallel processing;
- help in overcoming the constraints of fixed-size hardware;
- provision of a framework for the evaluation of different PE interconnection networks;
- provision of a theoretical basis for the design of a class of parallel hardware.

5.2.4. An Overview of the PDT Approach

The mapping of data on the memory of a processor array is represented in PDT terms by means of "mapping vectors". Each different mapping of an array of data, taken from a large class of possible mappings, is represented by a different mapping vector. The specification of a given data routing then requires only the specification of the initial and final mapping vectors. The form of the mapping vectors is such that code to perform the routing can be generated automatically from such a specification without explicit "programming".

In many cases, it is not necessary to give a complete description of the final mapping vector since all that is required is a mapping with a certain pairing of data. Mapping vectors have the property of explicitly indicating which data are paired (for important data pairings), making it easy to state what change is required to the mapping vector in order to achieve the desired pairings. The user of PDTs may state the simple change required to the mapping vector and the PDT system will automatically generate code to perform the implied data routing.

Thus, whether the data routing is specified in terms of a transformation between two mappings or in terms of a change required to achieve a certain pairing of data, the user always works in terms of mapping vectors (in "mapping vector" space) rather than in terms of the actual physical movement of data (in "physical" space). Specifications given in mapping vector space are processed by the PDT system to produce data routing code.

The PDT approach outline above gives a powerful tool for using processor arrays effectively. This is because the mapping vectors:

- concisely represent each mapping within a large class of important mappings;
- cater for multi-dimensional PE arrays;
- cater for multi-dimensional arrays of data;
- handle cases where the size and shape of data differ from that of the PE array;
- explicitly indicate how data is paired-up within each PE;
- represent each mapping as a data structure rather than as a procedure, in order to permit definitions of mappings to be manipulated at run-time.

Furthermore, code generated to perform data routing is able to make efficient use of the parallel data routing instructions and PE interconnection network provided by the hardware.

5.2.5. Implementation

Mapping vectors are concise, one-dimensional objects and any changes to them can be viewed as a permutation of the mapping vector elements (in fact "inversions" of mapping vector elements are also possible). Such permutations can be achieved by a suitable sequence of simpler changes to the mapping vector (for example, exchanging pairs of mapping vectors elements); these simpler changes are called "generators". The PDT system decomposes the overall transformation to the mapping vector into a sequence of generators which will lead to efficient realization of the transformation on the given hardware. Code is then generated to invoke these generators in the prescribed order.

Exchanges of pairs of mapping vector elements are good candidates for generators on processor arrays since any permutation of mapping vector elements can be achieved by a sequence of them, and because they are efficient to implement on an array of PEs having no more than local connectivity.

5.2.6 Summary

The task of using SIMD processor arrays effectively is dependent on having efficient means of mapping and routing data. Direct methods of handling data mapping and routing are available and they make use of the high level abstraction of working with arrays as complete objects rather than looping over individual array elements. A complementary method of handling the mapping and routing of data has been outlined here which reduces the complexity of the problem by working in the space of "mapping vectors" rather than in physical space. Software exists on the AMT DAP to implement this and has been used to produce efficient solutions to a variety of problems.

5.3. SOLUTION OF A LARGE SYSTEM OF EQUATION ON DAP USING A HYBRID GAUSS/GAUSS-JORDAN METHOD

5.3.1. Introduction

Effective use of parallel processors requires algorithms tailored to the degree of parallelism and structure of the computer. Often parallel algorithms are simple extensions of serial ones or involve a rediscovery of techniques well-known, but not normally used on serial machines.

An example of the latter arises in the solution of a system of equations. The normal serial method is Gaussian elimination. This involves taking each column (the "Pivot column") in turn and zeroising its below diagonal by subtracting multiples of the corresponding row (the "pivot row") from all those below it. The resulting upper triangular system is solved by back substitutions.

Gauss-Jordon elimination reduces the matrix to diagonal form by, at each step, subtracting multiples of the pivot row from those above it as well as below. The method is not normally used since it is slower; with an $n \times n$ matrix and one right-hand side Gauss-Jordan needs about $n^3/2$ add plus multiply operations whereas Gauss needs about $n^3/3$.

On DAP and with a matrix no larger than the DAP size, Gauss-Jordan elimination can be done in the same time as Gauss, and is faster overall since there is no back-substitution stage. The technique has been applied to matrix inversion using a method that progressively converts the input matrix into its inverse.[1] Full pivoting is used; finding the maximum of an array is very rapid on DAP and the necessary reordering of rows and columns is done at the end. Also the arithmetic operations on the pivot row and column are implemented as a subset of those on the rest of the matrix so there is negligible overhead for those special case computations.

The remainder of this discussion is concerned with applying these techniques to the solution of matrix equations where the matrices are larger than the DAP array, but small enough to be held entirely within DAP store.

5.3.2. Principles of the Equation Solver

The input matrix is considered as being partitioned into "sheets" that match the DAP in size, the sheets forming a two-dimensional matrix set. Thus, on a 64×64 DAP each sheet or matrix holds a 64×64 neighborhood of the large matrix. As before, each step of processing involves selecting one row of the large matrix as pivot row. Now this row extends across several sheets and the basic arithmetic involves taking each section of pivot row in turn and subtracting appropriate multiples of it from rows in its own sheet and in sheets in the same column of sheets.

The detailed form of the elimination will now be derived. First, likening sheets of the large matrix to individual matrix elements on a serial processor, it is clear that reduction to block upper triangular form is best. This can be regarded as Gaussian elimination applied to sheets. The general form of the final matrix is illustrated for a hypothetical 4×4 DAP and a 16×16 matrix (see Figure 24A).

In general, each of the elements A is non-zero and different from the others, as are B, C, D, F, G, H, K, L and U.

To make the large matrix upper triangular each of the diagonal sheets identified by A, F, K, U needs to be made upper triangular. However, as already noted it is as fast to make those sheets unit matrices by Gauss-Jordan elimination.

As a typical elimination step the matrix has the form shown in Figure 24B.

The next pivot column is identified by **. After completion of the forward elimination, the matrix has the form shown in Figure 24C. The right-hand side is also shown, corresponding transformations having been applied to it.

With this form of matrix, back-substitution is simple and rapid. Elements Z of the right-hand side are already equal to the corresponding result elements. To compute results corresponding to Y simply form a matrix each of whose rows is equal to vector Z and multiply element-by-element by matrix L. Then sum along the rows of the result and subtract from vector Y. Similarly results corresponding to vectors X and W are computed in turn using previously computed result vectors.

The usual column pivoting may readily be introduced in the elimination stage. This involves finding the position of the largest element in that part of

```
A A A A    B B B B    C C C C    D D D D
A A A A    B B B B    C C C C    D D D D
A A A A    B B B B    C C C C    D D D D
A A A A    B B B B    C C C C    D D D D

0 0 0 0    F F F F    G G G G    H H H H
0 0 0 0    F F F F    G G G G    H H H H
0 0 0 0    F F F F    G G G G    H H H H
0 0 0 0    F F F F    G G G G    H H H H

0 0 0 0    0 0 0 0    K K K K    L L L L
0 0 0 0    0 0 0 0    K K K K    L L L L
0 0 0 0    0 0 0 0    K K K K    L L L L
0 0 0 0    0 0 0 0    K K K K    L L L L

0 0 0 0    0 0 0 0    0 0 0 0    U U U U
0 0 0 0    0 0 0 0    0 0 0 0    U U U U
0 0 0 0    0 0 0 0    0 0 0 0    U U U U
0 0 0 0    0 0 0 0    0 0 0 0    U U U U
```

FIGURE 24A. 16 × 16 block upper triangular matrix.

```
                 *
                 *
1 0 0 0    B B B B    C C C C    D D D D
0 1 0 0    B B B B    C C C C    D D D D
0 0 1 0    B B B B    C C C C    D D D D
0 0 0 1    B B B B    C C C C    D D D D

0 0 0 0    1 0 0 F    G G G G    H H H H
0 0 0 0    0 1 0 F    G G G G    H H H H
0 0 0 0    0 0 1 F    G G G G    H H H H
0 0 0 0    0 0 0 F    G G G G    H H H H

0 0 0 0    0 0 0 J    K K K K    L L L L
0 0 0 0    0 0 0 J    K K K K    L L L L
0 0 0 0    0 0 0 J    K K K K    L L L L
0 0 0 0    0 0 0 J    K K K K    L L L L

0 0 0 0    0 0 0 S    T T T T    U U U U
0 0 0 0    0 0 0 S    T T T T    U U U U
0 0 0 0    0 0 0 S    T T T T    U U U U
0 0 0 0    0 0 0 S    T T T T    U U U U
                 *
                 *
```

FIGURE 24B. Intermediate matrix.

the pivot column on and below the diagonal. The corresponding row becomes the pivot row and is interchanged with the row having the same number as the pivot column.

1	0	0	0	B	B	B	B	C	C	C	C	D	D	D	D	W
0	1	0	0	B	B	B	B	C	C	C	C	D	D	D	D	W
0	0	1	0	B	B	B	B	C	C	C	C	D	D	D	D	W
0	0	0	1	B	B	B	B	C	C	C	C	D	D	D	D	W
0	0	0	0	1	0	0	0	G	G	G	G	H	H	H	H	X
0	0	0	0	0	1	0	0	G	G	G	G	H	H	H	H	X
0	0	0	0	0	0	1	0	G	G	G	G	H	H	H	H	X
0	0	0	0	0	0	0	1	G	G	G	G	H	H	H	H	X
0	0	0	0	0	0	0	0	1	0	0	0	L	L	L	L	Y
0	0	0	0	0	0	0	0	0	1	0	0	L	L	L	L	Y
0	0	0	0	0	0	0	0	0	0	1	0	L	L	L	L	Y
0	0	0	0	0	0	0	0	0	0	0	1	L	L	L	L	Y
0	0	0	0	0	0	0	0	0	0	0	0	1	0	0	0	Z
0	0	0	0	0	0	0	0	0	0	0	0	0	1	0	0	Z
0	0	0	0	0	0	0	0	0	0	0	0	0	0	1	0	Z
0	0	0	0	0	0	0	0	0	0	0	0	0	0	0	1	Z

FIGURE 24C. Matrix after forward elimination.

On DAP the relevant part of the pivot column will in general be spread across more than one sheet. In the present implementation, that part of the pivot column is first extracted from the various sheets in order to find the maximum in a single operation. The factors by which rows of the large matrix are multiplied are obtained by dividing the pivot column by the pivot element. This is done in a single matrix-scalar division operation on the extracted data.

5.3.3. Implementation Details

This section gives more detail of a DAP-FORTRAN demonstration program using the principles above. It can solve any size system subject to there being sufficient memory to hold the entire matrix. The right-hand side is combined with the n × n matrix as column (n + 1) and manipulated concurrently with the matrix; clearly if n is a multiple of the DAP dimension a small performance improvement could be achieved by processing it separately as a vector.

The following description follows the layout of the DAP program.

1. Initialize counts related to size of matrix.
2. Begin outer loop over columns of sheets.
3. Begin loop over columns within the sheet. Steps (2) and (3) identify a "current column". There is a correspondingly numbered "current row" and "current element".
4. Extract sections of the pivot column from the diagonal sheet and those below it; place in a matrix.

5. Find the position and value of the pivot element by examining the matrix formed in (4).
6. In the matrix of (4) place the diagonal (or current) element in the pivot element position and the value −1.0 in the diagonal element. Then divide the entire matrix by the pivot. Thus, the diagonal element contains −1.0/pivot.
7. Begin inner loop over columns of sheets beginning with the sheet-column that contains the column after the pivot column.
8. Extract the section of pivot row in the sheet-column selected at (7) and form a matrix each of whose rows is equal to this vector. Also copy the vector into the section of the current row.
9. In the main part of the matrix the arithmetic to be performed may be expressed informally as:

```
M = M-MATR(PROW) * MATC [PCOL/PIVOT] .... (a)
```

i.e., multiples of the pivot row are subtracted from each row. However, the pivot row (in the "current row" position) is to be divided by the pivot element:

$$PROW = PROW/PIVOT$$

or

```
M(Pivot Row) = - MATR(PROW) * [-1.0/PIVOT]
               .....(b)
```

The expression inside [ ] has been evaluated at step (6) for both cases (a), (b) and the MATR has been done at step (8). This step zeroises a section of the "current row" of the matrix, enabling step (10) to use expression (a) for case (b) also.

10. Loop over sheets containing the current row and those below it performing the operations of (9a) above. Each pass of the loop uses a different column selected from the pivot column matrix evaluated at step (6).
11. Loop to step (7) (next column of sheets).
12. Loop to step (3).
13. Loop to step (2).

At this point the elimination is complete and the matrix conceptually has the form of Figure 24C. In practice the diagonal and below-diagonal sheet will not be needed and they are left undefined.

14. Begin loop over sections of the result vector in reverse order beginning with the penultimate.

15. Loop along the sheet-row identified at (14) for the above-diagonal sheets of the matrix. Broadcast result selection using MATR, multiply element-by-element by sheet of matrix and accumulate.
16. Apply SUMC function to the matrix resulting from (15) and subtract the result vector from the right-hand side section giving the result section.
17. Loop to step (14).

5.3.4. Performance

Most of the time is spent in step (10) and most of this is highly parallel arithmetic. Thus, the method makes very effective use of DAP.

As might be expected the solution time is $0(m^3d)$ where the matrix is partitioned into (m × m) sheets each of size (d × d) to match the DAP array.

As an example, solution of a 501 × 501 matrix with one right-hand side took 9.5 s on a 64 × 64 DAP.

5.4. AN IMAGE UNDERSTANDING PERFORMANCE STUDY ON THE DISTRIBUTED ARRAY PROCESSOR

5.4.1. Introduction

A performance study has been carried out for the ICL Distributed Array Processor (DAP) on some aspects of image analysis. The algorithms studied were the application of a "Mexican Hat" shaped convolution, area filling, feature extraction and classification via a statistical classifier. Starting with a digitized image the goal is to process the image in such a way that objects of interest, within the field of view defined by the captured image, can be identified and their features recognized. The output from the entire process is the identification (by predefined classes) of these objects.

The Distributed Array Processor on which this study is based is the "Mil-DAP" version[7] of the larger machine initially available in 1980[1] (Goodyear's MPP[8] has a somewhat similar architecture). Previous work in low-level image processing on the ICL DAP includes that of Bird,[9] Reddaway,[10] and Marks.[11]

5.4.2. Overview

The main stages in the process are (1) image capture, covering the methods of inputting the image into the DAP in suitable format, and (2) difference of Gaussians (DOG) Channel application. The DOG, or Marr-Hildreth operator was postulated as the mechanism used by the eye to identify shapes of various sizes. We use a series of DOG operators of reducing size, and track the progress of "blobs" (positive valued points in the DOG convolution) through these channels. Thus it is possible to identify blobs in the finer channels which are part of a larger blob in the coarser channel. In practice we take one blob in the coarser channel and identify by looking for overlapping blobs in all the intervening ones. This involves techniques of area filling and labeling.[12] We then attempt to classify these children blobs by taking various measurements of "features". These features are used in a Mahalanobis distance classifier, a statistical method which relies on the prior calculation of covariance matrices for each set of features from some training examples. The application of this method has been discussed in Sabey.[13]

This paper is a detailed design study of a possible implementation of the application. It makes extensive use of the variable precision capability of the DAP, including single bit operation. When considering performance, it is necessary to consider bit length of operands and the precise quantity and distance of values to be shifted between PEs.

The DAP can be used in two modes, matrix or vector. In matrix mode we hold operands one per PE and perform 1024 operations simultaneously. In vector mode we map 32 bit items across PEs so that each item is shared between 32 PEs. In this mode 32 operations can be performed simultaneously, at a slightly faster rate (per set of operands) than in matrix mode.

The following timings are used in the performance calculations: Routing: between PE registers: 1 cycle per 2 places. Adds: 3 cycles per bit plane (of larger operand) generally, 2.5 cycles per bit plane when the result overwrites one of the operands. Multiplies: at the bit level these are a series of adds, so the cost is, asymptotically, 2.5 cycles times the product of the operand lengths. Multiplying by a scalar is much cheaper, since we can take advantage of the bit pattern in the scalar (e.g., skip zero bits). Multiplying by a power of 2 implies simply adjusting the start address of the item in PE store. MAX function: the maximum value from a set of matrix mode operands can be found at 7 cycles per bit plane.

5.4.3. Image Capture

The field of view is scanned twice for each image. Each scan is digitized into a raster of 512 horizontal by 256 vertical 8-bit gray-scale values (GSVs) or pixels. The image is input to the DAP via the Fast I/O (FIO) unit in the form of two interleaved scans to produce a complete image size of 512×512 pixels. The two extreme ways of mapping an oversize (compared with DAP array size) pixel array are "sheet" and "crinkled". In the case of sheet mapping, the image would be cut into local arrays each of the same size as the DAP, with the sheets stored in separate layers in the DAP. This form of mapping is appropriate where most of the computation relates to small regions of the image, since it incurs a high cost in operations which cross the boundaries between sheets. For this study we will use "crinkled" format. That is, each 16×16 area of the image is stored in row major order within a single PE (the PE's patch) such that corresponding positions in neighboring PEs differ by 16 pixels. The organization of the pixel values into this mapping is largely done as an asynchronous process in the FIO unit, independent of the DAP processor, with some cycle stealing to perform the store operations.

Timing: The size of each $1/2$ image scan is 128 kbytes. The FIO has two buffers, each of 64 kbytes, so we can read $1/4$ of a complete image from one of the buffers into DAP store in one operation. However, since we want to organize the pixels in crinkled format in the DAP store, successive interleaves will be stored in distinct layers in the DAP. Thus for each $1/4$ image bit plane we can transfer half of a DAP bit plane directly into the required position, stealing a cycle from the DAP processor for each transfer. For the first half of each DAP bit plane, the corresponding second half will be in the next $1/4$ image bit plane, i.e., in the other FIO buffer. It is possible to merge the two half bit planes in the buffer by retrieving the first $1/2$ bit plane, reading in the remainder, and storing the whole plane back into DAP store. For each whole DAP bit plane we have had to do one store to get the first $1/2$ in; one load to get it back again; and another store to copy the whole plane into DAP; i.e., 3 cycles.

The number of cycles stolen per image is thus $3 \times 256 \times 8 = 6144$, which at 150 ns per cycle takes 0.922 ms.

If the scanning device operates at 50 frames per second (12.5 kbytes/s), the load would be 0.31 Mcycles/s; i.e., a load factor of 4.6%, and well within the maximum transfer rate of the FIO which is 40 Mbytes/s.

5.4.4. DOG Convolutions

The purpose of this section of the process is to determine which parts of the picture represent objects of interest. This is done by applying a multi-channel DOG (difference of Gaussians) filter[14] to the pixel values previously stored. The convolution function consists of two Gaussian operators, an excitatory Gaussian with $\sigma = \sigma_e$ and an inhibitory Gaussian (subtracted) with $\sigma_i = 2\sqrt{\sigma_e}$. It can be expressed as:

$$DOG(\sigma) = (1/2\pi\sigma^2) [\exp(-x^2 + y^2)/2\sigma^2)$$
$$-0.5(\exp(-(x^2 + y^2)/4\sigma^2)] \qquad (12)$$

where σ varies from $8 \times \sqrt{2}$ to $\sqrt{2}$ by factors of $\sqrt{2}$. The different values of σ define the various channels; x and y are measured from the point under consideration and their max/min values are allowed to extend to 3σ in each direction. As σ decreases, the significant area covered decreases, the x and y decrease and the channel number increases from 1 to 7. Channel 1 is the coarsest channel; channel 7, the finest.

The DOG widths (3σ) (rounded up to next odd number) vary with channel number as follows:

Channel:	1	2	3	4	5	6	7
Width:	69	49	35	25	19	13	9

We can either apply this DOG function using Fourier transforms, or directly as discussed here. From Reference 8 it is clear that the two orthogonal components in x and y can be applied independently and successively thus:

$$DOG(\sigma) = (1/2\pi\sigma^2)[\exp(-x^2/2\sigma^2)\exp(-y^2/2\sigma^2)]$$
$$-(1/4\pi\sigma^2)[\exp(-x^2/4\sigma^2)\exp(-y^2/4\sigma^2)] \quad (13)$$

This represents four linear operations, i.e., the application of the two differently shaped Gaussians in the two dimensions X and Y. However, since successive channels differ in σ by a factor of $\sqrt{2}$ and the two Gaussians making up the DOG expression also differ in σ by a factor of $\sqrt{2}$ we can apply the Gaussians separately and use the partial results in two successive channels. This nearly halves the work provided there is space to store the intermediate results.

The output required from each DOG channel is a binary mask, set false wherever the value of the convoluted image has a negative value and true wherever the result of applying the convolution is zero or positive. Also, from the first DOG channel ($\sigma = 8\sqrt{2}$) the values of the convoluted image are required.

5.4.5. Direct Application of Convolutions

A direct application is more efficient than using FFTs when the extent of the convolution is small. To achieve this, we can scale the image and the DOG operator by averaging each 4×4 pixel square for the larger DOGs, and each 2×2 square for the middle ones. This keeps the size of the applied DOG operator to less than 20 pixels width. We will apply the two Gaussians separately and save intermediate results so that they may be used in successive steps. However, some points in the Gaussian operator that are initially used in a scaled image for only one result point will be used in a subsequent, less scaled channel for four result points.

Each result is obtained by accumulating the value at a point with some of its neighbors, having multiplied those neighbors by various amounts depending on the Gauss function. This can be done either by performing all the routing operations first to align all the values, and then applying the multiplies and adds; or by doing all the multiplies first and then the routing and accumulation. Although the first method reduces routing costs, it puts each number in two separate places for the multiply stage, thus almost doubling the arithmetic cost. We will perform the multiplies first, at the cost of shifting longer values for the summing process. The actual DAP space routing costs depend on how close the result pixel is to the edge of the PE's patch, and the packing density of image points to PEs. For example, with 8×8 patches in each PE and a DOG width of 19 (see DOG.2 below) the extreme edge and most internal result points are shown below:

PE M – 1	PE M	PE M+1	PE M+2
00000000	00000000	00000000	00000000
. .	. . . . *	. . . .	.

Shifts: $2 \times 1 + 7 \times 0 + 8 \times 1 + 1 \times 2$.

Total routing for edge result point = 10 @ 1 + 1 @ 2.

PE M-1	PE M	PE M+1	PE M+2
00000000	00000000	00000000	00000000
.....	....*...	......	

Total routing for inner result point = 11 @ 1.

Key: 0   the eight points per PE
      *   position of result point
          position of points to be added

Thus for the four different positions of result point, the total routing is 10 @ 1 + 1 @ 2 + 3 × (11 @ 1). Counting one cycle for each shift of length one or two, routing per result bit plane is 11 cycles.

Since the application of the Gaussians is symmetrical in the X and Y directions, the timings will be calculated for one dimension and we will multiply everything by a little more than two (to allow for increased numeric range — i.e., bit length) at the end.

The process involves the following steps:

1. Calculate the average over each 2 × 2 square of gray-scale values; i.e., reduce each set of four numbers to a single one. The original 256 points per PE are now reduced to 64 points. Each of these points requires 3 adds and a divide by 4, on 8-bit data. At 3 cycles per bit plane for an add (Note: dividing by 4 is a null operation), total is 64 × 3 × 8 × 3 = 4008 cycles.
2. Repeat (1) for each 2 × 2 square within the set of averaged values output by 1; number of points per PE now reduced to 16. Cost = 1152 cycles.
3. Apply the coarsest Gaussian operator to the output of 2 (scaled image 128^2 points) using precalculated weights. The width of the scaled operator is 19 pixels. Since we are eventually interested only in the relative values of the convolved values, the precalculated values can be scaled so that the central value is always a power of 2. This saves a multiply on these points and ensures that the DOG values all lie within a reasonable range. The direct method is applied in directions X and Y separately.
4. Apply the next coarsest Gaussian as above; scaled image size = 128^2, Gaussian width = 13.
5. Subtract the output of 4 from 3 giving the first DOG output. This is 16 matrix subtracts.
6. Apply the next coarsest Gaussian. For the second DOG channel, the inhibitory Gaussian is the same as the excitatory Gaussian in DOG.1. So we need only calculate the excitatory Gaussian for this one and subtract to produce the DOG. Here we apply a Gaussian of linear size 19, i.e., same operations as in (3). but this time there are 64 points per PE.

7. Subtract the output of 6 from the previous one; same as (5) but with 64 points.
8. Repeat (6) and (7) until the seven different channels have been completed. As the different grades of coarseness are applied, the scaling of the image will change so that generally the number of points in the scaled version increases from 128×128 up to 512×512 and the operator size always lies between 9 and 19.

    Each inhibitory Gaussian is larger than the corresponding excitatory, and we use each excitatory as the following channel's inhibitory. Thus to maintain precision at the finest channel, we apply the last excitatory Gaussian with the same extent as the inhibitory.
9. The output of each DOG channel is an array of convolved greyscale values, of which we are interested in the positive valued points. We can thus reduce the output to a 1-bit mask by simply setting a bit array to true wherever the DOG is positive, and false elsewhere.

For all the DOG Channels, operations are:

	Mults	Adds	Routing (@ 10 bits)		
DOG.1 G_i	9	18	10×15.5	$\times$	16 pts/PE
DOG.1 G_e	6	12	10×9	$\times$	16 pts/PE
DOG.2 G_e	9	18	10×11	$\times$	64 pts/PE
DOG.3 G_e	6	12	10×5.25	$\times$	64 pts/PE
DOG.4 G_e	9	18	10×4	$\times$	256 pts/PE
DOG.5 G_e	9	12	10×2.6	$\times$	256 pts/PE
DOG.6 G_e	4	8	10×1.3	$\times$	256 pts/PE
DOG.7 G_e	4	8	10×1.3	$\times$	256 pts/PE

Including the subtract steps (which count the same as an add) the total operation counts are 7088 multiplies, 15344 adds and routing of 37872 cycles.

In the case of the multiplies, the operands are an 8-bit GSV and a precalculable scalar value on the Gauss function which would be held with a precision of 9 or 10 bits to allow for rounding error. The length of the result also needs to be 10 bits. Since all elements in the array are to be multiplied by the same value we can use the fast scalar multiply method implemented by MacQueen,[15] which typically performs such multiplies well within 80 cycles. The operands of the adds go from 10 bits to 14 bits but we will conservatively count them all at 14 bits, which at 2.5 cycles per bit plane gives 35 cycles per add; giving a total operation count for the application of the DOGs in one dimension of 1142 kcycles.

If we allow 20% extra for the second dimension and adding the cycle counts from steps 1 and 3 gives a total of 2518 kcycles, which at 150 ns per cycle is 0.38 s.

5.4.6. Segmentation

In this part of the process we use the multiple DOG channels to select those areas in the picture which are of interest. First we find the blob containing the highest value in the coarsest DOG channel and then find all the blobs in successively finer channels which are overlapped by this blob, recursively. After the final step, a mask of interesting image segments is defined, which can be used in the computation of image features.

The amount of work in the propagation is reduced by splitting it into two levels: global and local. Continuation of a blob from one side of a patch to the other can be determined by ANDing bits together in the direction of scan, as discussed below. In detail the operations are:

1. Find the maximum value in DOG.7.
2. Using this as the initial seed point do one in-PE propagation up to the zero crossing in DOG.7. A logical mask indicates which pixels are a part of any blob, and another logical mask shows which pixels are parts of "interesting blobs" (i.e., connected to seed points). We AND each bit in the patch (area of image within the PE) in turn from the blob mask with its nearest neighbors in the "interesting" mask to propagate the interesting mask 1 bit at a time. The scanning process has a major and a minor direction (e.g., left and down), and in checking whether the bit is next to a marked one we only need to look "backwards" in these two directions. These directions determine the general drift of the propagated front, and it is necessary to perform two scans in opposing directions (e.g., left and down; right and up) to ensure that the edges of any convex shape are reached. A double scan of this form will also fill moderately reentrant shapes, and will be sufficient to ensure that at the edges of the appropriate 16×16 square, the pattern to be propagated into the next PE's patch is known. If the blob is locally more reentrant than this allows for, the other parts will normally be filled in the course of stage 7. Optimal use of DAP registers reduces the cycle count for a scan to 3.5 per bit plane.
3. In each 16×16 patch AND together each row and each column in the positive DOG bit map output from DOG.7 to produce two orthogonal 16-bit "scaled" components. Each bit in these scale vectors represents whether or not a negative DOG.7 value occurs anywhere in the corresponding row or column of the 16×16 patch.
4. For each of the four edges in all image patches, AND each bit with corresponding bits in neighboring PEs' scaled components, creating a propagation set in each successive neighboring PE up to the edge of the image. Each PE is to have two 16-bit propagation sets, one for the north-south and one for the east-west propagations. This step will assign values to one of these vectors in each of the four neighboring PEs.
5. We can now propagate at a global level; to north and south using scaled columns, and to east and west with the same method as the local one, ANDing scaled blob components with row/column propagation set bits instead of blob mask bits with "interesting" mask bits. When this is done,

unless the blob is very convoluted, we will have marked each PE which has some of its area within the required blob with appropriate values in the scaled vectors.

6. Within each PE we must now use the results of the global propagation as seeds for the local propagation step within each PE. The scale and propagation vectors are ANDed together into the corresponding position and the edges of the two-dimensional patches.
7. This is a repeat of step 1 but to take account of reentrant blobs it is prudent to repeat the double scan locally before checking if another global propagation step is needed.
8. To check if the process is complete we can perform another step such as that described in (7) and check for any change; but to check that no more PEs need be involved we must also ensure that all propagations across PE boundaries are complete, so when the scan reaches each edge point we must continue the scan one further point into the neighboring PE.
9. The process described above will fill all convex shapes and many convoluted shapes as well. Some very reentrant shapes, such as particularly well developed, thin spirals, and dumb-bell shapes in which the bar is long, thin and at an angle, may not be completely filled. To cope with some of these we can repeat steps 5 to 7 two or three times. If more than this is required then we can assume that such blobs are not likely to be interesting within the constraints of this study; i.e., that a "rogue blob" has entered the calculation, so we should stop the processing at this stage and discard such blobs. However, in order to identify which blobs are to be discarded, a mask growing process is required in "grow back" from the points identified in step 8 as being in incomplete blobs, to cover the previously grown mask area.

Timing

The costs in DAP cycles of the operations described above are as follows:

Operation	Cycles/bit plane	Cost
1. Find max in DOG.7	7	31256
2. In-pe prop'n	3.5	1792
3. 16-fold scaling	18/result	576
4. Initial prop'n	12	768
5. Global prop'n	12 (·2)	1536
6. Seed Gen (in-pe)	3	96
7. In-pe prop'n	3.5 (·2)	3584
8. Complet'n check (scan)	3.5	1792
Complet'n check (store prev.)	1	512
Complet'n check (ck neighbrs)	5	320

Step 1 is only done once. In a reasonably favorable case we will need to do steps 2 to 4 followed by 5 to 9 only once, when step 9 will indicate that the propagation is complete (note that each scan step already contains some redundancy).

In a worst case we can count the operations to include 3 performances of steps 5 to 8 and double the result to allow for grow-back (except that step 9 need only be done once); the total cycle is 44812 cycles.

This process is performed once for each of DOG channels 1 to 6. Total operation count for propagation through DOG channels is thus 300 kcycles; which, at 150 ns per cycle gives a time of 0.045 s.

5.4.7. Segment Labeling

This follows the scheme for mask growing from seeds, except that instead of propagating single bits, labels are used. The first task is to initialize as few points as possible in each blob as potential labels. This is done by applying the test shown below to ensure that a point on each top left hand corner is chosen; it also ignores isolated positive pixels.

a	b	c	For x:
			AND x with
d	x	e	NOT d AND NOT b
			AND (NOT c OR NOT e)
f	g	h	AND (g OR e)

The logical processing can be done in 11 cycles per bit plane. For the routing costs, since each move is one place in image space, only 1 in 16 need shifting between PEs; routing cost is thus 5/16 (*2) cycles per bit plane.

Having established which points in each blob are potential labels we must perform a "SUM LEFT" to give each potential label point a unique value. Clearly some blobs will generate multiple labels with this test, and when these labels are propagated over the entire blob we must ensure that a single label is consistently dominant. This is achieved by comparing the propagated values with values already in the blob, and only overwriting old values when the new one is smaller.

The differences between this section and the mask propagation step are the propagation of labels and the comparing process before each write. The size of the labels to be propagated can be minimized to the number of bits in the largest label found during the label initializing process. The maximum number of possible labels from this process is $1/6$ the number of pixels (i.e., a label bit length of 16 for the whole image, or 6 bits within a PE), but typically the size will be ten or fewer bits (for whole image).

Timing

The costs in DAP cycles for the above operations are as follows:

Operation	Cycles/bit plane	Cost
3 × 3 convolution	11	2816
3 × 3(routing)	10/16	160
SUM LEFT:		
(in pe, ave leng. 4)	12	3072
(ex pe, ave leng. 12)	36	360
(add pe start vals)	16 × 3	12288
label prop'n; as per mask prop'n, multiplying relevant parts by 10		327k

Total operation count for the segment labeling is 346 kcycles, which at 150 ns/cycle = 0.052 s.

5.4.8. Feature Generation

We now have a set of labeled segments, from which we can extract the features required by the classification process. This is a statistical classifier which, for the purposes of this study, will require eight features.

Studies are still in progress as to the best features to use, but taking a representative set to include blob contrast, perimeter length^2/area and the various moments (rotational invariant, skew invariant, etc.),[16] we can define a set of feature primitives, from which a large set of features may be calculated.

Most of the feature primitives require a calculation at each pixel, followed by a summing of the results over the area of the segment. The set of feature primitives we will use here are:

I. At each point, summed over blob:
   (a) Perimeter contribution (explained below)
   (b) 1 (for calculation of total area)
   (c) The gray-scale value (GSV) (for total weight)
   (d) x.GSV (for calculating x (see Reference 2))
   (e) y.GSV (for calculating y (see Reference 2))
   (f) x.y.GSV
   (g) x^2.GSV
   (h) y^2.GSV
   (i) x^3.GSV
   (j) y^3.GSV
   (k) $x.y^2$.GSV
   (l) $x^2.y$.GSV

II. For the whole blob:
   (m) Maximum GSV
   (n) Minimum GSV
   (o) Maximum value of x

(p) Minimum value of x
(q) Maximum value of y
(r) Minimum value of y

The values of x and y in the above are distances in the X and Y dimensions, respectively, of each pixel in the blob, measured from some local (to the blob) origin. We can reduce the total work by splitting x and y into global and local components. Thus, if i and j are addresses for a whole blob in x and y respectively, we can replace x_i by $(X_p + x'_m)$ and y_j by $(Y_q + y'_n)$, where X and Y are PE addresses within a blob and x', y' are pixel addresses within the PE's patch. For example:

$$\sum x_i^2 y_i \, GSV_{ij} = \sum (x_p^2 y_q \sum (GSV_{p+m,q+n})$$
$$+ 2x_p y_q \sum (x'_{p+m} GSV_{p+m,q+n})$$
$$+ x_p^2 \sum (y'_{q+n} GSV_{p+m,q+n})$$
$$+ 2x_p \sum (x'_{p+m} y'_{q+n} GSV_{p+m,q+n})$$
$$+ \sum (x'^2_{p+m} y'_{q+n} GSV_{p+m,q+n})) \qquad (14)$$

Apart from the primitives listed above, with the x and y replaced by x' and y', we also need to calculate the X and Y for each blob separately. First the addresses of the PEs in the DAP array are found, then we compute X_{min} and Y_{min} for each blob and subtract them from the PE addresses. This will typically reduce the bit length to less than 4 bits (we will use 4 for timing). The primitive feature values are calculated by summing individual values at each pixel, over each blob.

Timing
The costs in DAP cycles of the stages in the process are as follows:

Operation	Cycles/bit plane	Cost
SUM LEFT (for pe address)	n/a	323
Calculations at each pixel:		
perimeter contrib'n		
(add neighbors)	8.25	2112
(8bit GSV, x',y' 4bit)	96 (*2)	24.6k
x'^2GSV, y'^2GSV &		
x'y'GSV (mult above		
results by 4bit x',y')	144 (*3)	110.6k
x'^3GSV, x'^2y'GSV,		
x'y$'^2$GSV, y$'^3$GSV:		
(mult abv by 4bit)	192 (*4)	196.6k

Operation	Cycles/bit plane	Cost
Calculations over whole blob:		
Identify blob (MAX to 10-bit labels, broadcast & compare create new mask)	7 2 1.5	23.4k
X&Y, max&min	n/a	2232
x_{diff}, y_{diff}	15 (*2)	30

Sums within blob areas: within the PE values to be added are:

	Operand length	Result (sum) length	
Area	1	9	
Perimeter	2	10	
GSV (for weight)	8	16	
1st order moments	12	20	2 off.
2nd order moments	16	24	3 off.
3rd order moments	20	28	4 off.

The total operand length for the above 12 sums is 163 bits and for each of these there are 256 layers to be summed with PEs. The most efficient way to sum these is to first add pairs of numbers at the original bit length, and repeat this process adding 1 bit at each step, until we have one value per PE. For example, for GSVs the first 128 adds are on 8-bit values, the next 64 are on 9-bit values, and so on. For all 12 sums:

Step 1 involves 163 bits (*128 for the 256 layers)
Step 2 involves 175 bits (*64)
Step 3 involves 187 bits (*32)
Step 8 involves 247 bits (*1)

This gives an operation count at 2.5 cycles per bit plane for the in-PE adds of 111.3 kcycles.

Multiplies within PEs: the following multiplies are performed, one per PE:

Operation	Operand lengths	Cost
$X\Sigma GSV, Y\Sigma GSV$	4, 16	2×192
$X^2\Sigma GSV, Y^2\Sigma GSV$	4, 20	3×240
$X\Sigma xGSV, X\Sigma yGSV,$ $Y\Sigma xGSV, Y\Sigma yGSV$	4, 20	4×240
$X^2Y\Sigma GSV, XY^2\Sigma GSV$ etc.	4, 24	4×288
$X^2\Sigma yGSV, X\Sigma xyGSV$ etc.	4, 24	10×288

Total cost = 6096 cycles

We can now add the appropriate combination of elements to produce the value of each required primitive, per PE. Without going into detail, the cost of these (in-PE) adds is 2361 cycles.

It then remains to add these primitives across all PEs participating in the blob. The results of the sums are produced in horizontal (vector) mode, which is ideal for the classification stage. The adds required per PE to get the primitives are:

Operand length	Cost
2 @ 21 bits	2 @ 556
3 @ 25 bits	3 @ 1040
4 @ 31 bits	4 @ 1040
	Total = 8392 cycles

Since these results are all at least 31 bits long, we will convert them to floating-point format, at a cost of $2 \times 128 + 7 \times 200 = 1656$ cycles.

Finally we calculate max. and min. GSV at a cost of 7 cycles per bit plane = 28.7 kcycles.

To calculate the total operation count we need to multiply the per blob figures by 16 and add in all the per PE figures. This comes to 3080 kcycles, at 150 ns/cycle, time = 0.46 s.

5.4.9. Classification

Now that a full set of feature primitives for each blob has been calculated, we can use these values to generate a set of class expectation indices using the formula supplied by Radford.[14] This states that for class i, the expectation index G_i for the set of feature value x can be expressed in matrix notation as:

$$G_i(x) = -0.5 \cdot (x - m_i)^T \cdot S_i^{-1} \cdot (x - m_i) + K_i$$

where:
m_i is the set of mean feature values for class i
S_i is the covariance matrix for class i
K_i is a constant for class i

For the purposes of the performance study we assume a four class problem so each S is a set of four 8×8 matrices; and the calculation is to yield four expectation indices for each blob.

We can further assume that there will typically be about 16 or fewer blobs in this part of the process as a result of the blob overlapping and extraction computation.

The remainder of the computation will be in "feature space"; i.e. we remap the extracted primitives in store so that any dependent feature may now be calculated. The formula for a set of moment features are given in Reference 6, and once the appropriate ones have been obtained from the primitives we can apply formula for $G_i(x)$ above[14] to the feature vector of each blob. When all the G_i are calculated for a blob, $MAX(G_i)$ for all i is found, to identify the most likely class index for the blob.

Since the S matrices are 8×8, we can work in horizontal mode, with each blob's data occupying $1/4$ of the DAP layer. As there are 16 blobs to consider we will have to repeat each operation at four different layers in the DAP.

All data is already in the form of 32 bit floating point values in horizontal form as a result of the SUM operations in the last step.

Timing

The classification process, for each of the four layers, involves the following vector operations:

	Divide	Add/subtract
Contrast	1	
Aspect ratio	1	2
Peri.²/area	1	1

In calculating a sample set of moments, the following operations are also needed: 24 multiples, 18 vector-scalar multiples, 16 squares, 2 cubes, and 41 adds.

For the class index calculations, we use both vector and matrix mode, with the following costs: 1 vector add, 2 matrix multiplies, 1 mat.-scalar mult., 7 mat. adds.

Routing costs are 200 cycles for rotating features into matrix mode for operation of S^{-1} multiplies. Plus 90 cycles during process.

The total, when the factor of 4 for the 4 layers is included, comes to 178 kcycles which, at 150 ns/cycle, takes 0.027 s.

5.4.10. Conclusion

Taking the time figures for each section, the total processing time for the whole program for a single 512×512 image is:

0.00092	sec	Input image to DAP
0.38	sec	Direct DOG application
0.045	sec	Segmentation
0.052	sec	Blob labeling
0.46	sec	Feature extraction
0.027	sec	Classification
0.96 sec	Total	

DOG Convolution: A FORTRAN program has been implemented on a VAX11/780 for a direct DOG application, and Radford has reported[17] that a single complete DOG application of width 9 on an image size 256^2 takes 41 s.

We have done precisely this operation with DOG.3 above; including costs for the inhibitory Gaussian (DOG.2 G_e), it takes less than 0.01 s on the DAP, i.e., a speed-up of more than 4000.

Segmentation/Area Filling: This process is extremely data-dependent, but again using figures reported by Radford for a FORTRAN program on a VAX11/780 on the smaller image size (about 25 s for a three-pass blob, compared with under 5 ms for three global passes on the DAP), we can expect a speed-up of at least 5000.

Feature Extraction: Again, this algorithm is very data-dependent, especially on a serial processor. However, Bird has implemented a FORTRAN code on a VAX11/780 which extracts features from 512^2 pixel arrays at a conservative rate of about 3 ms per pixel.[9] Interpolating the figure of 0.46 s, which applies to 16 blobs of any size, a comparative figure for the DAP might be about 3.5 ms, i.e., a speed-up of over 1000.

We have shown that the DAP is a very appropriate and speedy tool for image processing. This is due both to the high parallelism and the flexibility of the bit-level operations.

REFERENCES

1. **Flanders, P. M., Hunt, D. J., Parkinson, D., and Reddaway, S. R.,** Efficient high speed computing with the distributed Array Processor, *Symp. on High Speed Computer and Algorithm Organization,* University of Illinois, Academic Press, New York, 1977, 113.
2. **Batcher, K. E.,** Design of a Massively Parallel Processor, *IEEE Trans. Comp.,* C-29, 153, 1980.
3. **Flanders, P. M.,** Fortran Extensions for a Highly Parallel Processor, Infotech State of the Art Report on Supercomputers, 1979.
4. **Metcalf, M. and Reid, J.,** *Fortran 8x Explained,* Oxford University Press, New York, 1987.
5. **Flanders, P. M. and Parkinson, D.,** Data mapping and routing for Highly Parallel Processor Arrays, *Future Computing System,* Vol. 2, No. 2, Oxford University Press, New York, 1987.
6. **Flanders, P. M.,** A unified approach to a class of data movements on an array processor, *IEEE Trans, Comp.,* C-31 (No. 9), 809, 1982.
7. **Simpson, P., Roberts, J. B. G., and Merrified, B. C.,** Mil-DAP: Its Architecture and Role as a Real Time Airborne Digital Signal Process, presented at the AGARD Conference, Lisbon, May 1985.
8. **Batcher, K. E.,** Architecture of the MPP, *IEEE Comp. Soc. Workshop on Computing Architecture for Pattern Analysis and Image Database Management,* 1983.
9. **Bird, P. A.,** private communication, August 1985.
10. **Reddaway, S. F.,** DAP and its application to image processing, *IEE Elec. Div. Colloq. on VLSI Modules for Image Processing,* 1983.
11. **Marks, P.,** Low level vision using an array processor, *Computer Graphics and Image Processing,* 14, 281, 1980.

12. **Haralick, R. M. and Shapiro, L. G.,** Image segmentation techniques, *Computer Vision, Graphics and Image Processing,* 29, 100, 1985.
13. **Sabey, J.,** A Facility for the Design and Assessment of Statistical Classifiers, presented at the 1983 RSRE IP Symposium.
14. **Radford, C. J.,** Definition of Further Algorithms for Possible Study, RSRE Paper, October 1984.
15. **MacQueen, K. S.,** Efficient Matrix-Scalar Multiplies on the DAP, in preparation.
16. **Bird, P. A.,** Moments in Pattern Recognition, Analysis Group Working Note AGWN/027, British Aerospace, December 1984.
17. **Radford, C. J.,** private communication, 1985.

6 THE GEOMETRIC ARITHMETIC PARALLEL PROCESSOR

6.1 GAPP DESIGN

6.1.1. Introduction

Massively parallel processors provide unique, high-performance solutions to a large class of problems. In the fall of 1981, Dr. Wlodzimierz Holsztynski applied his mathematical expertise to the research and development of solutions to Martin Marietta's image processing problems.

This research resulted in the invention of the Geometric Arithmetic Parallel Processor (GAPP)™ and a family of derivatives. The fundamental processing requirement for this system is to provide flexible processing power despite environmental and size constraints.

6.1.2. Background

The GAPP concept was first implemented as a medium scale integration (MSI) breadboard in 1982. This first system emulated a 6-by-12-cell array using programmable logic and discrete memory components to mechanize the cells. Later that year, this system demonstrated the execution of a simple pattern matching algorithm. The development of GAPP technology continued into 1983 with the commitment to develop a GAPP-based custom integrated circuit. NCR Corporation of Fort Collins, Colorado was licensed by Martin Marietta to design and build GAPP chips. The first design was a PLA-based approach that resulted in 3-by-6 cell chips. Prior to the completion of these parts, known as GAPP I, Martin Marietta began to improve the design of the basic cell toward higher cell density per chip (6-by-12 cells or 72 cells per chip). The first of these new chips, GAPP II, was delivered to Martin Marietta in late 1984. Two chip design and fabrication cycles were completed within 2 years. These chips were fabricated in 3 micron complementary metal oxide semiconductors (CMOS) using a double metal process. NCR has continued the process improvement and is now delivering parts from a 2 micron double metal CMOS process.

During chip design and development, Martin Marietta designed and built a prototype system. This system was designed to perform real-time (30 frames per second) video processing. The video source of primary interest and focus was a forward looking infrared (FLIR) sensor. This application was chosen because the FLIR is a major product line of Martin Marietta Electronics and Missiles Group in Orlando.

The desired result from this processor was the extraction of targets from each image and the rejection of all non-targets (clutter) in a tactical military scenario. In addition, the targets were classified by type. This GAPP system was designed as both a research tool and as an example of an automatic high speed processor. Packaged in a standard laboratory rack with considerable room for additional experimental pieces, the design included two GAPP arrays: the main array containing 41,472 processing cells or elements and the target array containing 4,608 cells. The system also contained two 29116 micro controllers, two 68000 single board micrcomputers, and an extended MIL-STD-1750A instruction set processor designed by Martin Marietta. The GAPP II chips and the system design came together in 1984 and has since been in continuous operation.

The merit of the GAPP computation approach was well recognized and additional versions of GAPP processors were authorized prior to completion of the first processor. A flightworthy helicopter system whose main array contained 51,840 processing cells was built and flown in 1986 and 1988. This system continues to serve as a testbed for various programs. The company also recognized that algorithms developed for GAPP-based systems would be dramatically improved by installing GAPP processors as peripherals in the image processing laboratories at Orlando. There are currently three GAPP systems attached to VAX-based systems in support of programs at Martin Marietta. The largest GAPP system's main array contains 82,944 processing elements. This is probably the largest array of processors ever constructed.

Martin Marietta and NCR jointly developed a peripheral processor for the NCR Tower Computers (also compatible with IBM PC-ATs and other compatible computers). Martin Marietta expanded the development so that the basic system is also compatible with SUN 3 systems. NCR continued its development system for their customers.

6.1.3. Cell Description
6.1.3.1. General

Martin Marietta intentionally designed the GAPP cell as simple as possible. This simplicity is the driving principle resulting from Dr. Holsztynski's work. Nothing should be included in the cell which is not involved in the computation clock cycle. This requirement keeps the cell structure small, allowing a large number of cells per chip. The cell consists of six active components: four 1-bit registers, a 1-bit full adder/subtractor (FAS), and 128 bits of memory. Additionally, multiplexers and data paths permit the movement of signals within the cell.

6.1.3.2. Registers

Three of the 1-bit registers, labeled North-South (NS), East-West (EW), and carry borrow (C) are connected to the inputs of the 1-bit FAS. Additionally, the NS register output is connected as an alternate input to the NS registers in the cells which exist geometrically to the north and south of this cell. Likewise, the EW register is connected as an alternate input to the EW registers in the cells which exist geometrically to the east and west of this cell. This is the nearest neighbor orthogonal connection of the fine grid array of GAPP cells. The contents of the C register are not available outside the cell in which it exists without passing through some other register.

6.1.3.3. Full Adder/Subtractor

The 1-bit full adder/subtractor (FAS) is the computational element of the cell. It implements a truth table.

The three 1-bit inputs come from the three previously-mentioned registers, NS, EW, and C. On every clock cycle the FAS automatically produces the result prescribed by the truth table. This truth table allows the construction of arithmetic and logical results, in a bit serial fashion, that are completely general. In principle, each cell can perform all arithmetic and logical operations with this element. The output labels represent respectively sum, carry, and borrow (SM, CY, BW).

6.1.3.4. RAM

The memory array in each cell is organized as a 1-by-128-bit static RAM. When a 7-bit address is supplied, along with read-write signals, 1-bit of data may be read from or written into every cell's addressed memory location.

6.1.3.5. Control/Clock

Each cell requires 20 bits of control/address information defining the activity required of the cell. The control section consists of 13 signals (the other 7 are associated with the RAM addressing) which primarily select data paths within the cell. There are five independent parallel groupings of data paths: one associated with each portion of the cell which can store data. Thus, the RAM and registers NS, EW, CM, and C can be manipulated in parallel. Additionally, each cell must receive a clock signal. All changes of state within the cell occur synchronously with the clock.

6.1.4. GAPP Chip Descriptions
6.1.4.1. Control

The control, address, and clock signals are common among all cells on the chip. Thus, every cell performs exactly as its neighboring cell. The only difference between activities are a function of the data content within each cells' registers and RAM. These data differences are crucial because a cell or group of cells, through the proper use of algorithms, can appear to be "turned off". The cells' ability to perform logical operations makes individual cell operations practical even in a SIMD control strategy.

6.1.4.2. Shift Register Groups

On a single GAPP chip, there are 6 sets of 12-bit NS shift registers and 12 sets of 6-bit EW shift registers. Since every processing element contains one NS register and one EW register, then these groups of NS and EW shift register form a geometric orthogonal arrangement across groups of cells. The contents of the EW registers to be transferred may use the NS registers and vice versa.

Similarly, the CM registers are organized as a group of shift registers, geometrically placed in parallel with the NS shift registers. The CM registers are unidirectional (from south to north), allowing images to be input from the south and output to the north.

6.1.4.3. Chip Performance/Mechanics

Each grouping of like-named registers, such as outputs from the FAS and RAM locations at the same address, can be thought of as planes of data. When an instruction is executed, every cell in the chip reacts in exactly the same way. Since each chip contains 72 cells, this has the effect of operating on a 72-bit 'word' within the chip for up to five planes (instructions involving NS, EW, C, CM and RAM) in one clock time. Usually one to three planes are moved at once.

The ends of each of the three groupings of shift registers (CM, NS, and EW) come to the edge of the chip. Both the CM and NS groups exit at the north and south edges while the EW group exits at the east and west edges. Each of these data groups may be thought of as input/output ports to the chip. In that sense, each chip has 6 ports; 2 bidirectional 6-bit ports (one at the northern and one at the southern edge for NS); 2 unidirectional 6-bit ports (one for output at the northern edge of CM and one for input at the southern edge of CM); and 2 bidirectional 12-bit ports (one at the western and one at the eastern edge for EW). Further, the system designer may choose to provide three simultaneously input (CMS, and E or W and N or S) and three simultaneously output (CMN, and E or W and N or S) paths on a given clock cycle. At a 10 MHz clock, each chip has an input/output bandwidth of 60 Mbytes/s, 30 MBytes/second input and 30 Mbytes/s output.

The data signals are deliberately pinned out of the chip package at four mechanical edges, providing relatively easy printed circuit board layout.

A cell requires 25 clocks to perform an 8-bit add ($3n + 1$, where n is the number of bits in each operand to be added). Each chip can be clocked at 10 MHz frequency. At this 100 ns rate, each cell can perform 400,000 8-bit adds per second. Each chip contains 72 cells each performing their add: thus the chip throughout is equivalent to 28.8 million 8-bit adds per second. The 8-bit add executes in 2.5 µs. As an example of an elementary image processing operation, a 3-by-3 neighborhood Sobel operator takes 54.6 µs or 18,315 Sobels per second per cell.

Each chip contains 84 pins for power, ground, clock, control, address, and data exchange. Each chip occupies about one square inch of board space and dissipates about one-half watt.

6.1.5. GAPP Array
6.1.5.1. Assembly

The assembly of an array is simple; each chip is connected to its logical neighbor (east connects to west and north to south). Clocks and control are distributed to every chip in the array. Practical limits exist and most are imposed by the choice of board housing, backplanes, bus standards, or system architecture. In the current design, arrays are modularized as 48-by-132 (6,336) cells on a single 9 µ board (14 3/8-by-16 in.).

6.1.5.2. Input/Output

In standard systems, input of data occurs via the CM south port and output occurs from the CM north port. This arrangement takes advantage of the CM plane, allowing for simultaneous input and output during computation. To obtain simultaneous input and output three conditions must exist. First, a result must be available at the start of the input/output operation. Second, a plane of input data must be available in the external world. Third, the algorithm currently running must require at least N clocks, where N is the size of the GAPP array in the north-south direction. To obtain free input/output, a result is loaded from RAM or registers into the CM plane in one clock cycle. Data in the CM register plane are shifted north one position for each clock. Simultaneously, a row of data are output into an appropriate buffer on the northern edge of CM. On the same clock, a row of data are input into the southern edge of CM. This operation continues for N clocks. During the N clocks, any other operation can occur within the array as long as it does not involve the CM plane. At the end of N clocks, data are written from CM into RAM or registers as directed by the program.

6.1.5.3. Sizes

The smallest size array is one chip. The required array size is tailored to the systems problem. In real-time image processing, the major parameters then determine size including input data rate, algorithm length (execution time), and array clock speed. For example, assume a 10 MHz clock speed machine accepting imagery arriving at 12 megapixels per second and that the algorithm requires 50,000 instructions or clocks (the equivalent of 2,000 8-bit adds for every pixel in the array).

The algorithm will require 5 ms to execute (50,000/ 10,000,000). The array

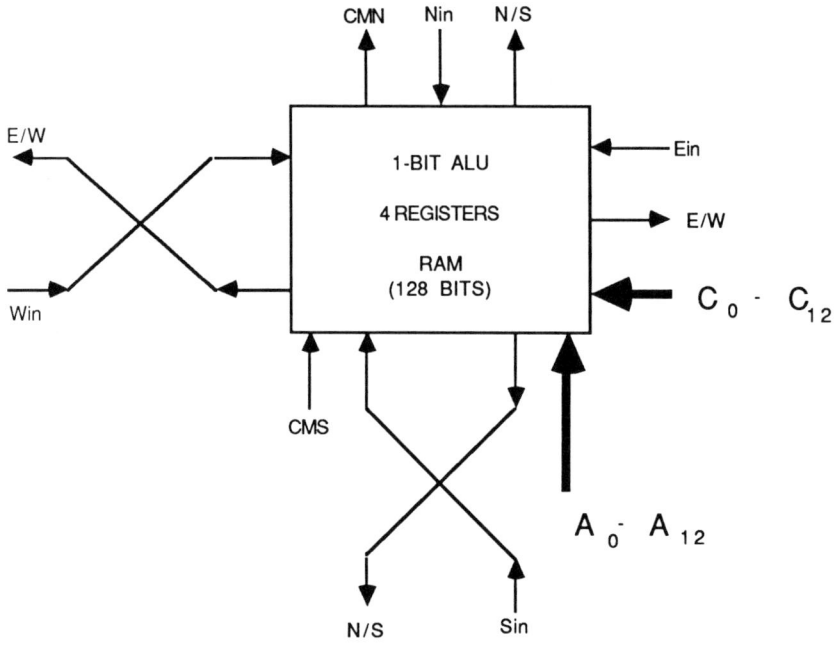

FIGURE 25. GAPP PE links to neighbors.

must contain at least 60,000 cells (5 ms-by-12,000) to maintain real-time rates without missing any data; this equates to about 833 chips. Using the 48-by-132 cell GAPP modules previously mentioned, a system containing ten modules will suffice (880 GAPP chips or 63,360 cells). The modular design approach can accommodate up to 24 GAPP modules, 2,112 chips, or 152,064 cells. At a 10 MHz clock frequency, a 24 module system would exhibit a computational throughput of 60 giga 8-bit adds per second. The largest GAPP systems to date contains 1,152 chips or 82,944 cells.

6.2. GAPP PROGRAMMING

Each processor element contains separate lines that link the cell to its neighbors and to the outside world. In addition to the North-South (NS) and East-West (EW) lines that pass data between cells, there are the CM South input (CMS) and CM North ouput (CMN) (Figure 25). There is also a complement of 22 external signal lines: 7 address lines (A_0 through A_6), 13 control lines (C_0 through C_{12}), one global output (GO), and one clock (CLK).

The chip's overall simplicity is reflected in the layout of a single processor element (Figure 26). Each of its four latches — CM, NS, EW, and C (referred to as the C register) —accepts data from up to eight possible sources, depending upon the setting of the control lines. C_0 and C_1 control the input to the CM

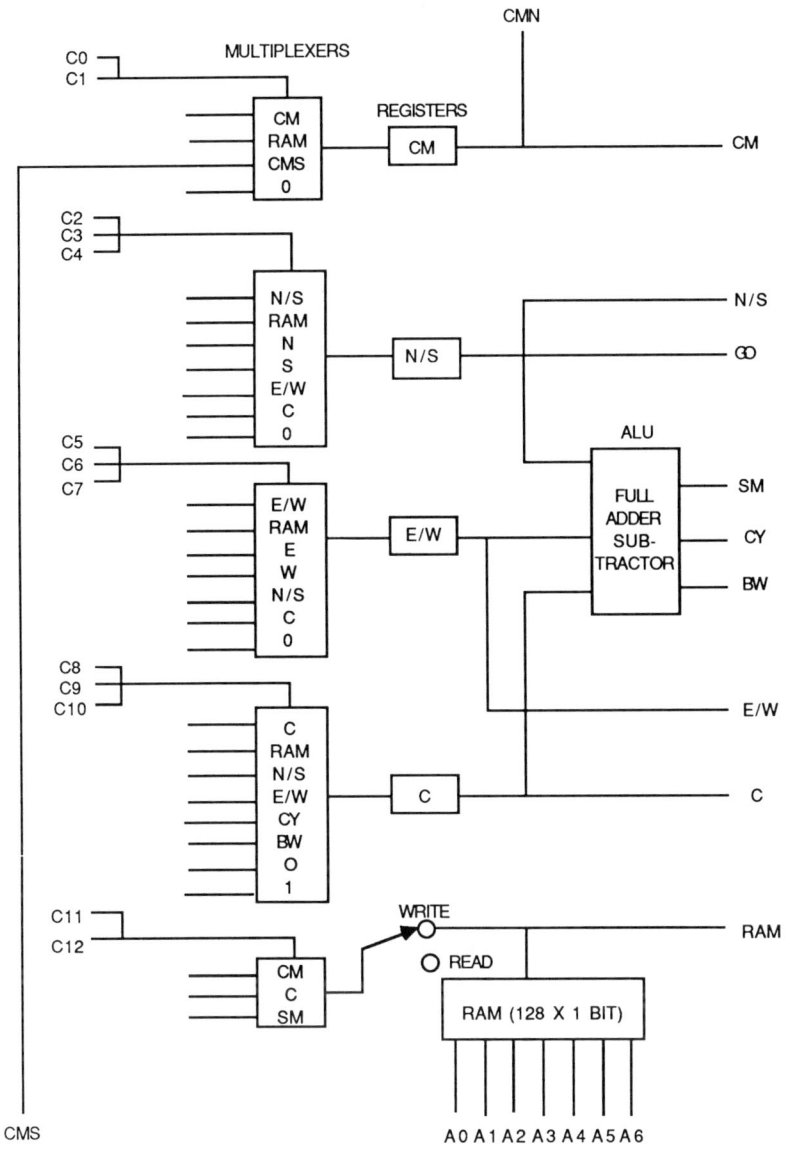

FIGURE 26. GAPP PE diagram.

latch; C_2 through C_4 govern the input to the NS latch; C_5 through C_7 manage the input to the EW latch; and C_8 through C_{10}, the input to the C register. Lines C_{11} and C_{12} handle reads to and writes from the 128-bit RAM.

Working from a truth table, the array performs additions and subtractions

TABLE 7A
GAPP Truth Table

Arithmetic Operations

Input			Output		
NS	EW	C	SM	CY	BW
0	0	0	0	0	0
0	1	0	1	0	1
1	0	0	1	0	0
1	1	0	0	1	0
0	0	1	1	0	1
0	1	1	0	1	1
1	0	1	0	1	0
1	1	1	1	1	1

TABLE 7B
GAPP Logic Operations

Logical operation	Output		Input	Input conditions
INV	SM	=	$\overline{NS}$	EW = 0, C = 1
	SM	=	$\overline{EW}$	NS = 0, C = 1
	SM	=	$\overline{C}$	NS = 0, EW = 1
AND	C	=	NS · EW	C = 0
	CY	=	EW · C	NS = 0
	BW	=	$\overline{NS}$ · EW	C = 0
OR	CY	=	NS + EW	C = 1
	BW	=	$\overline{NS}$ + EW	C = 1
	BW	=	EW + C	NS = 0
XOR	SM	=	NS + EW	C = 0
$\overline{XOR}$	SM	=	$\overline{NS + EW}$	C = 1

(Table 7A). The C, NS, and EW inputs to the multiplexers represent the contents of the C, NS, and EW registers, respectively. The summing output of the single bit ALU, SM, goes directly to the RAM and may also be simultaneously input to any of the four registers. The Carry and Borrow outputs (CY and BW, respectively) are open to the C register. A truth table is used as well to fulfill single- and dual-input logic functions — logical complement, exclusive-OR, exclusive-NOR, logical AND, and logical OR — on data in the NS and EW latches (Table 7B).

The chip is programmed with a sequence of instructions that, when compiled by an assembler, directs the appropriate control signals to every cell in

TABLE 7C
GAPP Instruction Set for the Systolic Array Processor

Register operation	Mnemonic	C_{12}	C_{11}	C_{10}	C_9	C_8	C_7	C_6	C_5	C_4	C_3	C_2	C_1	C_0	Description
CM	CM: = CM	X	X	X	X	X	X	X	X	X	X	X	0	0	Micro NOP
	CM: = RAM	X	X	X	X	X	X	X	X	X	X	X	0	1	Load CM from RAM
	CM: = CMS	X	X	X	X	X	X	X	X	X	X	X	1	0	Move from CMS into CM
	CM: = 0	X	X	X	X	X	X	X	X	X	X	X	1	1	Load zero into CM
NS	NS: = NS	X	X	X	X	X	X	X	0	0	0	X	X	X	Micro NOP
	NS: = RAM	X	X	X	X	X	X	X	0	0	1	X	X	X	Load NS from RAM
	NS: = N	X	X	X	X	X	X	X	0	1	0	X	X	X	Move from N into NS
	NS: = S	X	X	X	X	X	X	X	0	1	1	X	X	X	Move from S into NS
	NS: = EW	X	X	X	X	X	X	X	1	0	0	X	X	X	Move from EW into NS
	NS: = C	X	X	X	X	X	X	X	1	0	1	X	X	X	Move from C into NS
	NS: = 0	X	X	X	X	X	X	X	1	1	X	X	X	X	Load 0 into NS
EW	EW: = EW	X	X	X	X	X	0	0	0	X	X	X	X	X	Micro NOP
	EW: = RAM	X	X	X	X	X	0	0	1	X	X	X	X	X	Load EW from RAM
	EW: = E	X	X	X	X	X	0	1	0	X	X	X	X	X	Move from E into EW
	EW: = W	X	X	X	X	X	0	1	1	X	X	X	X	X	Move from W into EW
	EW: = NS	X	X	X	X	X	1	0	0	X	X	X	X	X	Move from NS into EW
	EW: = C	X	X	X	X	X	1	0	1	X	X	X	X	X	Move from C into EW
	EW: = 0	X	X	X	X	X	1	1	X	X	X	X	X	X	Load 0 into EW
C	C: = C	X	X	0	0	0	X	X	X	X	X	X	X	X	Micro NOP
	C: = RAM	X	X	0	0	1	X	X	X	X	X	X	X	X	Load C from RAM
	C: = NS	X	X	0	1	0	X	X	X	X	X	X	X	X	Move from NS into C
	C: = EW	X	X	0	1	1	X	X	X	X	X	X	X	X	Move from EW into C
	C: = Carry	X	X	1	0	0	X	X	X	X	X	X	X	X	Load C from Carry
	C: = Borrow	X	X	1	0	1	X	X	X	X	X	X	X	X	Load C from Borrow
	C: = 0	X	X	1	1	0	X	X	X	X	X	X	X	X	Load 0 into C
	C: = 1	X	X	1	1	1	X	X	X	X	X	X	X	X	Load 1 into C
RAM	Read	0	0	X	X	X	X	X	X	X	X	X	X	X	Read from RAM
	RAM: = CM	0	1	X	X	X	X	X	X	X	X	X	X	X	Load RAM from CM
	RAM: = C	1	0	X	X	X	X	X	X	X	X	X	X	X	Load RAM from C
	RAM: = Plus	1	1	X	X	X	X	X	X	X	X	X	X	X	Load RAM from Sum

the array. Up to five commands, one from each of the five groups that make up the overall set, can be executed simultaneously on every instruction cycle. The possible combinations of horizontally microcoded instructions results in nearly 6000 commands (see Table 7C).

Programming the array is well suited to tasks such as addition and multiplication. The algorithm for recognizing a 101 pattern demonstrates its ability to take care of both arithmetic and logical operations (Figure 27).

In the first step, the EW registers of a configuration of 8-by-5 processor elements are loaded with the test pattern found at RAM location 0 of each cell. Next, the entire test pattern is shifted one column to the left. In the third step, those data are loaded into the NS register and, simultaneously, the entire test

INPUT PLANE

X		X		X			
X	X		X	X			
	X	X		X			
X		X	X				

(a)

OUTPUT PLANE

1	0	1	0				
0	1	0	0				
0	0	1	0				
1	0	0	0				

(b)

FIGURE 27. GAPP algorithm to recognize 101 pattern.

pattern is shifted another column to the left into the EW register, and C is set to zero.

The NS registers thus contain data that represent the pattern before the second column shift, and the EW registers contain data that represent the pattern after the second-column shift. The C registers next receive the results from the borrow outputs of the ALUs of each processor element. Also in the fourth step, the EW register is reloaded with the original pattern. The Borrow outputs are equivalent to the logical expression:

$$BW = \overline{P_{M,N+1}} \cdot P_{M,N+2}$$

where $P_{M,N+1}$ is shifted-by-one version of the original pattern in cell row M, column N and $P_{M,N+2}$ is the shifted-by-two pattern. At the fifth step, the C register gets the Carry output that is logically equivalent to:

$$CY = P_{M,N+1} \cdot \overline{P_{M,N+1}} \cdot P_{M,N+2}$$

In the last step, RAM location 1 gets the final output. It is evident that only those processor elements that are at the left end of a 101 pattern will contain a 1.

Edge enhancement, a typical aspect of pattern recognition, well utilizes a high level of parallel processing. Since edge enhancement greatly amplifies noise within an image, additional filtering should be used to minimize noise effects.

Image enhancement using a technique known as dilation and erosion can reduce noise effects. The dilation operator compares each pixel to its neighboring pixels and replaces the pixel with the maximum value found within the neighborhood (forementioned pixel included). Erosion replaces a pixel with the minimum value found in the neighborhood. By applying these two operators one after the other, simple pixel noise will be removed. By applying one operator multiple times, following with an equal number of the other operator, other types of noise will be reduced. Each of these operations also reduces image details which may be of interest

Since an exact match between an image being viewed and an image stored in memory is rare, convolution and correlation are helpful functions for determining the similarities between the two. In the simple 101 pattern given earlier, perfect matches are simple to demonstrate. In the longer strings found in real-world pattern recognition, convolution and correlation adjust for minor discrepancies.

6.2.1. Convolution and Correlation

Convolution is employed in edge enhancement, for instance, to improve the quality of the image. In convolving a 3-by-3-pixel mask with an 8-bit grayscale, the mask is placed over every pixel in the image and the product terms in each 3-by-3-pixel window are summed.

Global broadcasting lets the system send a single portion of the 3-by-3-pixel mask to each of the cells within the array. To speed processing, the parts that make up an image plane can be shifted to the center of the convolution window before the required multiplication is performed. That increases efficiency because two 8-bit integers form a 16-bit product. If the shift were not carried out first, the product would have to be shifted bit-serially to the center of the window.

At the end of the convolution, the chip will have performed a total of nine global 8-bit multiplications, twelve shifts, and nine 16-bit additions. The approximate execution time for each multiplication is 25.2 μs; for a shift, 2.4 μs; and for addition, 4.9 μs. Thus a total of just 299.7 μs is expended on the operation.

A binary correlation mask using a binary image is conceptually identical to convolution, but bit-wide exclusive-OR operations are used instead of multiplications. The correlation creates a level of comparison, so that a decision threshold can be established to determine whether a match is close enough to meet system requirements. A score of 441 denotes a perfect match; 0 indicates an inverse image. Thresholds can be set at any level between to determine pass or fail.

Processing a 21-by-21-pixel binary correlation mask takes one exclusive-OR, 1536 shift operations, and 400 additions. The exclusive-OR and shift operation take 300 ns apiece, and additions take 1.6 μs. Total execution time is thus 1.3 ms.

Certain fundamental operations are common to both local and global algorithms. One such operation, or building block, is overflow detection, which is used for many tasks.

One approach to it conjoins a 1-bit field with each field to be operated upon. Adding a field of 3 bits and a field of 5 bits will probably cause an overflow if it is delivered to a 3-bit field, so a 1 will be placed into the overflow field. The resultant image provides useful information about the data being processed. For instance, the overflow bit may be used to generate a visual cue, like light or dark spots on the screen, to indicate which elements have overflowed. It can be used to interactively adjust the algorithm.

Among the other operations necessary for image processing are common arithmetic functions like addition, subtraction, and multiplication. Generally, images consist only of positive numbers representing the gray-scale value of the pixels. Image multiplication is needed for windowing or masking. A two-dimensional template representing a window may be shifted into the array and multiplied by the resident image. Any of these arithmetic operations may cause an overflow, which will be indicated if an overflow bit plane is used in the result field.

Register shifting is taken care of in the same manner as moving a contiguous section of memory on a standard machine. To shift upward in memory index, the highest numbered element in the block is shifted first, followed by the second highest, the third, and so on. Once again, overflow detection is needed to determine whether an element is shifted out of its field, since the program cannot write outside the field.

Translation, another basic operation, is one of the simplest for the chip to handle because of the relationship between neighboring processors. To shift toward the east a 1-bit field located at RAM address 12 within the processor array, simple execute:

```
EW: = RAM 12
EW: = W
C: = EW
RAM 12: = C
```

Here, overflow detection is not needed, since there is no possibility of an overflow taking place.

One basic task of image processing, thresholding, unites a number of the foregoing operations. Thresholding determines which pixel values are greater or less than a predetermined level. In an application that needs to zero (that is, ignore or turn into zeros) all the pixels with a gray-scale value of less than 20, the first step is to make a copy of the image's data base, which is destroyed as the task is carried out.

Since a 6-bit field can represent numbers from 0 to 63, adding 44 to every pixel will cause all those with values greater than 19 to overflow. The overflow bit plane must then be inverted to yield a zero overflow bit in every pixel where that occurred. If the inverted overflow bit is then ANDed with the original fields, all the pixels that overflowed will have their fields zeroed.

The entire task can be rapidly performed by using a global broadcast, which simultaneously places a given value (in this case, 44) in every processor element in the array. Obviously that is faster than moving the data through the array until it has reached each processor element. To place the binary value 101100 into RAM locations 21 to 26 in every element, the following instruction sequence would be executed:

```
C: = 1
RAM26: = C, C = 0
RAM25: = C, C = 1
RAM24: = C, C = 1
RAM23: = C, C = 0
RAM22: = C, C = 0
RAM21: = C
```

Another chore common to image processing, finding the maximum pixel value in an image, lends itself to the architecture of the systolic array. A

Program 1. Establishing the Highest Intensity Pixels

```
COMMENT: Initialize EW = 1
NS: = 0, EW: = 0, C: = 1
NS: = 0, EW: = C, C: = 0

COMMENT: Loop from MSB to LSB and deliver MAXVAL as bit serial
output on GO for n = 8 to 1 do
{
    NS: = RAMn, EW: = EW, C: = 0      (Read next bit from RAM into
                                       NS)
    NS: = NS, EW: = EW, C: = CY       (Form NS "and" EW)
    NS: = C, EW: = EW, C: = 0         (Send result to GO from NS)
    if GO = 1                         (Bit n of MAXVAL = 0 from NS)
    {
        EW: = EW                      (EW retains present value)
    }
    if GO = 0                         (Bit n of MAXVAL = 1)
    {
        EW: = NS                      (EW set to 0)
    }
}
```

number of algorithms could be used, depending on the desired objective. One takes advantage of the chip's Global Output (GO) line to furnish the value of the highest-intensity pixel (MAXVAL) within a O(k) interval (Program 1).

Once the algorithm is completed, the processor elements with the maximum intensity value will have a logic 1 stored in their EW registers. The same algorithm can also determine the value of the lowest intensity pixel (MINVAL) by first making a negative from the image, which is accomplished by simply inverting each bit of the pixel.

Counting the number of pixels that are displayed at maximum intensity is also done relatively simply and quickly with the array. Traditional processors would take $O(N \times N)$ operations, but an array-based binary tree approach performs a number of additions in parallel, hence requiring only $O(\log N)$ operations. Several pairs of numbers are added within all columns of an array, then pairs of these results are added in parallel. The resulting data flows upward through the block of arrays until the sum reaches the top processor element of each column.

At that point, a second algorithm sums the values in the rows until the total for the entire block is contained in the upper-left-hand processor element. Since translation operations cause data to shift into the edge of the array, these inputs must be set to zero so that the external data contributes zero to the sum. A binary-tree summation of a column of 64 numbers first assumes that the numbers are 8-bit pixel values. They are also assumed to reside in RAM locations 1 for the LSB to 8 for the MSB (Program 2). The partial sums are stored in RAM locations 1 through 14.

Program 2. Binary-Tree Summing

```
for m = 0 to 5 do
{
        c: = 0
        for n = n1 to (8 + m) do
        {
                NS: = RAMn, EW: = EW, C: = C
                NS: = NS, EW: = RAMn, C: = C
                for p = 1 to 2**m
                {
                        NS: = S, EW: = EW, C: = C
                {
                RAMn: = SM, C: = CY
        }
        RAM (M + 9): = CY
}
```

6.2.2. Straightforward Convolutions

Convolution is one of the most important jobs performed in image processing. It uses the previously described neighborhood algorithm to determine new values for pixels, thereby enhancing an image. Convolutions are put to work along the entire range of image processing from upgrading old photographs to improving the definition of edges in a robotic vision system.

Convolution is characterized by a high level of parallelism, so it is well suited to the systolic array. Typically, a template of new values is placed over the values of the camera image. Global broadcasting distributes the template. The objective is to move the sum outward in a spiral from the center of the template, which is the location of the new pixel value, to each of the matrix elements that reside under the template. At each matrix a location multiplication is performed, and the result is added to a traveling sum. The image resulting from this convolution is enhanced. Since all of the summations occur simultaneously, the parallel array processor handles the job at a good clip.

Histograms, which count the number of pixels containing particular gray-scale values, can make adjustments for changes in lighting, as well as let systems adjust to very light or very dark images. In that way they improve visual information at either end of the intensity spectrum.

The process is handled as quickly as the array's global-sum operation counts the processor elements. The elements to be counted are first identified by broadcasting a gray-scale value to every processor element and comparing it with the pixel value stored in each. Matches to the image stored in RAM locations 1 to 6 are determined by using a specific algorithm (Program 3). Various values are broadcast to create series of "bins", with different pixel levels sorted into the appropriate bins.

After this task is finished, every processor element that holds a pixel matching the broadcast pixel will have a logic 1 in RAM location 0. Before

Program 3. Sorting Pixels into Bins

```
NS: = 0, EW: = 0, C: = 1
NS: = 0, EW: = 0, C: = 1, RAM:0 = C    (Initialize RAM 0 = 1)
for n = 1 to 6 do
{                                       Broadcast bin bit n)
    NS: = 0, EW: = 0, C: = X            Where X is the value of
                                         bin bit n)
    NS: = RAMn, EW: = C, C: = 1         (Read bit n of image
                                         pixel)
    NS: = RAM 127, EW: = EW, C: = 1     (SM = 1 if NS matches
                                         EW)
    RAM 127: = SM
    NS: = NS, EW: = RAM 0, C: = 0       (Read RAM 0 and compare
                                         with RAM 127
    NS: = NS, EW: = EW, C: = CY         (CY = 1 if RAM 0 and
                                         RAM 217 were both 1)
    NS = NS, EW = EW, C = 1             (If all six bits match,
                                         then RAM 0 will
                                         continue to contain 1)
}
```

counting the number of pixels, a quick check for GO = 1 will indicate if there were any pixels at all which matched the broadcast value. By determining the number of pixels in the various bins, the system can figure out whether the image is dark or light or contains a variety of shades, making adjustments as necessary.

One of the most fundamental algorithms used in restoration consists mainly of adding successive frames of the same image (on a pixel-by-pixel basis) to yield a running average. Doing so improves the signal-to-noise ratio of the image, making it easier for the system to process and making it more visually pleasing to the operator overseeing the task on a display.

The actual code used to add two 8-bit images (see Program 4) assumes that the first is stored in RAM locations 0 to 7 (with the MSB at the highest location) and that the second is stored in RAM locations 8 to 15 (the MSB here is held in location 15). It also takes for granted that both words are positive. Simple extensions, though, allow negative numbers to be added or subtracted.

When the 25 instructions are finished, the two input images remain in the same RAM locations, while the sum of the images is stored in locations 16 to 24. The number of instructions needed to add an m-bit number to an n-bit number, where $n \geq m$, can easily be determined with the equation $3m + 2(n - m) + 1$. When both numbers are 8, as in the above example, the result is 25.

Another common algorithm, this one used in image enhancement is a finite-impulse-response filter. The equation:

$$Y(n) = \sum_{i=0}^{N-1} a(i)I(n-1)$$

Program 4. Double Vision: Adding Two 8-Bit Images

Line	Code	Line	Code
1	NS: = RAM 0; C = 0	14	EW: = RAM 12
2	EW: = RAM 8	15	RAM 20: = SM; C: = CY
3	RAM 16: = SM; C: = CY	16	NS: = RAM 5
4	NS: = RAM 1	17	EW: = RAM 13
5	EW: = RAM 9	18	RAM 21: = SM; C = CY
6	RAM 17: = SM; C: = CY	19	NS: = RAM 6
7	NS: = RAM 2	20	EW: = RAM 14
8	EW: = RAM 10	21	RAM 22: = SM; C = CY
9	RAM 18: = SM; C: = CY	22	NS: = RAM 7
10	NS: = RAM 3	23	EW: = RAM 15
11	EW: = RAM 11	24	RAM 23: = SM; C = CY
12	RAM 19: = SM; C: = CY	25	RAM 24: = C
13	NS: = RAM 4		

represents the output, Y(n) in terms of the input, I(n). It consists of both adds and shifts.

The objects being observed are generally well defined in contrast to the background, making it easy for the system to pull out the characteristics needed to recognize a pattern. When the data is received by the screening block, however, one of its key tasks is to replace the weak signals along the edges of the object with stronger signals.

A common technique used for edge enhancement calls for a Sobel filter, a two-dimensional finite-impulse-response filter with a thresholding algorithm. The Sobel filter takes an existing image and creates a new one comprising the magnitude and direction of all the strong edges of the object.

The filter works with neighborhood processing, determining the value of a pixel by examining those adjacent to it. With a 3-by-3-pixel grouping, consisting of pixels A through I, the equation that determines the Y axis values is

$$Y = (A + 2B + C) - (G + 2H + I)$$
$$= [(A + B) + (B + C)] - [(G + H) + (H + I)]$$

These X axis values are established by

$$X = (C + 2F + I) - (A + 2D + G)$$
$$= [(C + F) + (F + I)] - [(A + D) + (D + G)]$$

The code for processing the first equation (see Program 5) reveals the unusual aspects involved in programming the systolic array. The true keys to the chip's speed lies in the simultaneous computations it carries out. The first 25 instructions add each pixel to its neighbor to the west to form a new image.

Likewise, instructions 26 through 53 add this new image to the eastern

Program 5. Sobel-Filter to Establish Pixel Value

Line	Code	Line	Code
1	EW: − RAM 0; C: − 0	54	NS: − RAM 25
2	EW: − E; NS: − RAM 8	55	NS: − S; C: − 0
3	RAM 16: − SM; C: − CY	56	NS: − RAM 25; EW: − NS
4	EW: − RAM 1	57	NS: − N
5	EW: − E; NS: − RAM 9	58	RAM 35: − SM; C: − BW
6	RAM 17: − SM; C: − CY	59	NS: − RAM 26
7	EW: − RAM 2	60	NS: − S
8	EW: − E; NS: − RAM 10	61	NS: − RAM 26; EW: − NS
9	RAM 18: − SM; C: − CY	62	NS: − N
10	EW: − RAM 3	63	RAM 36: − SM; C: − BW
11	EW: − E; NS: − RAM 11	64	NS: − RAM 27
12	RAM 19: − SM; C: − CY	65	NS: − S
13	EW: − RAM 4	66	NS: − RAM 27; EW: − NS
14	EW: − E; NS: − RAM 12	67	NS: − N
15	RAM 20: − SM; C: − CY	68	RAM 37: − SM; C: − BW
16	EW: RAM 5	69	NS: − RAM 28
17	EW: − E; NS: − RAM 13	70	NS: − S
18	RAM 21: − SM; C: − CY	71	NS: − RAM 28; EW: − NS
19	EW: − RAM 6	72	NS: − N
20	EW: − E; NS: − RAM 14	73	RAM 38: − SM; C: − BW
21	RAM 22: − SM; C: − CY	74	NS: − RAM 29
22	EW: − RAM 7	75	NS: − S
23	EW: − E; NS: − RAM 15	76	NS: − RAM 29; EW: − NS
24	RAM 23: − SM; C: − CY	77	NS: − N
25	RAM 24: − C	78	RAM 39: − SM; C: − BW
26	EW: − RAM 16; C: − 0	79	NS: − RAM 30
27	EW: − W; NS: − RAM 16	80	NS: − S
28	RAM 25: − SM; C: − CY	81	NS: − RAM 30; EW: − NS
29	EW: − RAM 17	82	NS: − N
30	EW: − W; NS: − RAM 17	83	RAM 40: − SM; C: − BW
31	RAM 26: − SM; C: − CY	84	NS: − RAM 31
32	EW: − RAM 18	85	NS: − S
33	EW: − W; NS: − RAM 18	86	NS: − RAM 31; EW: − NS
34	RAM 27: − SM; C: − CY	87	NS: − N
35	EW: − RAM 19	88	RAM 41: − SM; C: − BW
36	EW: − W; NS: − RAM 19	89	NS: − RAM 32
37	RAM 28: − SM; C: − CY	90	NS: − S
38	EW: − RAM 20	91	NS: − RAM 32; EW: − NS
39	EW: − W; NS: − RAM 20	92	NS: − N
40	RAM 29: − SM; C: − CY	93	RAM 42: − SM; C: − BW
41	EW: − RAM 21	94	NS: − RAM 33
42	EW: − W; NS: − RAM 21	95	NS: − S
43	RAM 30: − SM; C: − CY	96	NS: − RAM 33; EW: − NS
44	EW: − RAM 22	97	NS: N
45	EW: − W; NS: − RAM 22	98	RAM 43: − SM; C: − BW
46	RAM 31: − SM; C: − CY	99	NS: − RAM 34
47	EW: − RAM 23	100	NS: − S
48	EW: − W; NS: − RAM 23	101	NS: − RAM 34; EW: − NS
49	RAM 32: − SM; C: − CY	102	NS: − N
50	EW: − RAM 24	103	RAM 44: − SM; C: − BW
51	EW: − W; NS: − RAM 24	104	NS: − 0; EW: − 0
52	RAM 33: − SM; C: − CY	105	RAM 45: − SM
53	RAM 34: − C		

neighbors, forming a second image. Instructions 54 to 105 add each pixel in that new image to its neighbor to the north and finally add each pixel in this third image to its southern neighbor. The resulting data, which is determined for every pixel in the image, can have as many as 10 bits, as well as a + or − sign. The latter denotes whether the edge gradient is changing from black to white or white to black. The fact that each of these values is computed in parallel results in fast throughput, since the value for the middle row is shifted upward, where it becomes the bottom row of the 3-by-3-pixel grid being processed in an adjacent processor element. The value also is shifted downward to the adjacent processor element, where it becomes the top row for the grid being processed there — a 3-by-3-pixel group centered around H.

The following Sobel operations are performed within individual cells of the array. First, a reasonable approximation of the magnitude of the gradient vector is computed by adding the absolute values of X and Y. To obtain that result, the sign bit plane for X determines whether to invert it and add one to it to form the absolute value. Processing the absolute value of a number takes $3m + 8$ instructions, where m equals the number of bits.

Next, the direction of the gradient to the nearest 45 line is determined. This must be done so that data can be processed by the extractor block and is accomplished by using the signs of both X and Y to determine which quadrant contains the gradient. Once the direction is established, another process, consisting of $18m + 76$ instructions, is performed to bring the vector in line with the nearest 45 angle.

In thresholding, the final step, the value is compared with a predetermined constant. The result of this operation is used to pass or reject the direction vector, a step necessary to ascertain that it is valid data and not simply noise. If the vector information is below the threshold, the resulting word consists of 8 zeros. If it is valid, the location of a single 1 in the 8-bit word will denote which of eight directions the vector lies closest to. Thresholding requires $m + 13$ instructions, with another 18 instructions needed to properly place the 1 in valid words.

6.3 GAPP SYSTEM

6.3.1. Introduction

Before the GAPP was invented at Martin Marietta-Orlando, image processing techniques requiring the many benefits of systolic array processing were conceptualized but found impractical. Until recently, however, a general purpose, personal computer size, real-time image processing workstation did not exist. The system discussed in this section processes and displays real-time results of pre-programmed GAPP algorithms at RS-170 video rates and is called the GAPP Peripheral Processor $(GP^2)^{TM}$.

The architecture used to implement the final GP^2 system evolved from past Martin Marietta image and signal processor designs with an emphasis on small

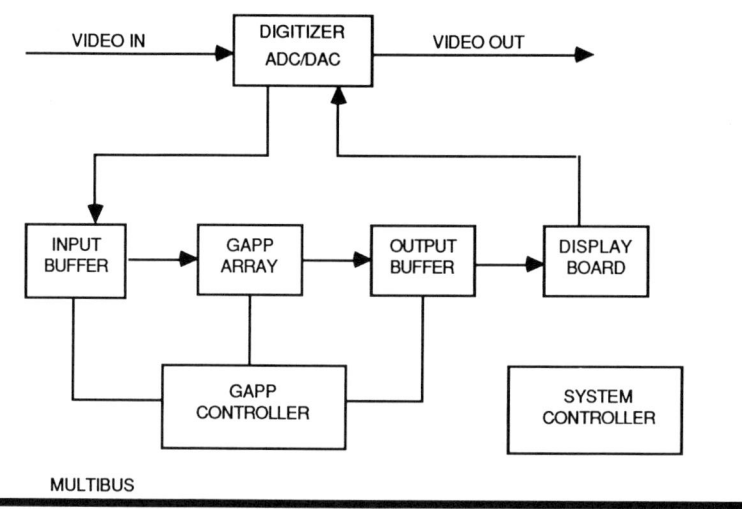

FIGURE 28. GP^2 Functional block diagram.

size, low costs, and display of results in real time. The system is designed to accept Forward Looking Infrared (FLIR), camera or recorded analog data, in a standard RS-170 video format. The system processes this data and provides analog Red/Green/Blue (RGB) or composite signal out to drive a monitor. Processed results are delayed by two frames, but no frames are omitted in processing. Also no data compression takes place.

6.3.2. Architecture

The GP^2 is divided into several functional blocks as shown in Figure 28. The system comprises a System Controller (SC) or an interface to a host computer, a Digitizer, a GAPP Controller (GC), an Input Buffer (IB), a GAPP Array, an Output Buffer (OB), and a Display Board (DB). These major parts of GP^2 are addressed individually as they occur in the data path. Data formats and communication rates are standard to external devices, but some unique implementations are internally present in the GP^2.

System Controller. The SC is used to: (1) control the interface to the user, (2) compile developed algorithms into GAPP code, (3) initialize the boards in the system for the mode of interest, and (4) run diagnostic routines to determine system health. Most of the boards in the system have interface circuitry that allow communication to the SC via the Multibus™. The SC can be a single board computer that resides in the Multibus chassis (in GP^2), or it can be a personal computer that is linked to GP^2 by a Multibus-to-PC bus repeater card. The SC is used exclusively in editing and compiling GAPP algorithms. However, after loading the compiled GAPP code to the GC for real-time operation and initializing the rest of the GP^2 system, the SC becomes unnecessary.

Digitizer. The Digitizer board serves two major functions to the GP^2 system: (1) to digitize and interface with incoming analog video data and (2) to convert a stream of output digital data into an analog video signal. The digitizer can connect up to four separate standard video sources at once and select one of these sources to process. Some acceptable inputs to the GP^2 are FLIR, standard video camera, and recorded imagery on tape or video disk that adhere to the normal RS-170 raster-scan video format. Video is digitized and then passed to the IB at a 10 mega-pixel per second rate. After this data is processed in the parallel GAPP Array processor, the DB extracts resultant data stored in the OB and passes it back to the Digitizer for re-conversion to a RS-170 RGB video signal. The GP^2 system additionally can provide a composite National Television Standard Committee (NTSC) video output to drive a display.

GAPP Controller. The GC is configured as a 256K instruction microcode memory that sequences through GAPP Array commands and addresses. It not only controls the GAPP Array but also synchronizes IB and OB operations. Each GC instruction is 32 bits wide: 20 bits for GAPP command and addresses, 2 bits for input/output control, 4 bits for GC sequencer control, and 6 bits for Constant RAM control. The GC additionally contains a command register, multiple vector registers, status register, and a program counter. These registers, along with part of the microcode memory, provide system input and output discretes to the entire GP^2 signal processing system for control and status purposes. The Constant RAM is memory accessible over the Multibus that allows the GC to inject dynamic constants into the GAPP Array during execution of the program.

Input Buffer. The IB receives video data from the Digitizer, converts the data into a form usable by the GAPP Array, and stores the reformatted image data in one of two blocks of memory. The GP^2 is configured to process four subframes the size of the GAPP Array (96 × 108 pixels) in real time. The subframes are removed from four quadrants evenly arranged about the midpoint of the input image frame, with some vertical subframe overlap, so that the area processed appears to be the central 192 by 200 pixel block in an input image frame. These subframes are processed sequentially by the GAPP Array after being loaded sequentially to an IB memory block. Each block of memory on the IB accommodates 128 pixels in the scan direction at a maximum input rate of 10 MHz. It also handles up to 512 rows of 8-bit pixels orthogonal to the scan direction.

The IB receives data as a series of pixels in a raster scan fashion. These pixels must be converted into bit planes prior to being input to the GAPP Array. An image bit plane is a two-dimensional rectangle of bits of common significance extracted from an image of pixels. Conversion to bit planes is accomplished by inputting the pixels into a Corner Turning Array (CTA), a 12-by-18 arrangement of Processing Elements (PEs) contained on the IB. Data output from the orthogonal periphery of the CTA is then mapped into a local memory bank called the image RAM.

The IB contains dual blocks of image RAM. This allows the IB to reformat the current field/frame of imagery with one block of RAM, while the other block of previously reformatted data may be simultaneously and asynchronously input into the GAPP Array under the direction of the GC. Because the IB image RAM is accessible via the Multibus, the debug and integration of this system element is simplified, as well as all succeeding hardware downstream from the IB.

The IB outputs information to the GAPP Array as a series of bit-rows, sequentially shifting the bit-planes through the south periphery of the GAPP Array, until an entire image is input. An image bit-row is a one-dimensional row of bits of common significance extracted from a row of pixels in the IB.

GAPP Array. After the GAPP Array receives image data from the IB, algorithms loaded to the GC manipulate the GAPP Array to process the image data in an SIMD parallel fashion. The GAPP Array is constructed of 144 identical, interconnected GAPP chips. Each GAPP chip contains a 6×12 array of interconnected PEs. The GAPP chips are arranged to provide a systolic array of 16 by 9 devices (or 10,368 processing elements) on a single GAPP Array board. Each pixel in an image will be input into a specific GAPP PE, and that PE will operate on that pixel.

The GAPP Array is capable of implementing typical arithmetic/logic functions plus a global OR function. The global OR is essentially 1 output pin per chip (or 1 output pin per 72 PE units) that will be active if there are 1 or more cells of the 72 cells in the chip that are active in the bit-plane of interest. The global OR outputs from all of the GAPP chips on the GAPP Array are combined and fed back to the GC. This information can then be used to make decisions in GAPP code execution. After the algorithm is completed, the results are passed from the GAPP Array to the OB.

Output Buffer. The OB receives and stores image and information data from the GAPP Array under the direction of the GC. The OB has dual blocks of image RAM that allow the GAPP array to write processed image data into one block of image RAM, while the other block, which contains previously output data, is being accessed. The OB can accommodate 128 pixels in the scan direction, and it can handle up to 963 rows of 17-bit pixels orthogonal to the scan direction. The OB receives data from the GAPP Array as a series of bit-rows, 108 bits wide. The information contained in the OB is, therefore, in bit-plane format. A port on the OB for accessing processed bit-plane data, called the SCAN bus, is used by the DB to allow this bit-plane data to be reformatted into pixels and displayed on a monitor.

Display Board. The DB accesses the image RAM in the OB and retrieves image bit-plane data. Next, the bit-plane data is reformatted into pixel data and then passed to the Digitizer for display on a monitor.

The bit-plane data from the OB is input to a CTA on the DB. This array (12×18 PEs) reformats the bit-plane data into pixel data. Next the pixel data is stored in a line buffer. When filled, the line buffer represents one video scan line of data processed by the GAPP Array. The DB tracks the Digitizer's synchronization signals and appropriately inserts the data from the line buffer into the video field for display. The final result obtained from GP^2 is the processed image. However, depending on the desired application, the DB's function can be altered to perform some enhanced image processing or control functions based on the image results obtained from the GAPP Array.

Peak performance figures for the GP^2 with one completely populated array board have been measured as follows:

8 bit addition	4.1 billion additions/s
8 bit multiply, 16 bit result	382.0 million multiplications/s
8 bit 3 × 3 convolution	95.0 million convolutions/s
8 bit 5 × 5 convolution	33.0 million convolutions/s
Binary 21 × 21 correlation	6.7 million correlations/s

The following algorithm performance figures include data I/O instructions and code to process a 192 × 200 image at real-time video rates. More image data can be processed by adding hardware with no penalty in processing time.

Algorithms	# of Instructions	Execution Time (10 MHz clock)
Low pass filter implemented by 3 × 3 convolution	12922	1.29 ms
Moving target indicator	7118	.72 ms
Sobel operator	8734	.87 ms
Template matching 5 × 5 image pixels	6553	.66 ms

The GP^2 executes at an internal system clock of 10 MHz and results in the execution of a maximum of 330,000 instructions during a video frame time.

The 1-bit system wide global OR function decreases instruction counts by an order of magnitude for many applications. The min/max function, a function that will find the minimum/maximum values in an image, was reduced from over 1000 instructions to less than 100 instructions. The global OR data can also be used for data dependent jumping and looping within an algorithm.

7 THE CONNECTION MACHINE

The Connection Machine (CM) is a product of Thinking Machines Corporation, Cambridge, Massachusetts. It was founded in June 1983, about the same time as the Japanese government was initiating its Fifth Generation computer project to exploit new computer architectures in its drive for worldwide industry leadership. European countries were beginning the Esprit and Alvey projects along similar lines. In America the federal government launched a major effort within the Defense Advanced Research Projects Agency (DARPA) in October of that year. This effort is termed the Strategic Computing Initiative.

TMC President and Founder Sheryl Handler was joined by W. Daniel Hillis, whose MIT thesis served as the basis for the CM, Marvin Minsky, Donner Professor of Science at MIT, Nobel Laureate Richard Feynman, late of Cal-Tech and Vice President Richard Clayton, formerly Vice President for Computer Systems Development at Digital Equipment Corporation among others to employ parallel processing in data intensive applications for science and engineering.

The company's first product, the 1000 MIPS CM-1, used 65,536 (64K) single-bit processors. The first commercial installation, a 16K sized version, was at MRJ, Inc., a Perkin-Elmer subsidiary, in August 1986. The 2500 MFLOPS CM-2 was introduced in April 1987. The CM-2 had 16X more memory (64K bits per processor rather than 4K), a faster clock and floating-point hardware.

The CM was initially conceived as an engine for artificial intelligence applications, and may yet find that role. To date its primary use has been for engineering applications using a more algorithmic approach, but the long range goal is to build what Hillis calls an amateur system — a machine with common sense.

The company received $4.7M from the DARPA Strategic Computing Initiative to develop the first machine. Over the years the SCI (now the Strategic

Computing Program) has given rise to a number of important parallel architectures including the Butterfly from BB & N and the WARP from Carnegie-Mellon University, as well as numerous advances in other computing technologies.

Hillis concentrated his design on interprocessor communication. He settled on a hypercube connectivity. In this scheme there are 2^N nodes (N integer). Each node has N directly connected nodes (nearest neighbors) and the longest path between any two nodes has N steps through other nodes. In this context a single node machine is a zero dimensional cube, a machine with two connected nodes is a one dimensional cube, four nodes connected as a square is a two dimensional cube, and in three dimensions eight nodes form the traditional cube, and so forth. The term hypercube is used because 65,536 nodes corresponds to a 16-dimensional cube. This interconnect design turns out to be an excellent compromise between the mesh connectivity arrangement of the Illiac IV and some other SIMD machines, and a fully connected design which for 65,536 processors would be an impractical number of interconnects.

A list of some of the sites with a Connection Machine installed is shown on Table 8. The second column indicates the size machine at each site.

7.1. THE CM SYSTEM

At the heart of any large computational problem is the data set: some combination of interconnected data objects, such as numbers, characters, records, structures, and arrays. In any application these data must be selected, combined, and operated upon.

At the heart of the Connection Machine Model CM-2 system is the parallel processing unit, which consists of thousands of processors, each with thousands of bits of memory. These processors can not only process the data stored in their memory, but also can be logically interconnected so that information can be exchanged among the processors.

One way to view the relationship of the CM-2 parallel processing unit to the other parts of the system is to consider it as an intelligent extension to the memory of the front-end computer. The data parallel data objects are stored by assigning each one to the memory of a processor. Then the operations can be specified to operate simultaneously on any or all data objects in this memory.

Data objects are left in the Connection Machine memory during execution of the program and are operated upon in parallel at the command of the front end. This model differs from the serial model of processing data objects from a computer's memory one at a time, by reading each one in turn, operating it, and then storing the result back in memory before processing the next object.

The flow of control is handled entirely by the front end, including storage and execution of the program and all interaction with the user and/or programmer. The data set, for the most part, is stored in the Connection Machine memory. In this way, the entire data set can be operated upon in parallel

TABLE 8
Some Connection Machine Sites

MRJ	2—16K
MIT-AI LAB	16K
MIT-MEDIA LAB	16K
RCA	16K
NRL	16K
DOW JONES	2—32K
YALE	16K
DARPA (SAIC)	16K
UNIV OF MARYLAND	16K
WHITNEY-DEMOS	16K
SRC	64K
UNIV. OF SYRACUSE	32K
NASA AMES	16K
UCSD ISI	16K
UCLA	16K
LOS ALAMOS	64K
NOSC	8K
SANDIA	16K
UTRC	16K
BOSTON UNIV.	16K
TASC	8K
ARGONNE	16K
NCAR	8K
LOCKHEED AUSTIN	32K

through commands sent to the Connection Machine processors by the front end. The front end can also operate upon data stored in individual processors in the Connection Machine, treating them logically as memory locations in its virtual memory.

There are several direct benefits to maintaining program control on the front end. First, programmers can work in an environment that is familiar. The front end interacts with the Connection Machine parallel processing unit using an integrated command set, and so the programming languages, debugging environment, and operating system of the front end remain relatively unchanged. Second, a large part of the program code for any application pertains to the interfaces between the program, the user, and the operating system. Since the control of the program remains on the front end, code developed for these purposes is useful with or without the Connection Machine parallel processing unit, and only the code that pertains specifically to the data residing on the Connection Machine processors needs to use the data parallel language extensions. Finally, parts of the program that are especially suited for the front end, such as file manipulation, user interface, and low-bandwidth I/O, can be done on the front end, while the parts of the program that run efficiently in parallel, namely the "inner loops" that operate on the data set, can be done on the Connection Machine. In this way, the individual strengths of both the serial front end and the Connection Machine processors can be exploited.

In general, the Connection Machine system appears to be an extension of the front-end system. The data parallel hardware looks like intelligent memory; the data parallel software extends the front end's capabilities to allow the direct execution of parallel operations.

7.1.1. Overview

The Connection Machine system implements data parallel programming constructs directly in hardware. The system includes 65,536 physical processors, each with its own memory. Parallel data structures are spread across the data processors, with a single element stored in each processor's memory. When parallel data structures have more than 65,536 data elements (the normal case), the hardware operates in virtual processor mode, presenting the user with a larger number of processors, each with a correspondingly smaller memory.

Communication among elements of a parallel data structure is implemented by a high-speed routing network. Processors that hold interrelated data elements store pointers to one another. When data is needed, it is passed over the routing network to the appropriate processor.

Scalar data is held in a front-end processor. The front end also controls execution of the overall data parallel program. Program steps that involve parallel data are passed over an interface to the Connection Machine parallel processing unit, where they are broadcast for execution by all the processors at once.

The Connection Machine front end provides the programming environment for the system. Programs can be stored on front-end disks. Network communications links are most effectively implemented on the front end as well.

High-speed transfers between peripheral devices and Connection Machine memory take place through the Connection Machine I/O system. All processors, in parallel, pass data to and from I/O buffers. The data are then moved between the buffers and the peripheral devices. Connection Machine high-speed peripherals include the DataVault mass storage system and the Connection Machine graphics display system.

The Connection Machine system software is designed to utilize existing programming languages and environments as much as possible. The languages are based on well-known standards; the extensions to support data parallel constructs are minimal so that a new programming style is not required. The CM-2 front-end operating system (either UNIX or Lisp) remains largely unchanged.

The *Lisp and CM-Lisp languages are data parallel dialects of Common Lisp (a version of Lisp currently being standardized by ANSI technical committee X3 J13). *Lisp gives programmers fine control over the CM-2 hardware while maintaining the flexibility of Lisp. CM-Lisp is a higher-level language that adds small syntactic changes to the language interface and creates a very powerful data parallel programming language.

The C* language is a data parallel extension of the C programming language (as described in the draft C standard proposed by ANSI technical committee X3 J11). C* programs can be read and written like serial C programs; the extensions are unobtrusive and easy to learn.

The assembly language of the CM-2 is Paris. This is the target language of the high-level language compilers. This language logically extends the instruction set of the front end and masks the physical implementation of the CM-2 processing unit.

7.1.2. The Connection Machine

The Connection Machine Model CM-2 is a computing system that provides both development and execution facilities for data parallel programs. Its hardware consists of a parallel processing unit, from one to four front-end computers, and an I/O system that supports mass storage and graphic display devices (see Figure 29). The user interacts with the front-end computer; all program development and execution takes place within the front end. Because the front-end computer runs standard serial software, the user sees a familiar system environment with additional languages and utilities.

The central element in the system is the CM-2 parallel processing unit, which contains:

- thousands of data processors
- an interprocessor communication network
- one or more sequencers
- an interface to one or more front-end computers
- zero or more I/O controllers and/or framebuffers

A parallel processing unit may contain 64K, 32K, or 16K data processors. (Here, and throughout this document, "K" stands for 1024. Thus 64K means 65,536; 32K means 32,768; 16K means 16,384; 8K means 8,192; and so on.) Each data processor has 64K bits (8 kbytes) of bit-addressable local memory and an arithmetic-logic unit (ALU) that can operate on variable-length operands. Each data processor can access its memory at a rate of at least 5 Mbits/s. A fully configured CM-2 thus has 512 Mbytes of memory that can be read or written at about 300 Gb/s. When 64K processors are operating in parallel, each performing a 32-bit integer addition, the CM-2 parallel processing unit operates at about 2500 MIPS. (This figure includes all overhead for instruction issuing and decoding.) In addition to the standard ALU, the CM-2 parallel processing unit has an optional parallel floating-point accelerator that performs at 3500 MFLOPS (single precision) or 2500 MFLOPS (double precision).

One of the most important requirements of general purpose data parallel computing is the ability of the data elements to communicate information among themselves in patterns that vary according to the problem and with time. The CM-2 system provides two forms of communication within the parallel processing unit. The more general mechanism is known as the router, which

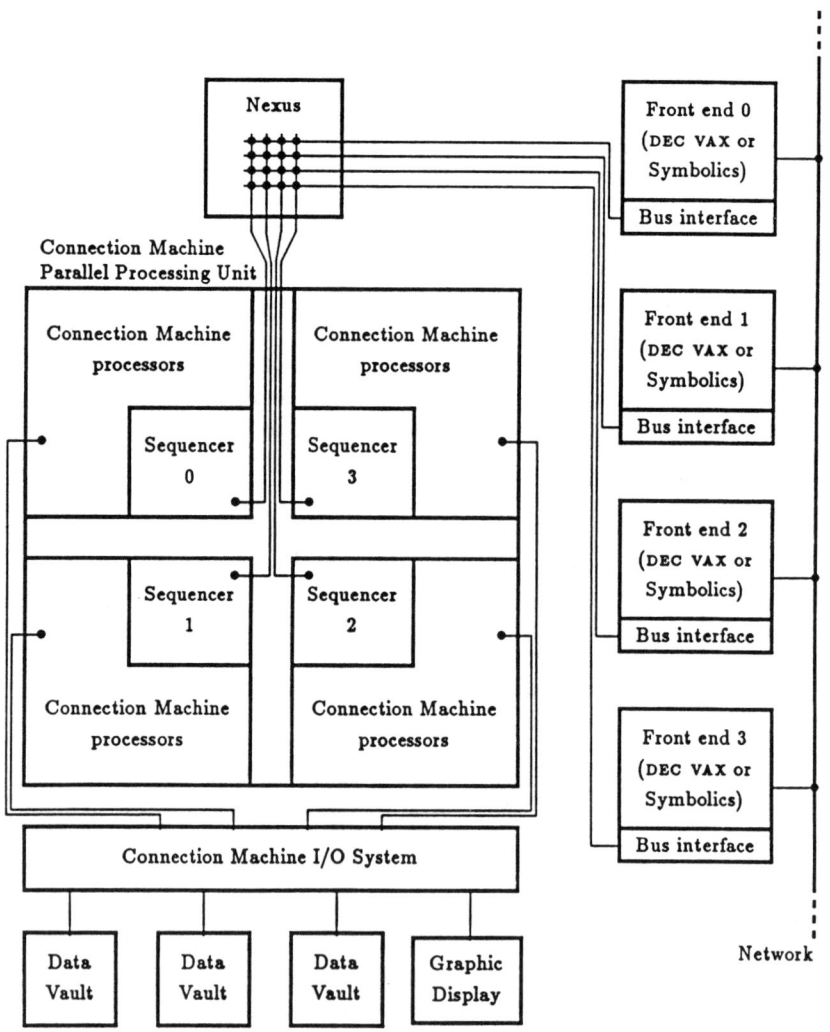

FIGURE 29. Connection machine model CM-2 System organization.

allows any processor to communicate with any other processor. One may think of the router as allowing every processor to send a message to any other processor, with all messages being sent and delivered at the same time. Alternatively, one may think of the router as allowing every processor to access any memory location within the parallel processing unit, with all processors making memory accesses at the same time; in effect, the router allows the local memories of the data processors to be treated as a single large shared memory. The messages (or accessed fields, if you will) may be of any length. The throughput of the router depends on the message length and on the pattern of accesses; typical values are 80 million to 250 million 32-bit accesses per second.

The CM-2 parallel processing unit also has a more structured, somewhat faster communication mechanism called the NEWS grid. In the CM-1 and some other fine-grained parallel systems, communication can take place over a fixed two-dimensional grid. The CM-2, however, supports programmable grids with arbitrarily many dimensions. Possible grid configurations for 64K processors include 256×256, 1024×64, 8×8192, $64 \times 32 \times 32$, $16 \times 16 \times 16 \times 16$, and $8 \times 8 \times 4 \times 8 \times 8 \times 4$. The NEWS grid allows processors to pass data according to a regular rectangular pattern. For example, in a two-dimensional grid each processor could receive a data item from its neighbor to the east, thereby shifting the grid of data items one position to the left. The advantage of this mechanism over the router is merely that the overhead of explicitly specifying destination addresses is eliminated; for many applications this is a worthwhile optimization.

The parallel processing unit is designed to operate under the programmed control of a front-end computer, which may be a Sun 4, a Symbolics 3600 Lisp machine or a DEC VAX 8000 series computer with a BI bus. The front end provides the program development and execution environment. All Connection Machine programs execute on a front end; during the course of execution the front end issues instructions to the CM-2 parallel processing unit. In effect, the CM-2 parallel processing unit extends the instruction set and I/O capabilities of the front-end computer. The set of instructions that the front end may issue to the parallel processing unit is called Paris. It is designed for convenient use by front-end programs, and includes not only such operations as integer arithmetic, floating-point arithmetic, and interprocessor communication, but also such powerful operations as vector summation, matrix multiplication, and sorting.

The data processors do not handle Paris instructions directly. Instead, Paris instructions from the front end are processed by a sequencer in the parallel processing unit. The task of the sequencer is to break down each Paris instruction into a sequence of low-level data processor and memory operations. The sequencer broadcasts these low-level operations to the data processors, which execute them at a rate of several million per second.

To increase the flexibility of program development and execution, the CM-2 processing unit may be divided into as many as four sections. Depending on the configuration, a section will have either 8K or 16K data processors. For example, a parallel processing unit with 64K data processors will be divided into four sections of 16K data processors; a processing unit with 32K data processors could consist of either two 16K sections or four 8K sections.

Each section can be treated as a complete parallel processing unit in itself; in particular, each section contains its own sequencer, router, and NEWS grid. Sections may also be ganged; when this is done, their sequencers are also ganged and behave as a single sequencer, their routers cooperate as a single router, and their NEWS grids cooperate to form a single grid. A programmable, bidirectional switch called the Nexus allows up to four front-end computers to be attached to a single parallel processing unit. The front ends need not all be of the same type. Under front-end software control, the Nexus can connect any front end to any section or valid combination of sections in the CM-2 parallel processing unit. For example, in a CM-2 system with 32K data processors (in four 8K sections) and four front ends, one could assign one section to each front end for software testing; or one could gang all four sections to be controlled by any one front end for a production run; or one could assign 8K sections to each of two front ends, gang the other two sections to give 16K data processors to a third front end, and use the fourth front end for purposes unrelated to the parallel processing unit. The Nexus can be reconfigured in seconds; once this is done, data and instructions flow between the front end and the sequencers without visible intervention by the Nexus.

For every group of 8K data processors there is one I/O channel. (A section with 8K processors therefore has one channel; a section with 16K processors has two channels.) To each I/O channel may be connected either one high-resolution graphics display framebuffer module or one general I/O controller supporting an I/O bus to which several DataVault mass storage devices may be connected. The front end controls I/O transfers in exactly the same manner that it controls the data processors, by issuing Paris instructions to the sequencer. The sequencer can then send low-level commands to the I/O channels and interrogate channel status. Data is transferred directly and in parallel between the I/O devices and the data processors, without being funneled through the sequencers.

The Connection Machine Model CM-2 parallel processing unit contains thousands of data processors. Each data processor contains:

- an arithmetic-logic unit (ALU) and associated latches
- 64K bits of bit-addressable memory
- four 1-bit flag registers
- optional floating-point accelerator
- router interface
- NEWS grid interface
- I/O interface

The data processors are implemented using four chip types. A proprietary custom chip contains the ALU, flag bits, router interface, NEWS grid interface, and I/O interface for 16 data processors, and also contains proportionate pieces of the router and NEWS grid network controllers. The memory consists of commercial RAM chips. The floating-point accelerator consists of a custom floating-point interface chip and a floating-point execution chip; one of each is required for every 32 data processors. A fully configured parallel processing unit contains 64K data processors, and therefore contains 4096 processor chips, 2048 floating point interface chips, and 2048 floating-point execution chips, and half a gigabyte of RAM.

A CM-2 ALU consists of a 3-input, 2-output logic element and associated latches and memory interface. The basic conceptual ALU cycle first reads two data bits from memory and one data bit from a flag; the logic element then computes two result bits from the three input bits; finally, one of the two results is stored back into memory and the other result into a flag. One additional feature is that the entire operation is conditional on the value of a third flag; if the flag is zero, then the results for that data processor are not stored after all.

The logic element can compute any two boolean functions on three inputs; these functions are simply specified (by the sequencer) as two 8-bit bytes representing the truth tables for the two functions.

This simple ALU suffices to carry out, under control of the sequencer, all the operations of the Paris instruction set. Consider, for example, addition of two k-bit signed integers. First the virtual processor context flag is loaded into a hardware flag register (which is then used as the condition flag for all remaining ALU operations). Next a second hardware flag is cleared for use as a carry bit. next come k iterations of an ALU cycle that reads one bit of each operand from memory and also the carry bit, computes the sum (a three-way exclusive OR) and carry-out (a three-input majority function), and stores the sum back into memory and the carry-out back into the carry flag. These cycles start with the least significant bits of the operands and proceed toward the most significant bits. The last of the k cycles stores the carry-out into a different hardware flag, so that the last two carry-outs may be compared to determine whether overflow has occurred. Arithmetic is therefore carried out in a bit-serial fashion; at about half a microsecond per bit, plus instruction decoding and other overhead, a 32-bit add takes about 21 μs. With 64K processors all computing in parallel, this produces an aggregate rate of 2500 MIPS (that is, 2.5 billion 32-bit adds per second). All other Paris operations are carried out in like fashion.

The ALU cycle is broken down into subcycles. On each cycle the data processors can execute one low-level instruction (called a nanoinstruction) from the sequencer and the memories can perform one read or write operation. The basic ALU cycle for a two-operand integer add consists of three nanoinstructions:

LOADA: read memory operand A, read flag operand, latch one truth table
LOADB: read memory operand B, read condition flag, latch other truth table
STORE: store memory operand A, store result flag

Other nanoinstructions direct the router, NEWS grid, and floating–point accelerator, initiate I/O operations, and perform diagnostic functions.

Interprocessor communication is accomplished in the CM-2 parallel unit by special-purpose hardware. Message passing happens in a data parallel fashion; all processors can simultaneously send data into the local memories of other processors, or fetch data from the local memories of other processors into their own. The hardware supports certain message-combining operations: that is, the communication circuitry may be operated in such a way that processors to which multiple messages are sent receive the bitwise logical OR of all the messages, or the numerically largest, or the integer sum.

Each CM-2 processor chip contains one router node, which serves the 16 data processors on the chip. The router nodes on all the processor chips are wired together to form the complete router network. The topology of this network happens to be a boolean n-cube, but this fact is not apparent at the Paris level. For a fully configured CM-2 system, the network is a 12-cube connecting 4096 processing chips. Each router node is connected to 12 other router nodes; specifically, router node i (serving data processors 16i through 16i + 15) is connected to router node j if and only if $|i - j| = 2^k$ for some integer k, in which case we say that routers i and j are connected along dimension k.

Each message travels from one router node to another until it reaches the chip containing the destination processor. The router nodes automatically forward messages and perform some dynamic load balancing. For example, suppose that processor 117 (which is processor 5 on router node 7, because 117 = 16 × 7 + 5) has a message M whose destination is processor 361 (which is processor 9 on router node 22). Since $22 = 7 + 2^4 - 2^0$, this message must traverse dimensions 0 and 4 to reach its destination. In the absence of congestion, router 7 forwards the message to router 6 ($6 = 7 - 2^0$, which forwards it to router 22 ($22 = 6 + 2^4$), which delivers the message to the processor 361. On the other hand, if router 7 has another message that needs to use dimension 0, it may choose to send message M along dimension 4 first, to router 23 ($23 = 7 + 2^4$), which then forwards the message to router 22, which then delivers it.

The algorithm used by the router can be broken into stages called petit cycles. The delivery of all the messages for a Paris send operation might require only one petit cycle if only a few processors are active, but if every processor is active then typically many petit cycles are required. It is possible for a message to traverse many dimensions, possibly all 12, in a single petit cycle, provided that congestion does not cause it to be blocked; the message data is forwarded through multiple router nodes in a pipelined fashion. A message that cannot be delivered by the end of a petit cycle is buffered in whatever router node it happens to have reached, and continues its journey during the next petit cycle. If petit cycles are regarded as atomic operations, then the router may be viewed as a store-and-forward packet-switched network. Within a petit cycle, however, the router is better regarded as a circuit-switched network, where dimension wires are assigned to particular messages whose contents are then pumped through the reserved circuits.

Each router node has a limited ALU, distinct from those for the data processors. During each petit cycle, each router node checks to see if its buffers hold several messages that are all going to the same processor. If so, the messages are combined. This may be done by taking the numerically greatest, summing them, taking the bitwise logical OR, or by arbitrarily discarding all but one. Other combining functions are implemented in terms of these. For example, combining with bitwise logical AND is performed by inverting the original message data, sending it with OR-combining, and re-inverting received messages. (Such tricks are implemented by the sequencer, transparently to the Paris user.) This hardware support for combining accelerates such Paris instructions as send-with-logand, send-with-s-add, and send-with-u-max. The combining hardware also combines read requests during execution of the Paris get instruction, so that a value fetched once from a processor can be returned to many requestors in a single petit cycle.

Each router node also contains specialized logic to support virtual processors. When a message is to be delivered by a router node, it is placed not only within the correct physical processor, but in the correct region of memory for the virtual processor originally specified as the message's destination.

In addition to the bit-serial data processors described above, the CM-2 parallel processing unit has an optional floating point accelerator that is closely integrated with the processing unit. There are two possible options for this accelerator: Single Precision or Double Precision. Both options support IEEE standard floating-point formats and operations. They each increase the rate of floating-point calculations by more than a factor of 20. Taking advantage of this speed requires no change in user software.

The hardware associated with each of these options consists of two special purpose VLSI chips, a memory interface unit and a floating-point execution unit, for each pair of CM-2 processor chips.

As an example of the operation of the floating-point accelerator, consider the execution of a two-operand floating-point instruction such as f-add-2 or f-multiply-2. Execution proceeds in five stages; each stage is generally comprised of 32 nanoinstruction cycles (one cycle for each of the 32 data processors on the two CM-2 processor chips).

1. The first operand for each 32 data processors is transferred from memory to the interface chip.
2. The first operand is transferred from the interface chip to the floating-point execution chip. (The floating-point execution chip is capable of storing 32 values of a given precision.) Simultaneously, the second operand is transferred from memory to the interface chip.
3. The second operand is transferred from the floating-point interface chip to the floating-point execution chip, where the operation is performed. At the end of this stage, the floating-point execution chip contains the 32 results.

4. The results are transferred from the floating-point execution chip to the interface chip.
5. The results are transferred from the interface chip to memory.

If the virtual processor ratio is N, this process is pipelined so as to require only 3N + 2 stages instead of 5N stages.

The Connection Machine I/O structure allows data to be moved into or out of the parallel processing unit at aggregate peak rates as high as 320 Mbytes/s for a system with multiple I/O controllers. Input/output is done in parallel, with as many as 2K data processors able to send or receive data at a time. All transfers are parity checked on a byte-by-byte basis.

The data processors send and receive data via I/O controllers, which interface through an I/O channel to Connection Machine data lines. These I/O controllers, in turn, operate under the control of the parallel processing unit sequencers. There may be as many as four sequencers in a fully configured system. A maximum I/O configuration for a 64K processor Connection Machine system includes eight I/O channels, each of which permits input and output operations for a set of 8K physical processors.

An I/O controller treats its 8K physical processors as two banks of 4K. Each CM-2 processor chip contains 16 data processors and has 1 I/O line, so each bank of 4K processors is implemented on 256 chips and has 256 I/O lines. A bank can therefore pass 256 bits in parallel at a time to its associated I/O controller. Each sequencer controls a bank switch that determines which bank is active.

I/O controllers store data internally in 288-bit chunks (256 data bits plus 32 parity bits). Parity is checked each time data is transferred between a controller and the data processors. Each controller has the ability to store 512 of these 288-bit chunks in its own internal memory. Data transfers between I/O controllers and data processors proceed under control of a Connection Machine sequencer. Two I/O controllers may be active simultaneously on each sequencer.

A Connection Machine I/O bus runs from each I/O controller to the devices it controls. This bus is 80 bits wide (64 data bits, 8 parity bits, and 8 control bits). The I/O controller multiplexes and demultiplexes between 256-bit processor chunks and 64-bit I/O bus chunks. The controller also acts as arbitrator allocating bus access to the various devices on the bus.

Since standard peripheral devices do not operate at the speeds that the Connection Machine system itself can sustain, it is often desirable to place multiple devices on multiple buses. For example, each of eight disk units could interface to several sections of data processors via several I/O controllers, each disk reading and writing data in parallel with the others. In this way, up to eight times the aggregate transfer rate of a single disk unit is achieved. Alternatively, devices may be interfaced to a single bus, interfaced in turn to I/O controllers in all sections of the parallel processing unit, allowing data to be moved directly between that device and any part of the processing unit. Typical configurations use a mix of these techniques. Some devices are connected to multiple controllers. Others connect to just one controller, and the Connection Machine router is used as necessary to move data to its final destination in the parallel processing unit.

7.1.2.1. Performance Specifications

The specifications below assume a fully configured Connection Machine model CM-2 system with 64K data processors and eight I/O channels. Specifications for floating-point performance assume the use of a floating-point accelerator.

General Specifications
Processors	65,536
Memory	512 Mbytes
Memory bandwidth	300 Gbits/s

The memory bandwidth is the maximum sustained transfer rate of data to or from memory.

Input/Output Channels
Number of channels	8
Capacity per I/O controller	40 Mbytes/s
Total I/O controller transfer rate	320 Mbytes/s
Capacity per framebuffer	1 Gbit/s

Each I/O channel may support either one general-purpose I/O controller or one framebuffer module. The total I/O controller transfer rate assumes simultaneous use of eight I/O controllers.

Typical Application Performance (Fixed Point)
General computing	2500 MIPS
Terrain mapping	1000 MIPS
Document search	6000 MIPS

These numbers indicate the averaged performance of the machine on applications for which it is well matched. The numbers are based on actual measurements that include all overhead in the sequencer, the operating system, the front-end user code, and inefficiencies of I/O transfers and algorithm design. The terrain mapping application, for example, cited as running at 1000 MIPS, does indeed run approximately 1000 times faster than the same application on a serial computer rated at 1 MIPS.

Interprocessor Communication
Regular pattern of 32-bit messages	250 million/s

Random pattern of 32-bit messages	80 million/s
Sort 65,536 32-bit keys	30 ms

The amount of time required to deliver messages depends on the pattern. A fully loaded random pattern is the worst case that has currently been measured. Sparse message patterns are faster, as are patterns with regular structure, such as grids, trees, or shuffles. The sort time is given here because it is a communication-intensive benchmark.

Variable Precision Fixed Point

64-bit integer add	1500 MIPS
32-bit integer add	2500 MIPS
16-bit integer add	3300 MIPS
8-bit integer add	4000 MIPS
64-bit move	2000 MIPS
32-bit move	3000 MIPS
16-bit move	3800 MIPS
8-bit move	4500 MIPS

These numbers indicate the performance of the machine running repeated cycles of the same instructions. The rates include the worst case for all overhead associated with virtual processors. For applications using large numbers of virtual processors per physical processor, the performance will be higher, especially when operating on small fields.

Double Precision Floating Point

4K × 4K matrix multiply benchmark	2500 MFLOPS
Dot product	5000 MFLOPS

The 4K × 4K matrix multiply benchmark starts with two matrices; approximately 16,000,000 elements each are distributed to the machine. The result is the matrix product. The number includes all communications overhead. The dot product rate is for multiplying two vectors, approximately a hundred elements each, stored within each processor in the optimal format, using the Paris f-vector-dot-product operation. This gives an indication of high rates that can be achieved for short periods of time. It is unusual to sustain such rates over the course of a computation. All double precision rates assume the machine is equipped with a double precision floating-point accelerator.

Single Precision Floating Point

Addition	4000 MFLOPS
Subtraction	4000 MFLOPS
Multiplication	4000 MFLOPS
Division	1500 MFLOPS

4K × 4K matrix multiply benchmark   3500 MFLOPS
Dot product   10,000 MFLOPS

Single precision rates are for a CM-2 equipped with either a double precision or a single precision floating-point accelerator. The rates for addition, subtraction, multiplication, and division assume the use of two-address, unconditional Paris instructions with a virtual processor ratio of 32 (2048K virtual processors), and include all instruction issuing and decoding overhead.

Parallel Processing Unit Physical Dimensions

Size	56" × 56" × 62"
Weight	2600 lbs.

These dimensions are for the parallel processing unit only and do not include the front-end computer(s), the high-resolution graphics display monitor, or the DataVault mass storage system.

Parallel Processing Unit Environmental Requirements

Power dissipation	28 kW
Power input	Four 30-amp 3-phase 110/208 V
Operating temperature	70°F ± 5°F
Operating relative humidity	50% ± 10%

These figures are for the parallel processing unit only and do not include the front-end computer(s), the high-resolution graphics monitor, or the DataVault mass storage system.

7.1.3. DataVault

The DataVault is the Connection Machine mass storage system. It combines very high reliability with very fast transfer rates for large blocks of data. The DataVault holds 5 Gbytes of data, expandable to 10 Gbytes. It transfers data at a rate of 40 Mbytes/s. Eight DataVaults, operating in parallel, offer a combined data transfer rate 320 Mbytes/s and hold up to 80 Gbytes of data.

Each DataVault unit stores its data in an array of 39 individual disk drives. Data is spread across the drives. Each 64-bit data chunk received from the Connection Machine I/O bus is split into two 32-bit words. After verifying parity, the DataVault controller adds 7 bits of Error Correcting Code (ECC) and stores the resulting 39 bits on 39 individual drives. Subsequent failure of any one of the 39 drives does not impair reading of the data, since the ECC code allows any single bit error to be detected and corrected. Although operation is possible with a single failed drive, three spare drives are available to replace failed units until they are repaired. The ECC codes permit 100% recovery of the data on the failed disk, allowing a new copy of this data to be reconstructed and written onto the replacement disk. Once this recovery is complete, the data base is considered to be healed.

The DataVault supports job staging and data base storage. New jobs may be loaded onto the DataVault from external devices such as magnetic tape drives. Once in the DataVault, a maximum-size 512-Mbyte memory image may be loaded in under 15 s. This same 512-Mbyte memory image may be loaded in less than 2 s on a system with eight DataVaults operating in parallel. Running jobs may use the DataVault for file storage, opening and accessing files as needed.

All DataVault operations take place under the control of a file server, which is a standard microcomputer. File server commands include creating files, as well as opening, reading, writing, and determining status. Commands to be executed by the file server are passed to it over the Connection Machine I/O bus. Commands such as "open" or "status" that do not involve data transfers are completed by the file server, and a completion message is returned via the I/O bus to the front end.

The file server supports "read" and "write" commands that can specify a field of any size. The data in this field is then transferred between each Connection Machine processor and the DataVault.

In systems with multiple DataVaults, a single master file server controls the file creation and deletion process, although the file itself may be spread across multiple units. Each file server that has a portion of the file maintains a file of disk block locations that allows files to be mapped into disk blocks. These files are not stored on the DataVault itself. They are stored redundantly on two independent file server disks to prevent a single medium failure from blocking access to the file. Two write operations are performed each time the information is changed. When a file is opened, the block location information is moved to the file server's main memory for faster access during subsequent reads and writes. File space is allocated in blocks of 32K bytes.

Off-line storage devices (such as magnetic tape) interface directly to the file server minicomputer. New data may be loaded into the DataVault without involvement of the rest of the Connection Machine system. Dumping of DataVault information to magnetic tape for backup also occurs without involving the rest of the system.

Data transfers move information between parallel variables in Connection Machine memory and DataVault files. A single read or write moves a specified number of bits (which could correspond to a single parallel variable or to a series of parallel variables that are contiguous in memory) into or out of each Connection Machine virtual processor.

Reading and writing of data are very similar operations. Here, the process of writing data will be described under the assumption that no errors occur.

A write operation is initiated by the front end. The front end issues a write instruction to the appropriate sequencer, which in turn activates the necessary I/O controllers. The request is received by the DataVault file server, which translates the logical file request into a series of physical disk addresses.

Data from Connection Machine memory is moved to the I/O controllers, with parity checked for each byte, and stored in the 288-bit × 512 buffer memories on those controllers. When the buffers are sufficiently full, the I/O controller signals its readiness to send data to the DataVault. At this point, the Connection Machine processors are free to proceed with other tasks.

Data in the I/O controllers is split into 64-bit units. Eight parity bits are added and the resulting 72-bit unit is sent on a Connection Machine I/O bus to the DataVault.

Parity of data arriving at the DataVault is checked twice, by two independent sets of logic. If both parity checkers agree that the data is valid, the 64 bits of data are split into two 32-bit words. For each 32-bit word, two independent ECC circuits generate 7 ECC bits for the data. As long as both units generate the same code, the resulting 39 bits are split up and each bit is sent to one of 39 disk buffers. As these buffers fill up, the data is written out to the individual disks.

When all data has been moved from Connection Machine memory through the I/O controllers and the disk buffers and physically written on the disks, a signal is returned to the front end that the transfer is complete.

Data being read into the Connection Machine memory from the DataVault follows the same path as for writing, but in reverse order), through the disk buffers, the I/O bus, and the I/O controller buffers. The data coming off the disks is checked by two independent ECC circuits. Errors are checked for, corrected, and logged, and the data is written to the I/O bus.

A transfer status may indicate that a single disk drive is failing and that the ECC has been required to correct the data. At this point, system operation should be interrupted to verify that, in fact, a drive has failed. If it has, it must be switched out of the array and a spare drive switched in. Switching and sparing is done automatically by the DataVault. To assure integrity of the data, the information stored along with each 32 bits of data allows this reconstruction. Regeneration of this data takes about 10 min, after which the data is again protected against the failure of another drive.

Repairs or replacement of the failed drive allows it to return to active use. Restoration of data at this point is very straightforward. It is only necessary to copy the contents of the spare drive that has been used in the interim. Once this transfer is completed, the repaired drive may be returned to active status. The spare drive is again marked as unused, and the data base is fully healed.

7.2. PROGRAMMING

The data parallel style of programming associates a processor with every element of a program's data. There are very few differences between a data parallel program and a conventional serial program. In both cases, a single sequence of instructions is used, with a serial control structure. The Connection

Machine system provides parallel processing without requiring the application programmer to indicate synchronization explicitly in programs.

Because the data parallel and serial programming styles are similar, they utilize the same languages. The languages currently supported for the Connection Machine system are C* (pronounced "see-star"), *Lisp (pronounced "star-lisp") and CM-Lisp (prounounced "see-em-lisp,). Each of the three languages are very close to the corresponding serial language specification, but in each case extends it by adding a new data type; very little new syntax is added, the power of parallelism arising instead from extending the meaning of existing program syntax when applied to parallel data. CM-Fortran is used at some sites.

There are some broad themes common to any data parallel programming language that are useful to keep in mind when examining a language description:

Establishing Parallel Data Structures. Data parallel programs can be expressed in terms of the same data structures used in serial programs. The difference is that the individual elements of a composite data structure, such as an array, are spread across processing elements, so that each data element has an associated processor. Since each processing element has its own dedicated memory, the task of associating data elements with processing elements is simply the task of assigning memory locations across processors. This assignment is done by the compilers when the array is first declared or created. In C*, the data types in a declaration implicitly specify whether a data structure is parallel. In *Lisp and CM-Lisp, data structures are created dynamically, and different creation operations are used by the programmer to indicate creation of serial or parallel data structures.

Establishing Linkages Among Data Elements. During the execution of a program, data from different problem elements are used together. Data parallel programs use pointers or array subscripts to establish connections between processors and hence between their data elements. An array of pointers, itself a parallel data structure, establishes an arbitrary pattern of intercommunication. If the required patterns are regular and local, such as processors sharing data with their nearest neighbors, then no explicit array of pointers is needed because each processor can easily calculate the address of its neighbors as needed.

Establishing Scalar Data. Some data is not parallel. For example, it is wasteful to place a copy of a constant in every processor's memory since the constant can be efficiently broadcast as needed from a central point. For this reason, scalar data (whether constant or variable) is declared as such and stored in the front end.

Operations on Parallel Data Structures. In a data parallel program, a single operation can affect all the elements of a parallel data structure at once, since each data element has its own processor. The same operation in a serial program must be expressed as a loop, with the basic operation applied sequentially to all the elements of the array. Some parallel operations are totally local to individual processing elements. The required data elements are all in the processing element's memory and the result is to be stored there. Other parallel operations have implicit communications cycles imbedded in them since some or all of the required data resides in other processors' memories.

Operations on Mixed Data. Operations that use both scalar and parallel data typically involve replication or reduction. If a scalar value participates in an operation that yields a parallel result (such as adding a constant to every element of an array), the scalar value is replicated by broadcasting it to all processors at once. If parallel data participates in an operation that yields a scalar result (such as finding the sum of all of the elements of an array), a reduction operation is used; given one processor for each data element, such an operation can be completed in time logarithmic in the number of data elements, by organizing the operations on the data into a balanced binary tree. (This organization is carried out by the underlying language implementation.) As with the sequential programming style, data parallel programmers do not need to do anything special when mixing scalar and parallel data.

Conditions. Data parallel programs implement conditionals by limiting the impact of operations to a certain subset of processing elements, and hence to a subset of the elements of a parallel data structure. The if ... then operation first tests a specified condition in all elements of a parallel data structure and then performs the indicated operations only in processors where the conditional was true. As in serial programs, conditionals may be nested in very general ways.

7.2.1. The *Lisp Language

The *Lisp language is an extension of Common Lisp for programming the Connection Machine in a data parallel style. It is intended for people who wish to write Connection Machine programs in Lisp that map simply onto Connection Machine hardware features. It supports primitives that correspond directly to the operation of the hardware, and also allows the users to build their own abstractions on top of those primitives. The language is a fully compatible extension of the Common Lisp standard.

Because the primitives of the language correspond very closely to the instruction set of the Connection Machine, it is possible to write code that executes very efficiently.

The parallel primitives of *Lisp support a model of the Connection Machine in which each processor executes a subset of Common Lisp, with a single thread of control residing on the front-end computer. For most Common Lisp functions, *Lisp provides a corresponding parallel function that can operate on all processors, or some selected subsets, simultaneously. In addition, the language provides Lisp-level operators for communicating between processors, both through pointers and in regular patterns. Sequential Common Lisp code, running on the front end, can be freely intermixed with the parallel code executed on the Connection Machine.

Most *Lisp functionality corresponds directly to underlying Paris instructions. As a result, the execution speed of a *Lisp program is predictable and easily computed by hand, and direct calls to Paris instructions and special purpose microcode blend in naturally with *Lisp code.

*Lisp provides a safe programming environment. The run time system will signal an error when the user causes a computation to overflow or when a pointer is used illegally. All user type declarations are continuously verified during the execution of the application. This error checking may be turned off for better performance.

The *Lisp implementation consists of an interpreter and a compiler. Both are written in Common Lisp and are transportable to any front-end computer that supports Common Lisp.

*Lisp supports all of the standard Common Lisp data types, including symbols, fixed and floating-point numbers, and arrays. It also supports an additional parallel data type called a pvar (parallel variable). A pvar is a first-class Lisp data type that has value for each processor in the machine. It is similar to an array, except that it is also possible to access and update its elements in parallel.

There are two ways of viewing a pvar. In one model, each processor is simultaneously running the same Common Lisp program, and the pvar represents a variable that exists in all processors and gets operated upon simultaneously in all processors. In the other model, the pvar represents an array whose size is the same as the number of processors in the machine. The elements of the array are located in consecutive processors.

The individual elements of a pvar may contain different types of data. *Lisp supports data of type integer, float, boolean (t and nil), and pointer. The integer and float types may be of any size supported by Paris. Although integers of any size from 1 to 128 bits are guaranteed to work, most operations work for sizes up to the amount of memory available in a processor.

Like Common Lisp, *Lisp supports run-time type checking, so a *Lisp program requires no declarations. If desired, programmers may insert type declarations to improve performance. *Lisp adheres to the standard Common Lisp declaration syntax.

The Connection Machine supports two different types of communication between processors. One is general pointer-based addressing, and the other is local communication on an n-dimensional grid. For general addressing, each processor is assigned a single number between zero and the number of processors in the machine. For grid addressing, each processor is assigned a vector address. *Lisp provides functions for communication in both modes.

The standard functions for reading or writing the contents of a pvar in a single processor are pref and pref-grid. The Common Lisp macro setf is used in combination with pref and pref-grid to store data from the front end into a pvar of a processor.

For example, (setf (pref my-pvar 10) 123.4) will store the quantity 123.4 into processor 10 of the pvar my-pvar. Thereafter (pref my-pvar 10) will return 123.4.

Similarly, (setf (pref-grid my-pvar 5 7) 111) will store 111 into pvar my-pvar at grid location (5,7) (assuming of course that the processors are configured as a two-dimensional grid). Thereafter (pref-grid my-pvar 5 7) will return 111.

All the functions in this section operate only on active processors.

The assignment operator is called *set. It takes a destination pvar and a pvar expression whose value is to be stored into that destination.

For example, (*set pvar1 pvar2) will store the contents of pvar2 into pvar1 in all active processors.

The function !! accepts a scalar and returns a pvar that contains the scalar in all active processors.

The statement (*set pvar1 (!! 5)) will store the quantity 5 into pvar1 for all active processors.

The functions +!!, –!!, *!!, and /!! perform the same operation as the Common Lisp functions +, –, *, and /, but in all active processors.

The statement (*set pvar1 (+!! pvar1 (!! 1))) will increment the value of pvar1 in all active processors.

There are many other *Lisp functions for manipulating other types of data. For example, the functions and!!, not!!, and or!! return boolean (t or nil) quantities in active processors. After the statement

```
(*set boolean-pvar (and!! (=!! pvar1 (!! 5))
                          (>!! pvar2 (!! 100))))
```

is executed, boolean-pvar will be t in all processors where pvar1 contains 5 and pvar2 is greater than 100.

Most Common Lisp functions have parallel equivalents in *Lisp. Typically, the user thinks of the name of a Common Lisp function and appends the characters "!!" to the function name to arrive at the parallel version. The characters "!!" are meant to represent the mathematical symbol , which means "parallel". Some of these functions are:

```
mod!!   ash!!   round!!   integerp!!
max!!   min!!   if!!      eql!!
ldb!!   dpb!!   byte!!    numberp!!
```

There are also parallel functions in *Lisp that do not have a corresponding Common Lisp equivalent. For example, (*set pvar1 (self-address!!)) will set pvar1 to the send address of each processor, and (*set pvar1 (self-address-grid!! (!! 1))) will set pvar1 to the first component of each processor's vector grid address.

All basic *Lisp functions will compute values only in active processors. Pvars in inactive processors are always left unmodified. Some of the *Lisp macros for manipulating the current set of active processors are *all and *when. The *all construct activates all processors for the block of *Lisp code in its body; the *when construct subselects, for the duration of the block of code in its body, all already active processors that satisfy a predicate.

```
(*all (*set pvar1 (!! 10)))      ;store 10 in all
                                  processors pvar1
(setf (pref pvar1 100) 0)        ;set processor 100
                                  pvar1 to 0
(*when (/=!! pvar1 (!! 0))
  (*set pvar1 (1-!! pvar1))      ;decrement non-zero
                                  values
```

One may nest *when and *all statements to any depth.

One of the primary strengths of the Connection Machine lies in its communication abilities. The basic functions for using the communication system are pref!! and pref-grid!!. Whereas pref and pref-grid allow the front-end computer to serially read or write the data in a pvar in a single processor, the !! versions allow each active processor to simultaneously read/write the value of a pvar in any processor. Even if two or more processors attempt to read the data of a single processor, they all receive the same correct data. (This is supported by the Connection Machine router hardware.)

The following two pieces of code have equivalent effects, although they achieve these effects in different ways:

```
(*all (*set pvar2 (pref!! pvar1 (!! 23))))
(*all (*set pvar2 (!! (pref pvar1 23))))
```

Note that the second form freely mixes serial and parallel code.

Although the previous example used pref!! to access the data of a single processor, it may also be used to access data in any processor. For example, the statement

```
(*all (*set pvar2 (pref!! pvar1 (random!! (!! 100)))))
```

causes every processor to make a pseudo-random choice from the first 100 elements of pvar1 and store the fetched value in pvar2.

Some other standard *Lisp routines that use the communication network especially efficiently are sort!! and enumerate!!.

Some *Lisp functions reduce the contents of a pvar in all active processors to a single value, which is then returned to the front-end computer. Examples of this class of functions are *min, *sum, and *logior. For example, (*all (*sum (!! 1))) will sum together the quantity 1 in all processors in the Connection Machine. The results will be the number of processors in the particular Connection Machine system being used (actually, the number of virtual processors into which the system has been configured).

7.2.2. THE C* LANGUAGE

C* is an extension of the C programming language designed to support programming in the data parallel style, in which the programmer writes code as if a processor were associated with every data element. C* features a single new data type (based on classes in C++), a synchronous execution model, and a minimal number of extensions to C statement and expression syntax. Rather than introducing a plethora of new language constructs to express parallelism, C* relies on existing C operators, applied to parallel data, to express such notions as broadcasting, reduction, and interprocessor communication in both regular and irregular patterns. While C* effectively allows the processing of large arrays of data, it preserves the interchangeability of arrays with pointers, a feature central to the C language. C* relies on pointers for interprocessor communication.

Just as the C language assumes an abstract machine model with certain interesting abstract properties (sequential execution, uniform address space, meaningful pointer arithmetic), so C* assumes a certain abstract machine model. The C* model is an extension of the plain C model. They share such important features as a uniform address space and meaningful pointer arithmetic. C* extends C by having many processors instead of just one, all executing the same instruction stream. The C* model may be summarized as providing the programmer with lots of processors of an otherwise conventional nature, operating within a uniform address space in a synchronous execution mode.

C* assumes a synchronous model of computation, in which all instructions are issued from a single source, a distinguished processor called the front end. All the other processors are called data processors. At any time, the data processors that are executing the instructions stream sent out from the front end are called the "active set". The local memory of an idle processor does not change, unless another processor writes it.

The layout of memory within each data processor is conventional. Except for the fact that no code is stored in the memory of a data processor, memory is laid out exactly as for a C program in a conventional sequential computer. One end of memory is used to hold statically allocated variables (storage classes static and extern), and the other end is used as a stack area for the allocation of automatic variables (storage class auto).

Processor memory layout can be informally described as a record structure, that is, a C struct. (The C language is very good at describing arbitrary memory layouts.) When there are many processors, as in the C* machine model, different processors may have different memory layouts because they may hold different kinds of data for different purposes. If we think of a data processor's memory layout as being a record structure, then we might as well say that a processor's memory really does belong to such a structure type, and we can distinguish groups of processors by that type. In C* a structure type that describes the memory of a data processor is called a domain. The layouts of 26 different processors might be described as follows:

```
domain employee {
  double salary;
  employee_type type;
  char *name;
  int knowledge;
};
domain part {
  int part_number;
  double price;
  vendor *supplier;
  char *description;
};
domain book {
  char *title, *ISBN;
  int content;
  employee *owner;
};
domain employee Fred;                    /* Processor 0 */
domain employee Sally;                   /* Processor 1 */
domain part grommet;                     /* Processor 2 */
domain employee George;                  /* Processor 3 */
domain part wing_nut;                    /* Processor 4 */
domain book my_novel;                    /* Processor 5 */
domain employee programmer[20];          /* Processors 6-25 */
```

In C*, all code is divided into two kinds: serial and parallel. Code that belongs to a domain is parallel, and may be executed by many data processors at once. Other code is serial, and is executed by the front end as if it were ordinary sequential C code. The two types of code are distinguished by syntactic context: code may belong to a domain (and therefore be parallel) only as the body of a member function of the domain or as the substatement of a selection statement that selects the domain. Once the context is established, however, the two types of code are written using the same syntax; parallel code, taken out of context, looks exactly like ordinary sequential C code.

In C*, all data is also divided into two kinds: scalar and parallel. These are described in the language using two new keywords, mono and poly; they are used somewhat like the storage class keywords extern, static, and auto, but describe an independent attribute. In certain situations they may sensibly be used in the same way as the const and volatile keywords of proposed ANSI standard C. Some example declarations:

```
mono int total;
poly int salary;
poly extern float coefficients[10];
mono int *poly x;  /* A poly pointer
                      to a mono integer. */
poly static struct foo x[20];
poly auto double all_the_day;
```

The mono or poly attribute may be omitted, and usually is, just as the storage class is often omitted in ordinary C code. Within parallel code, the default is poly; within serial code, the default is mono. (The declaration of poly variables is in fact forbidden within serial code, and so the keyword poly is required only in pointer-declaration contexts and casts.)

Scalar (mono) data resides in the memory of the front end, and parallel (poly) data resides in the memory of the data processors. Note that poly data is only potentially parallel; it is processed in parallel only if referred to by parallel code. It is possible for the front end, executing serial code, to access poly data in a sequential manner. Similarly, serial data may be processed by many data processors at once if the front end will first broadcast copies.

Domains differ from classes in that member declarations for domains can use the storage class keywords auto, register, static, and extern. In particular, different files can declare different members of a domain, and the extern keyword can mark members that are defined in one file but referenced in another. (In contrast, a C++ class may not be declared in such a piecemeal fashion.) Note that auto variables in member functions are allocated within each instance, on per-processor stacks. This is all consistent with the fact that the memory of each data processor is organized in the same way as for a sequential C program.

For convenience, the C* language includes maximum and minimum operators, which are really arithmetic operators. The minimum operator <? and the maximum operator >? may be applied to pointers as well as to numeric data. By themselves these operators are relatively unimportant, but the assignment operators <?= and >?+ have great utility in C*.

In C* most assignment operators may be used as unary operators. This unary use of existing binary operators is introduced purely for convenience, as an abbreviation for a frequently used and otherwise rather awkward idiom.

Instead of adding new operators for parallel computation, C* takes advantage of the compile-time distinction between scalar (mono) and parallel (poly) data, and extends existing operators, through overloading, to operate on parallel data. These extended interpretations allow us to express various interesting patterns of communication:

reading: fetching one value from a particular data processor to the front end

writing: storing a value from the front end into a particular data processor
replication: broadcasting a value from the front end to all data processors
reduction: combining values from all data processors to produce one result
permutation: interprocessor communication (in both regular and irregular patterns)

All these patterns of communication are achieved by using the standard C operators and by adding two rules to the usual rules of C evaluation:

Replication Rule—A scalar value is automatically replicated where necessary to form a parallel value.

As-If-Serial Rule—A parallel operator is executed for all active processors as if in some serial order.

The Replication Rule requires that when a binary (or ternary) operator combines mono and poly data, the mono value is replicated before you do the operation. A mono value is also replicated if passed as an argument to a function whose corresponding formal parameter is poly. In other words, replication occurs automatically wherever necessary.

The As-If-Serial Rule is more subtle; it facilitates parallelism by imposing a sequential semantics (while permitting a parallel implementation). The following code segment illustrates the point:

```
double total_salary;
...
{
    total_salary = 0;
    [domain employee].{
        total_salary += salary;
    }
}
```

The second assignment (the one within the selection of domain employee) will first replicate the variable total_salary as an lvalue; then the processor for every employee will attempt to perform the += operation on its own salary and that same lvalue. The As-If-Serial Rule is a simple way of stating the guarantee that, from the programmer's point of view, the processors do not interfere with each other. The net effect is that every employee's salary value has been added into total_salary exactly once. This, then, is how reduction is expressed in C*. The other C assignment operators may be used in a similar manner.

The other three patterns of communication, namely reading, writing, and permutation, arise naturally from the fact that addressing and the use of pointers in C* is perfectly as general as in C. To put it another way, the language restrictions that one might fear would be imposed because of implementation considerations are not imposed after all.

The communication pattern of reading is expressed quite simply. Within serial code one might write, for example,

```
strcmp(programmer[2].name, "Jane Jetson");
```

As in the example introduced earlier, programmer is an array of 20 employees, and so the elements of this array are instances of the class employee, residing in the memories of the data processors. The front end can refer to the name component of programmer number 2 simply by referring to programmer[2].name in the natural way.

Writing is expressed in exactly the same manner; for example, because the name component is public and writable, one can change an employee's name in the obvious way:

```
programmer[2].name = "Jane Eyre";
```

Permutation is also achieved through the natural use of C pointers. Any parallel processor can have pointers into the memory of any other processor. Therefore, if x is some poly variable of type T, and p is a poly variable of type "pointer to T", then the statement

```
*p = x;
```

means "send message x to processor p" (or more precisely to a specific variable within a processor, both being indicated by p); all active processors do this in parallel.

The use of an explicit pointer variable p allows any topological communications pattern to be expressed. The space required for such a pointer may be eliminated in cases where the pattern is sufficiently regular that it may easily be computed "on the fly". Here the ability of the C language to express address arithmetic is valuable; every processor can obtain a pointer to itself (by referring to the variable this) and then perform arithmetic on that pointer, allowing all kinds of relative addressing. For example,

```
x = (this+1)->x;
```

causes all x values to be shifted downward by one processor (every processor fetched the x value from the processor one above it).

Parallel Statements

C* adds only one new type statement to C, the selection statement, which is used to activate multiple processors.

All of the standard C statement types may be used in C* in both serial and parallel code. The treatment of control flow in parallel code satisfies the following design goals:

- As long as processors do not interact, the program behaves as if each processor were executing its own code independently. It is as if each of the parallel processes were executing ordinary serial C code.
- When processors do interact, the interactions are completely predictable, deterministic, and repeatable. This is achieved without ever requiring the programmer to write explicit synchronization code.

The format of a selection statement is as follows:

```
[domain tag] . statement
```

A selection statement activates all instances of a specified domain and then executes a substatement. (As with the switch statement, the substatement may be any statement but in practice it is typically a block.) On completion of the substatement, the instances activated by the selection statement are deactivated.

Within the substatement, the keyword this is bound to the primal parallel value: for each active instance, this is a pointer to that very instance. Because writing the name of a member variable memvar is equivalent to writing this → memvar, all references to such a variable also constitute parallel values.

The selection statement is the means by which serial code initiates parallel execution. The selection statement is also used within parallel code; in this case all instances active just before execution of this statement become inactive, and on completion of the statement the same instances become active again.

In parallel code, the expression in an if statement is treated as a poly value, so that each active domain instance has its own value for the test. (If the expression is not poly, then one may regard the parallel if statement either as behaving like an ordinary serial if statement or as first casting the value of the expression to be poly, thereby replicating it. These two points of view are equivalent.)

For the statement

```
if ( expression ) statement
```

the statement is executed with only those instances active whose test value was non-zero.

For the statement

```
if ( expression ) statement else statement
```

the first substatement is executed with only those instances active whose test value was non-zero, and then the second substatement is executed with only those instances active whose test value was zero.

On each iteration of the statement

while (expression) do statement

the expression is calculated as for an if statement. Instances that calculate the value zero became inactive; instances that calculate a non-zero value execute the substatement and then loop. The while loop completes if and when the active set becomes empty. At that time each individual processor has executed the substatement some number of times, and each may have executed it a different number of times, depending on the data being processed.

If the processors do not interact during the course of the loop, then it is as if each processor executes the while statement independently, each iterating the appropriate number of times, and then all processors become resynchronized when all have completed.

If the processors do interact, then their interactions are predictable; for example, all processors that execute the substatement as many as three times will be executing it for the third time together.

The C* compiler for the Connection Machine computer system is implemented as a translator in ordinary C code that is then compiled by an ordinary C compiler for the front-end computer. The C* compiler parses the C* source code, performs type and data flow analyses, and then translates parallel code into a series of function calls that invoke Connection Machine Paris operations. The use of the front end's usual C compiler allows all the programming tools associated with the front-end programming environment to be applied to C* program.

7.2.3. The Paris Language

The instructions that the front end may issue the parallel processing unit constitute a language called Paris (from the phrase "parallel instruction set"). It is the lowest-level protocol by which the front-end computer directs the actions of Connection Machine processors.

Paris is intended primarily as a base upon which to build higher-level languages for the Connection Machine system. It provides a large number of operations similar to the machine-level instruction set of an ordinary computer. Paris supports primitive operations on signed and unsigned integers and floating point numbers, as well as message-passing operations, I/O commands, and facilities for transferring data between the Connection Machine processors and the front-end computer.

Paris instructions direct the handling of data by the Connection Machine processors. Control instructions, such as subroutine calls, **if-then-else** conditionals, and while loops are not a part of the Paris instruction set. The control structure for an application loop are not a part of the Paris instruction set. The control structure for an application is provided by the front-end computer. A program that is "written in Paris" must actually be written in some ordinary sequential language for the front end, such as C, Fortran, Pascal, or Lisp.

The Paris user interface consists of a set of functions, subroutines, and global variables. The functions and subroutines direct the actions of the Connection Machine processors, and the variables allow the user program to find out such information about the Connection Machine system as the number of processors available and the amount of memory per processor.

As a simple example, here is a bit of C code that repeatedly causes every processor whose floating point z field is greater than 1.0 to be divided by two; the loop is terminated when no processor has a z value greater than one.

```
while (CM_f_gt_constant(z, 1.0, 23, 8),
       CM_global_logior(CM_test_flag, 1)) {
  CM_f_divide_constant_2(z, 2.0, 23, 8);
}
```

The functions whose names begin with "CM_" are Paris operations: CM_f_gt_constant causes every processor to compare a field to a common, broadcast constant, storing a bit reflecting the result in its "test" flag; CM_f_divide_constant similarly causes every processor to divide a floating-point field by a common constant; and CM_global_logior takes a bit field (in this example, a 1-bit field, namely the test flag) from every processor, and returns to the front end the result of a many-way bitwise inclusive-OR operation. The while constant is an ordinary C while loop, and is not a part of the Paris language proper.

Several different versions of the user interface are provided, one for each front-end programming language in which Paris is to be embedded. These interfaces are functionally identical; they differ only in conforming to the syntax and data types of one language or the other. Here is what the preceding example would look like if embedded in the Lisp language:

```
(do ()
    ((progn (CM:f-gt-constant z 1.0 23 8)
            (zerop (CM:global-logior CM:test-flag 1))))
  (CM:f-divide-constant-2 z 2.0 23 8))
```

This example of Lisp code uses a Lisp control structure, do, that is nearest in function to the C while statement. (It is actually a do-until statement, and the Lisp function zerop is used here to invert the sense of the result of CM:global-logior.) However, it would be appropriate to Lisp programming style to use a recursive function instead of to express such a loop:

```
(defun loop ()
  (CM:f-gt-constant z 1.0 23 8)
  (unless (zerop (CM:global-logior CM:test-flag 1))
    (CM:f-divide-constant-2 z 2.0 23 8)
    (loop)))
```

This example underscores the point that the control structure of the program may be written in any programming language (even the assembly language of the front-end computer, if necessary), and in any style suitable to that programming language. Paris merely extends that language by providing for the parallel processing of data.

Paris presents the user an abstract machine architecture that is very much like the physical Connection Machine hardware architecture, but with two important extensions: the virtual processor abstraction and a much richer instruction set.

The virtual processor abstraction (on which most higher-level software depends) is supported at the Paris level. When the Connection Machine system is initialized for a particular application, the number of virtual processors required by the application may be specified. If this number exceeds the number of available physical processors, then the local memory of each processor is split up into as many regions as necessary, and for every Paris instruction the processors are automatically time-sliced among the regions. For example, if an application should need to process a million pieces of data, it might request $V = 2^{20}$ virtual processors. Assume the available hardware to have $P = 2^{16}$ physical processors each with $M = 2^{16}$ bits of memory. Then each physical processor would support $V/P = 16$ virtual processors; this ratio V/P, usually denoted N, is called the virtual processor ratio, or VP ratio. In this example each virtual processor would have $M/N = 2^{12}$ bits of memory and would appear to execute code at about $1/N = 1/16$ the speed of a physical processor.

The time taken to perform a move depends on the length of the field to be moved and also on the number of virtual processors in use. If each physical processor is simulating N virtual processors, then issuing a single move instruction causes each physical processor to execute N move operations. This will take N times as long as if virtual processors were not in use, but also does N times as much work, so the Mips measurement is about the same. Indeed, the use of virtual processors usually increases the measured Mips rate, because the instruction needs to be decoded by the sequencer only once for N execution, and so the decoding overhead may be amortized.

Each virtual processor has some local memory and also a number of 1-bit flags. Most of the flags are condition codes such as overflow and float-inexact. The context flag, however, controls conditional execution; for most Paris operations a processor executes the operation if its context flag is 1, but does not participate if its context flag is 0. Processors whose context flag is 1 are said to be active; the set of active processors is called the current context. A few operations are unconditional, being executed by all processors regardless of the values of their context flags. (It is important, for example, that there be a way to set all context flags to 1 unconditionally!)

Most Paris operations deal with fields in the local memories of the Connection Machine processors. A field is specified by two quantities: the address of its first bit, and its length in bits. Uninterpreted bit fields (as processed by such operations as move, send, and logand) may be of any length. The length of an unsigned integer may range from 0 to 128 bits, and the length of a signed integer may range from 2 to 128 bits. (Some very simple arithmetic operations, such as addition, subtraction, and comparisons, are not limited to 128 bits.) Floating-point operations are available in a variety of precisions, including 32-bit, 64-bit, and 80-bit formats.

Nearly all operations are memory-to-memory; for example, the signed integer addition operation can add the value of one memory field into another memory field (two-address mode) or can replace a memory field with the sum of two other fields (three-address mode). The flags are addressed as if they were 1-bit memory fields.

Many operations come in several forms, differing from each other in up to three categories:

- *Addressing modes.* The operations s-add-2 and s-add-3 both perform signed integer addition, but the one takes two addresses and a length and the other takes three addresses and a length. The operation s-add takes three addresses and three lengths, allowing the three fields involved to be of different sizes. Anything s-add-2 can do, s-add-3 can do by duplicating one address operand; anything s-add-3 can do, 2-add can do by triplicating the length operand. The concise addressing modes improve performance by reducing total instruction size; the front end has fewer operands to send to the sequencer, and the sequencer has fewer operands to decode.
- *Conditionalization.* Most operations are executed only by active processors, but some are executed unconditionally by all processors. For example, the operation move copies one memory field to another for processors in the current context, but the operation move-always copies one memory field to another in all processors, regardless of the current context.
- *Immediate operands.* The operation s-add-2 adds one memory field into another in all active processors; the operation s-add-constant-2 adds an immediate quantity, sent from the front end as part of the instruction, into a memory field in all active processors. Note that the word "constant" in the instruction name is a relative term. The immediate operand is constant in being the same for all the data processors, but need not be constant within the front-end program; the front end may calculate the value to be sent to the sequencer.

7.3. APPLICATION

Diverse applications have been implemented on the Connection Machine. Here a sampling is provided from the fields of physics, operations research, image processing, text search, neural nets and orbit analysis.

7.3.1. Physics

Three physics application examples are provided. The first presented is electromagnetic scattering using the method of moments. The second and third, thermal diffusion and the wave equation, are treated together.

NUMERICAL COMPUTATION OF ELECTROMAGNETIC SCATTERING ON THE CONNECTION MACHINE USING THE METHOD OF MOMENTS

With the advent of fine grained parallel processors such as the Connection Machine, the computation time for many numerical algorithms can be reduced significantly over a comparable serial implementation. A parallel implementation of the method of moments on the Connection Machine is examined for the triangular surface patch technique as applied to perfect conductors using the Electric Field Integral Equation (EFIE)[1] and to homogeneous dielectric bodies using the Combined Field Integral Equation (CFIE).[2]

Introduction

The amount of local memory in each processor and the general machine architecture are crucial parameters in the design of algorithms for the CM or any other parallel processor. The processor memory limit dictates the number of global and temporary variables that each processor can manipulate. The interprocessor connection scheme has a great influence on the layout of the individual values stored in the variables in the processor memory. If the data is layed out in such a fashion as to conform to the interprocessor connection scheme, the amount of time spent on interprocessor communication can be minimized. These tradeoffs will be discussed below in terms of developing a parallel algorithm for the EFIE and CFIE formulations of the method of moments.

Method of Moments

The Method of Moments (MOM) has proven to be a very effective technique for computing the surface currents and the scattering cross-section for bodies whose size is on the order of a few wavelengths. For the case of closed bodies, the surface of the object can best be described by using small triangular surface patches which are the simplest geometric shapes which can be used to patch a closed surface. This method was first introduced by Rao et al.,[1] and it makes use of a set of edge centered basis functions which describe the current flow over the surface of the object by a finite sum as follows:

$$J(r) = \sum J_i f_i(r) \qquad (15)$$

In a radiating system with sources, the potential from which the radiated fields can be derived is written in terms of a Green's function integral of the surface current or charge density over the surface of the conducting object. The scattered electric field at some point, r, outside the body can be derived from the potential using:

$$E_s(r) = -j\omega A(r) - \nabla\varphi(r) \qquad (16)$$

By setting the above expression to the negative of the tangential incident field on the surface at the conductor, an analytic expression for the surface currents is obtained:

$$t \cdot E_{inc}(r) = t \cdot [-j\omega A\{J(r)\} - \nabla\phi\{J(r)\}] \qquad (17)$$

where t is the unit vector tangential to the surface at the point r.

Although the integral expression in Equation 17 is a valid expression for every point on the surface of the object of interest, the N unknowns J_i must still be found. The N equations required to find the J_i can be obtained by taking moments of Equation 17 using the N triangular patch basis functions and the inner product defined by:

$$< f(r), g(r) > = \int f(r) \cdot g(r)\, ds \qquad (18)$$

This leads to the expressions which will be solved:

$$< f_m(r), E_{inc}(r) > = < f_m(r), [-j\omega A\{J(r)\} - \nabla\phi\{J(r)\}] > \qquad (19)$$

where the tangential unit vector has been eliminated since the vector basis functions are all tangential to the surface. This system can be written in terms of a matrix equation which can be solved for the coefficients in the expression for the surface currents J_i.

Computing the electromagnetic fields scattered from a dielectric body is more complicated than the case of a perfect conductor, since the electromagnetic fields can penetrate the interior of the object. The tangential fields (E and H) at the surface of the dielectric object must be continuous across that surface. To be able to adequately describe the fields in both regions, a fictious magnetic surface current density is introduced. This current density gives rise to an electric vector potential and a magnetic scalar potential. The continuity of the fields across the surface yields an equation for the incident electric field and an equation for the incident magnetic field in terms of the surface current densities.

The scattered fields in the two regions are derived from

$$\begin{aligned} E_{sm} &= -j\omega A_m - \nabla\phi_m - \frac{1}{\varepsilon_m} \nabla \times F_m \\ H_{sm} &= -j\omega F_m - \nabla\psi_m - \frac{1}{\mu_m} \nabla \times A_m \end{aligned} \qquad (20)$$

where the potentials are defined in terms of Green's function integrals. If the scattered fields are written in terms of their respective potential functions, the field continuity at the surface can be expressed as:

$$t \cdot E_{inc}(r) = t \cdot [E_{s1}\{J(r), M(r)\} + E_{s2}\{J(r), M(r)\}]$$
$$t \cdot H_{inc}(r) = t \cdot [H_{s1}\{J(r), M(r)\} + H_{s2}\{J(r), M(r)\}] \quad (21)$$

The triangular patch basis function can now be used to take moments of Equations 21 when the surface currents are written in the form of Equation 15 resulting in a matrix equation which is divided into four blocks — two due to the terms involving the electric surface currents and two due to the terms involving the magnetic surface currents. The full expression becomes:

$$\begin{bmatrix} Z & C \\ D & Y \end{bmatrix} \begin{bmatrix} J \\ M \end{bmatrix} = \begin{bmatrix} V \\ I \end{bmatrix} \quad (22)$$

where the V source terms are derived from the incident electric fields and the I terms are derived from the incident magnetic fields.

Connection Machine Implementation

The implementation of the triangular surface patch MOM code on the CM presents some interesting problems since not only must the overall algorithm be considered, but the data layout across the processors must also be considered. The code can be separated into three distinct components: the patcher, the matrix filler, and the matrix solver. Each of the three components has a different data layout which is specific to the computations that must be performed. In some components, the data is arranged so as to minimize the amount of time spent on interprocessor communication, while in others, the data is arranged so as to make the conceptual data handling as simple as possible. The latter layout sacrifices some speed for conceptual simplicity.

The patcher is a relatively small pre-processor which will take the parameters of some simple objects — spheres, cubes and parallelepipeds — as inputs, and outputs a list of the patches in order of patch number. In addition to assigning patches to the surface of an object, the patcher also assigns each of the edges in a patch a unique number and direction flag which defines the direction of the current flow. The edge numbers and direction flags are assigned self consistently such that two adjacent patches have the same edge number and opposite direction flags assigned to the common edge. The patch data is then converted into an edge centered data format in which each processor holds the location of the three vertices for the two patches which are adjacent to some particular edge.

The matrix fill routine computes the individual elements that will eventually make up the entire MOM matrix. This is done on a row by row basis. The CM is filled with as much row data as will fit across the processors, computes the matrix elements and then moves (sends) the elements to the appropriate location within the machine. Each of the elements requires the computation of a Green's function integral which is given by:

$$\int f_m(r) \cdot O \int O' f_n(r) \, G(r - r') \, ds' \, ds \qquad (23)$$

where O and O' are some vector differential operators. Since each basis function integral is an integral over two patches, the edge data is spread across four processors. The individual processors perform one patch pair surface integral. The results of the four computations are summed and the total is put into the i – j element of the matrix. In the case of EFIE, one pass through the edge data is required to fill the matrix, while, for the CFIE case, one pass is required to produce the Z and Y blocks in Equation 22; another pass is required to produce the C and D blocks. The four blocks are then assembled into the full matrix equation.

The matrix fill code must also calculate the source terms which appear on the right-hand side of the matrix equation. The individual terms are given in integrals which have the form of:

$$\int f_m(r) \cdot F_{inc}(r) \, ds \qquad (24)$$

where F is either incident field (E or H) and the above surface integral must be computed over two patches. The integral for the ith component of the right-hand side is computed in processors $4j$ and $4j + 2$. The two patch integrals are summed and the total is moved (sent) to the appropriate location for the right-hand side. The EFIE case requires only the integral for the incident E field, while CFIE requires the integrals for both the incident E and H fields.

The matrix equation is solved by using a simple Gaussian elimination scheme with partial pivoting. The matrix data is stored in a chip by chip basis where each chip in the CM is a block of 16 processors. The 16 processor block holds one row of the matrix. The remaining memory space is filled with right-hand side terms corresponding to that particular row. The basic data layout is such that the elements which, in this case, are complex are stored in a contiguous memory space. The source elements are stored in blocks which are adjacent to the memory area for the matrix elements. This complicated data layout allows the matrix reduction to take place on a chip by chip basis which is much faster than a simpler processor layout scheme, since the interprocessor communication is absolutely minimized.

The total order of the algorithm can be approximated from the orders of the individual parts. The patcher forms such a small part of the overall package that its contribution can be neglected. The matrix fill performs its function on a row by row basis computing as many rows of the matrix as will fit across the processors of the machine. The fill algorithm is therefore proportional to the number of rows in the matrix, N. The matrix solver also performs its reduction on a row by row basis which leaves the solver proportional to N as well. This gives the overall algorithm order of N.

TABLE 9
EFIE MOM Execution Time

Number of patches	Number of edges	Total CM time (min)	Total VAX time (min)
48	72	.347	4.65
120	180	.757	37.68
160	240	1.11	77.79
280	420	2.56	293.3
500	750	5.98	1385*

* This figure is an estimate.

Results

The numerical predictions made for the surface currents by the CM code have been validated against the predictions made by the code for the VAX which was used to produce the results in Reference 2. The surface currents were also validated against analytic expressions for the surface fields for a perfectly conducting sphere and a homogeneous dielectric sphere. The surface current validation was used, since the far field (scattering cross-section) computation will wash out many of the numerical errors present in the surface current.

The execution time for the CM algorithm is significantly reduced over that of a VAX 8300 especially for very large problems. This is primarily due to the fact that the CM algorithm scales approximately as N, whereas the VAX algorithm scales as N^2 — the number of elements in the matrix — for the fill routine and as N^3 for the matrix inversion routine. The current algorithm also makes the CM or any other parallel processor a viable and attractive alternative to supercomputers for certain classes of problems. The execution time for the EFIE algorithm are summarized in Table 9.

In conclusion, the run time for a moment method algorithm on a parallel processor decreases the execution time by a significant amount for very large problems since the algorithm order is approximately N. These gains can become drastic as the problem size becomes very large. The overall MOM algorithm agrees very well with results which are predicted analytically which also makes it an overall useful method. With improved execution times and with larger and larger parallel processors, the applicability of the MOM method can now be extended to larger and larger bodies.

PARALLEL IMPLICIT METHODS FOR NUMERICAL PHYSICS
USING THE CONNECTION MACHINE

A fast parallel algorithm for tri-diagonal matrix inversion for the Connection Machine is presented and applied to the three-dimensional thermal diffusion and wave equations.

Introduction

The numerical solution to the thermal diffusion and wave equations have been known for quite some time and can be found in any standard text (see, for example, Reference 3). The numerical stability of the solution has also been well studied, and it has been found that a time implicit method can be used to ensure stability for arbitrarily large time steps. In this section, parallel solutions to the three dimensional thermal diffusion and wave equations are presented. The numerical solutions utilize an alternating dimension implicit (ADI)[4] formulation. They also utilize a very efficient tri-diagonal matrix inversion algorithm, which takes into account the gains in computation speed that can be achieved by using a parallel architecture. The summary of the timing results are presented with a discussion of the tradeoffs that can be made to influence the execution time.

Implicit Methods for Thermal Diffusion and Wave Motion

The most general form of the thermal diffusion equation can be written as follows:

$$\frac{\partial \psi}{\partial t}(\vec{x}, t) = \nabla \circ K(\psi) \nabla \psi(\vec{x}, t) \tag{25}$$

where $\psi(x,t)$ is the local temperature and $K(\psi)$ is the diffusion coefficient which can be temperature dependent. Assuming that K is independent of temperature, Equation 25 reduces to

$$\frac{\partial \psi}{\partial t}(\vec{x}, t) = K \nabla^2 \psi(\vec{x}, t) \tag{26}$$

In three dimensions, the above equation can be rewritten as

$$\frac{\partial \omega(x,y,z,t)}{\partial t} = K \left[\frac{\partial^2 \psi(x,y,z,t)}{\partial x^2} + \frac{\partial^2 \psi(x,y,z,t)}{\partial y^2} + \frac{\partial^2 \psi(x,y,z,t)}{\partial z^2} \right] \tag{27}$$

which can be discretized using

$$\psi^{\alpha}_{\lambda \upsilon \nu} = \psi(\lambda \Delta x, \mu \Delta y, \nu \Delta z, \alpha \Delta t) \tag{28}$$

where the space of the problem has been covered with a grid whose segments have length Δx in the x direction, Δy in the y direction, and Δz in the z direction and where α, λ, μ, and ν are integers. In discretized form Equation 27 becomes

$$\frac{\psi_{klm}^{\mu+1} - \psi_{klm}^{\mu}}{\Delta t} = K \left[\frac{\psi_{(k+1)lm}^{\alpha} - 2\psi_{klm}^{\alpha} + \psi_{(k-1)lm}^{\alpha}}{\Delta x^2} + \frac{\psi_{k(l+1)m}^{\beta} - 2\psi_{klm}^{\beta} + \psi_{k(l-1)m}^{\beta}}{\Delta y^2} + \frac{\psi_{kl(m+1)}^{\gamma} - 2\psi_{klm}^{\gamma} + \psi_{kl(m-1)}^{\gamma}}{\Delta z^2} \right]$$
(29)

The fully implicit formulation applied to the above equation ($\alpha = \beta = \gamma = m + 1$) would yield a matrix which has seven bands. This adds an additional degree of difficulty in the solution of Equation 29 which can be avoided by making one dimension implicit at a time.[4] In an ADI formulation in three dimensions, one time step is actually three time steps each of which has a different dimension implicit, which leads to the equations:

$$\frac{\psi_{klm}^{\mu+1/3} - \psi_{klm}^{\mu}}{\frac{\Delta t}{3}} = K \left[\frac{\psi_{(k+1)lm}^{\mu+1/3} - 2\psi_{klm}^{\mu+1/3} + \psi_{(k-1)lm}^{\mu+1/3}}{\Delta x^2} + \frac{\psi_{k(l+1)m}^{\mu} - 2\psi_{klm}^{\mu} + \psi_{k(l-1)m}^{\mu}}{\Delta y^2} + \frac{\psi_{kl(m+1)}^{\mu} - 2\psi_{klm}^{\mu} + \psi_{kl(m-1)}^{\mu}}{\Delta z^2} \right]$$
(30)

$$\frac{\psi_{klm}^{\mu+2/3} - \psi_{klm}^{\mu+1/3}}{\frac{\Delta t}{3}} = K \left[\frac{\psi_{(k+1)lm}^{\mu+1/3} - 2\psi_{klm}^{\mu+1/3} + \psi_{(k-1)lm}^{\mu+1/3}}{\Delta x^2} + \frac{\psi_{k(l+1)m}^{\mu+2/3} - 2\psi_{klm}^{\mu+2/3} + \psi_{k(l-1)m}^{\mu+2/3}}{\Delta y^2} + \frac{\psi_{kl(m+1)}^{\mu+1/3} - 2\psi_{klm}^{\mu+1/3} + \psi_{kl(m-1)}^{\mu+1/3}}{\Delta z^2} \right]$$
(31)

and

$$\frac{\psi_{klm}^{\mu+1} - \psi_{klm}^{\mu+2/3}}{\frac{\Delta t}{3}} = K\left[\frac{\psi_{(k+1)lm}^{\mu+2/3} - 2\psi_{klm}^{\mu+2/3} + \psi_{(k-1)lm}^{\mu+2/3}}{\Delta x^2} + \frac{\psi_{k(l+1)m}^{\mu+2/3} - 2\psi_{klm}^{\mu+2/3} + \psi_{k(l-1)m}^{\mu+2/3}}{\Delta y^2} + \frac{\psi_{kl(m+1)}^{\mu+1} - 2\psi_{klm}^{\mu+1} + \psi_{kl(m-1)}^{\mu+1}}{\Delta z^2}\right] \quad (32)$$

Equations 30, 31 and 32 can be expressed as a system of matrix equations each of which has the form:

$$\begin{bmatrix} 1 & 0 & 0 & 0 & . & . & & & & \\ -k & 1+2k & -k & 0 & . & . & . & & & \\ 0 & -k & 1+2k & -k & . & . & . & & & \\ 0 & 0 & -k & 1+2k & . & . & . & & & \\ 0 & 0 & 0 & -k & . & . & . & & & \\ . & . & . & . & . & & & & & \\ . & . & . & . & . & & & & & \\ . & . & . & . & . & & & & & \\ & & & & -k & 1+2k & -k & 0 \\ & & & & 0 & -k & 1+2k & -k \\ & & & & 0 & 0 & 0 & 1 \end{bmatrix} \begin{bmatrix} \psi^{\mu+1/3} \\ \psi^{\mu+1/3} \\ \psi^{\mu+1/3} \\ \psi^{\mu+1/3} \\ \psi^{\mu+1/3} \\ \\ \\ \\ \psi^{\mu+1/3} \\ \psi^{\mu+1/3} \\ \psi^{\mu+1/3} \end{bmatrix} = \begin{bmatrix} R^\mu \\ R^\mu \\ R^\mu \\ R^\mu \\ R^\mu \\ \\ \\ \\ R^\mu \\ R^\mu \\ R^\mu \end{bmatrix} \quad (33)$$

where

$$k = \frac{K\Delta t}{3\Delta x}, \frac{K\Delta t}{3\Delta y}, \frac{K\Delta t}{3\Delta z} \quad (34)$$

and

$$R_{mi}^u = \psi_{mi}^u + \frac{NK}{3}\left[\nabla^2 - \partial^2 m\right]\psi_{mi}^u \quad (35)$$

The above system can be solved by a simple tri-diagonal matrix inversion algorithm.

The numerical solution to the wave equation,

$$\frac{\partial^2 \psi}{\partial t^2} = C^2 \nabla^2 \psi \quad (36)$$

can be obtained in similar fashion to the numerical solution of the thermal diffusion equation. If Ψ is discretized in three dimensions Equation 28 yields:

$$\frac{\Psi_{klm}^{\mu+1} - 2\Psi_{klm}^{\mu} + \Psi_{klm}^{\mu-1}}{\Delta t^2} = C^2 \left[\frac{\Psi_{(k+1)lm}^{\alpha} - 2\Psi_{klm}^{\alpha} + \Psi_{(k-1)lm}^{\alpha}}{\Delta x^2} + \frac{\Psi_{k(l+1)m}^{\beta} - 2\Psi_{klm}^{\beta} + \Psi_{k(l-1)m}^{\beta}}{\Delta y^2} + \frac{\Psi_{kl(m+1)}^{\chi} - 2\Psi_{klm}^{\chi} + \Psi_{kl(m-1)}^{\chi}}{\Delta z^2} \right] \quad (37)$$

The above equation can be put into a three-dimensional ADI formulation as follows:

$$\frac{\Psi_{klm}^{u+1/3} - 2\Psi_{klm}^{u} + \Psi_{klm}^{u-1/3}}{\Delta t^2} = \frac{C^2}{9} \left[\frac{\Psi_{(k+1)lm}^{u+1/3} - 2\Psi_{klm}^{u+1/3} + \Psi_{(k-1)lm}^{u+1/3}}{\Delta x^2} + \frac{\Psi_{k(l+1)m}^{u} - 2\Psi_{klm}^{u} + \Psi_{k(l-1)m}^{u}}{\Delta y^2} + \frac{\Psi_{kl(m+1)}^{u} - 2\Psi_{klm}^{u} + \Psi_{kl(m-1)}^{u}}{\Delta z^2} \right] \quad (38)$$

$$\frac{\Psi_{klm}^{u+2/3} - 2\Psi_{klm}^{u+1/3} + \Psi_{klm}^{u}}{\Delta t^2} = \frac{C^2}{9} \left[\frac{\Psi_{(k+1)lm}^{u+1/3} - 2\Psi_{klm}^{u+1/3} + \Psi_{(k-1)lm}^{u+1/3}}{\Delta x^2} + \frac{\Psi_{k(l+1)m}^{u+2/3} - 2\Psi_{klm}^{u+2/3} + \Psi_{k(l-1)m}^{u+2/3}}{\Delta y^2} + \frac{\Psi_{kl(m+1)}^{u+1/3} - 2\Psi_{klm}^{u+1/3} + \Psi_{kl(m-1)}^{u+1/3}}{\Delta z^2} \right] \quad (39)$$

$$\frac{\Psi_{klm}^{u+1} - 2\Psi_{klm}^{u+2/3} + \Psi_{klm}^{u+1/3}}{\Delta t^2} = \frac{C^2}{9} \left[\frac{\Psi_{(k+1)lm}^{u+2/3} - 2\Psi_{klm}^{u+2/3} + \Psi_{(k-1)lm}^{u+2/3}}{\Delta x^2} + \frac{\Psi_{k(l+1)m}^{u+2/3} - 2\Psi_{klm}^{u+2/3} + \Psi_{k(l-1)m}^{u+2/3}}{\Delta y^2} + \frac{\Psi_{kl(m+1)}^{u+1} - 2\Psi_{klm}^{u+1} + \Psi_{kl(m-1)}^{u+1}}{\Delta z^2} \right] \quad (40)$$

The above system of equations can be put in the form of Equation 33 where, in this case,

$$k = \frac{C^2 \Delta t^2}{9\Delta x^2}, \frac{C^2 \Delta t^2}{9\Delta y^2}, \frac{C^2 \Delta t^2}{9\Delta z^2} \qquad (41)$$

and

$$R^u_{mi} = 2\psi^u_{mi} - \psi^{u-1/3}_{mi} + \frac{C^2 \Delta t^2}{9}\left[\nabla^2 - \partial^2 m\right]\psi^u_{mi} \qquad (42)$$

which can again be solved by a simple tri-diagonal matrix inversion.

Parallel Tri-Diagonal Matrix Inversion

Since each of the processors in a parallel machine can be performing operations simultaneously, the standard serial algorithm for performing a tri-diagonal matrix inversion is very inefficient. In 1978, Heller[5] presented an efficient algorithm for performing a tri-diagonal matrix inversion on a parallel machine more commonly known as cyclic reduction.

Consider the arbitrary tri-diagonal matrix equation:

$$\begin{bmatrix} b_1 & c_1 & 0 & 0 & \cdots & & & & \\ a_2 & b_2 & c_2 & 0 & \cdots & & & & \\ 0 & a_3 & b_3 & c_3 & \cdots & & & & \\ 0 & 0 & a_4 & b_4 & \cdots & & & & \\ 0 & 0 & 0 & a_5 & \cdots & & & & \\ \vdots & & & & & & & & \\ & & & & a_{m-2} & b_{m-2} & c & 0 \\ & & & & 0 & a_{m-1} & b & c_{m-1} \\ & & & & 0 & 0 & a & b_m \end{bmatrix} \begin{bmatrix} u_1 \\ u_2 \\ u_3 \\ u_4 \\ u_5 \\ \vdots \\ u_{m-2} \\ u_{m-1} \\ u_m \end{bmatrix} = \begin{bmatrix} v_1 \\ v_2 \\ v_3 \\ v_4 \\ v_5 \\ \vdots \\ v_{m-2} \\ v_{m-1} \\ v_m \end{bmatrix} \qquad (43)$$

where the vector **v** is known, and the solution **u** is sought. Since each processing element in a parallel machine can manipulate its data simultaneously with the other processors, the symbolic routine shown below allows each of the processors to manipulate a row of the matrix while performing the inversion in $\log_2(m)$ steps.

```
do i = 1,m step i
    a_k ← a_k/b_k
    c_k ← c_k/b_k
```

```
        u_k ← u_k/b_k
        b_k ← 1
        if k - i > 0
            a_k ← -a_k * a_{k-i}
            b_k ← b_k - a_k * c_{k-i}
            u_k ← u_k - a_k * u_{k-i}
        else
            a_k ← 0
        end if
        if k + i < m
            c_k ← -c_k - a_k * c_{k+i}
            b_k ← b_k - c_k * u_{k+i}
        else
            c_k ← 0
        end if
continue
```

where the subscripts refer to the address of the processor. This routine has the effect of using the rows 2^i above and 2^i below the rows in question to eliminate the a_k and c_k terms until there is one a_k term left in the lower left hand corner of the matrix and until there is one c_k term in the upper right hand corner of the matrix which are eliminated using the first and last rows, respectively.

Parallel Methods for Thermal Diffusion and Wave Motion for the Connection Machine

Since the self-address of each processor can be used to identify the sequential points in a one-dimensional grid, one dimensional problems are very easy to conceptualize on the Connection Machine. In three dimensions, this process could be very cumbersome if the problem could only be constructed as a grid of small one-dimensional segments. However, a grid can be constructed to simplify the implementation of problems in three dimensions. The DUNEWS-grid (Down-Up-North-East-West-South)[6,7] is imposed upon the hypercube in such a fashion as to optimize the communication between neighboring processors; the nearest neighbors in the grid structure are also nearest neighbors on the hypercube.

Using the structure provided by the Connection Machine, the thermal diffusion and wave equation can easily be put on the Connection Machine by simply assigning one point of the problem to one processor on the machine:

$$P(DUNEWS)(i,j,k)) = \psi(i\Delta x, j\Delta y, k\Delta z) \tag{44}$$

This method takes care of assigning the field values to the processors of the machine, but the geometry of the problem must be translated into the matrices of Equation 43. This can be achieved by allowing one processor to represent one row of the matrix, i.e., the matrix reductions are performed along the individual segments of the grids. Where more than one matrix needs to be inverted at one time, the individual matrix reductions are done simultaneously, maximizing use of the Connection Machine's resources.

In the case of thermal diffusion, each processor in the Connection Machine must hold the value of the local temperature, Ψ and the matrix coefficients, a, b and c. The algorithm becomes:

```
if ((self-address-3d-grid(i)    1) and (self-address-
3d-grid(i)    N))

    b ← 2k_i+1.0
    a ← -k_i
    c ← -k_i
else
    b ← 1.0
    a ← 0.0
    c ← 0.0
end if
do j = 1, n step j
    a ← a/b
    c ← c/b
    ψ ← ψ/b
    b ← 1.0
    if (self-address-3d-grid(i) - j > 0)
        a ← -a * a (self-address-3d-grid(i) - j)
        b ← b - a* c (self-address-3d-grid(i) - j)
        ψ ← ψ - a* ψ (self-address-3d-grid(i) - j)
    else
        a ← 0.0
    end if
    if (self-address-3d-grid(i) + j < N(i))
        c ← c * c (self-address-3d-grid(i) + j)
        b ← b - c * a (self-address-3d-grid(i) + j)
        ψ ← ψ - c * ψ (self-address-3d-grid(i) + j)
    else
        c ← 0.0
    end if
continue
```

where k_i is given by Equation 34, and self-address-3d-grid (i) gives the processor's location in the x-y-z grid.

The algorithm for the wave equation is similar to the one discussed above. In the Connection Machine, each processor now holds the current wave amplitude, Ψ^m, a previous wave amplitude, Ψ^{m-1}, and the matrix coefficients, a, b, and c. The reduction proceeds exactly as above except that the right hand side is replaced with

$$\psi \leftarrow 2\psi^n - \psi^{n-1} \tag{45}$$

and the matrix coefficients are assigned the appropriate values using the k values of Equation 31 instead of Equation 34.

Results

By using a simple sinusoid as an initial condition for both the thermal diffusion and wave equation, the accuracy of the numerical solution can be compared to the exact analytical result. The execution time of the algorithm can also be compared to the scaling of the execution time of the same problems run on a serial machine.

The question of accuracy of the numerical algorithm on the Connection Machine raises many interesting issues. Since the architecture of the individual processors is somewhat flexible, the precision is not restricted to 32 bits or 64 bits. Any arbitrary precision can be used, and the precision can be tailored to the problem such that a tradeoff can be made between the accuracy and the execution time. Using a 64-bit word, the comparison of the numerical solution to the analytic solution yielded an error of approximately 100 parts per million. For the ADI formulation, the error in the numerical solution varied very little over the range from 32 bits to 64 bits of precision.

Since the execution time of the algorithm requires the inversion of a system of tri-diagonal matrices, the total execution time scales as $\log_2(N)$, where N is the total number of points in the problem.

$$\begin{aligned} T &= \log_2(N1) + \log_2(N2) + \log_2(N3) = \\ &\quad \log_2(N1 * N2 * N3) = \log_2(N) \end{aligned} \tag{46}$$

This variation was found to be exact when only the execution time for the Connection Machine was taken into account; i.e., the time taken for the front-end program control and the time for the I/O was subtracted out of the total execution time.

The best algorithm for tri-diagonal matrix inversion — the heart of the algorithm — on a serial machine is one where the diagonal elements are used to eliminate the terms of the lower diagonal, and then the last value is used to back substitute to solve for the individual terms of the solution. However, this algorithm scales as N,

$$T = N1 * N2 * N3 = N \tag{47}$$

since the reduction loop must be executed for the total number of implicit segments in the problem.

Conclusion

Since the execution time for a serial machine scales as N, and the execution time on the Connection Machine scales as $\log_2(N)$, an order of magnitude or more increase in the execution time for many numerical simulations can be achieved. The size of the problem that can be handled depends upon the amount of storage in each processor and the number of available processors. A tradeoff can also be made between the execution time and the desired accuracy, since many problems do not require 64 or even 32 bits of precision to get a reasonable estimate of the physics of the problem. This also gives the potential of solving problems numerically that have not been attempted in the past due to long execution times or lack of storage space.

7.3.2. Operations Research

Optimal flow through a network with nonlinear costs is the first application, drawn from a multitude of possible examples, from the field of operations research. Another example, the 0/1 Knapsack problem, is also presented.

NONLINEAR NETWORK OPTIMIZATION ON A
MASSIVELY PARALLEL CONNECTION MACHINE

Introduction

Nonlinear programming problems with network constraints arise in a wide range of engineering, management, statistical, economic and other applications. Such problems are typically characterized by their large size that may extend to thousands of variables and constraints. Researchers are able to solve efficiently very large network problems by specialized algorithms that capitalize on the underlying graph structure. Progress has been made in the realm of linear network problems, nonlinear problems and multicommodity problems — both linear and nonlinear. Existing literature is quite extensive and interested readers are referred to the survey by Dembo et al.[8] for the discussion of applications, solution methodologies and software availability. As a result of the activities discussed in the above reference we can now solve on a routine basis problems with a few thousands of variables and constraints. Problems still exist, however, that are beyond the reach of current technology; such examples include models for air-traffic control, real-time dispatching of freight trains, personnel management in military applications and the multicommodity network problems that arise from studies in transportation.

The development of parallel computer systems provides an additional motive for attacking the large models that arise in these applications. Already progress has been made on several instances of the problem. Chang et al.[9] and Chen and Meyer[10] discuss the solution of generalized networks and nonlinear multicommodity problems on the CRYSTAL multicomputer. Zenios and Mulvey[11,12] discuss the use of network optimization software on vector and shared memory supercomputers and discuss computational experiences using (primarily) the CRAY series of vector supercomputers. Other areas of computational mathematical programming have been investigated in the context of parallel computing, including the multiple-cost-row linear program by Phillips and Rosen,[13] linear complementarity by Mangasarian and De Leone,[14] combinatorial optimization by Kindervater and Lenstra[15] and others. The volumes by Meyer and Zenios[16] and Mangasarian and Meyer[17] serve as general references.

While progress has been made in the context of optimization using vector supercomputers and small to medium scale parallel systems, the domain of massively parallel computing is virtually unexplored. The interface of this domain with network optimization has been investigated by Bertsekas and collaborators over the last few years. Motivated by the ideas of chaotic relaxation of Chazan and Miranker[18] they proposed a distributed iterative algorithm for convex network problems — Bertsekas and El Baz.[19] A modification to the Bertsekas-El Baz scheme was proposed by Zenios and Mulvey[20] and was evaluated empirically on a simulated parallel environment. Results in the later study were very encouraging with the parallel algorithm approaching linear speedup factors. Similar ideas were extended to the linear network flow problem by Bertsekas and Eckstein[21] and to the assignment problem by Bertsekas.[22] A survey of possible applications of distributed asynchronous algorithms and the underlying convergence theories can be found in Bertsekas et al.[23] The dissertation by Goldberg[24] discusses the design and implementation of parallel algorithms for (linear) network flow problems.

In this section we present the computational study of a relaxation algorithm for strictly convex network problems on a massively parallel computer. This is, to our knowledge, the first attempt to evaluate any of the ideas of massively parallel network optimization with commercially available multiprocessors. We discuss the implementation of distributed relaxation for strictly convex network problems on a Connection Machine CM-1 configured with 16,384 processing elements. Two related versions of the algorithm generated quadratic transportation problems. Internal tactics of the implementation that are suitable for a parallel environment are discussed. We also compare the performance of the distributed algorithm on the CM-1 with the performance of a commonly used sequential algorithm — the primal truncated Newton (PTN) — as implemented on large mainframes (IBM 3081-D), vector supercomputers (IBM 3090-600) and shared memory vector multiprocessors (Alliant FX/8). We show that distributed algorithms on parallel machines have the potential of solving efficiently several of the applications that are considered unsolvable by current standards.

Massively Parallel Network Optimization

Let $G = \{N,E\}$ be a directed network with N denoting the set of nodes and

E the set of edges (arcs). The (primal) nonlinear optimization problem over the network is defined as follows:

[NLN]
$$\text{Minimize}_{x} \quad F(x) = \sum_{(i,j) \in E} f_{ij}(x_{ij})$$
$$\text{subject to} \quad A \cdot x = 0$$
$$\ell \leq x \leq u \tag{48}$$

where $F(x)$ is a strictly convex continuously differentiable real-valued function, $x = \{x_{ij} | (i,j) \in E\}$ is the vector of decision variables indicating flows through the arcs of the network, A is a node-arc constraint matrix, and l and u are bounds on the decision variables. The constraint matrix A is characterized by the presence of at most two non-zero entries in every column. In the case where all entries are +1 or -1 the network problem is termed pure; whenever the entries are arbitrary it is defined as generalized. Readers who are interested in the field of network optimization and its applications are referred to the survey paper by Dembo et al.[8] In the following section we describe a dual relaxation algorithm for solving [NLN].

Distributed Relaxation Algorithm

The relaxation algorithm for [NLN] falls in the category of dual coordinate descent algorithms for monotropic optimization. It has been discussed in the form presented here by several authors, including primarily Bertsekas[25] and Bertsekas and Tseng,[26] Stern,[27] Ohuchi and Kaji[28] and Zenios and Mulvey.[29] Bertsekas et al.[30] develop convergence theories for a broad class of relaxation algorithms, including the one presented here. We summarize now the algorithm defining first the dual problem:

$$[\text{DNLN}] \quad \text{Minimize}_{p} \quad F^*(p) = \sum_{(i,j) \in E} f_{ij}^* (p_i - p_j) \tag{49}$$

where f^* and F^* denote the convex conjugate to f and F, respectively, and $p = \{p_i | i \in N\}$ is the vector of dual variables. The dual variables are unrestricted in magnitude and sign and the optimality conditions for problems [NLN] and [DNLN] are the well known: (1) primal feasibility, (2) dual feasibility, and (3) complementary slackness.

$$x_{ij} = u_{ij} \text{ if } p_i - p_j > \left. \frac{\partial f_{ij}(x_{ij})}{\partial x_{ij}} \right|_{x_{ij} = u_{ij}} \tag{50}$$

$$x_{ij} = l_{ij} \text{ if } p_i - p_j < \left. \frac{\partial f_{ij}(x_{ij})}{\partial x_{ij}} \right|_{x_{ij} = l_{ij}} \tag{51}$$

$$x_{ij} = \hat{x}_{ij} \text{ if } p_i - p_j = \frac{\partial f_{ij}(x_{ij})}{\partial x_{ij}}\bigg|x_{ij} = \hat{x}_{ij} \text{ for } l_{ij} \le \hat{x}_{ij} \le u_{ij} \quad (52)$$

We define also the deficit of a node as

$$d_i = \sum_{(i,j)\varepsilon E} x_{ij} - \sum_{(k,i)\varepsilon E} x_{ki} \quad (53)$$

and note that the partial derivative of the dual objective $F^*(p)$ with respect to p_i is equal to the deficit d_i of node i. The relaxation algorithm iterates on the price of one node at a time (p_i) satisfying complementary slackness conditions and primal feasibility for node i while minimizing the dual function along the ith coordinate direction. The dual function is minimized along this direction when the deficit d_i becomes zero. The algorithm can now be formalized.

The Relaxation Algorithm
Step 0: Set $k \leftarrow 0$. Choose starting prices $p^o = \{p_1^o, p_2^o, ..., p_n^o\}$. Compute a vector x^o satisfying conditions (50) to (52). Choose an arbitrary node j and set $p_j^k = p_j^o$ for all k.
Step 1: Set $k \leftarrow k + 1$. Compute the deficit vector d^k. Find a node i j with deficit d_i^k. If no such node exists then terminate with solution (x^k, p^k).
Step 2: (Coordinate-wise minimization). If $d_i^k < 0$, increase the dual variable p_i^{k-1} until the deficit — corresponding to the vector x that satisfies conditions (50) to (52) — becomes zero. For positive deficit d_i^k the variable p_i^k decreases. Let p_i^k be the solution and return to Step 1.

Our description leaves open several implementation details, like the initialization phase of Step 0, the rule for choosing node i in Step 1, and the procedure used in the minimization phase of Step 2. The paper by Zenios and Mulvey[29] discusses the implementation of this algorithm for large scale problems. Its convergence is established in Bertsekas et al.[30]

A distributed implementation of this algorithm executes Steps 1 and 2 concurrently for multiple nodes. To formalize the discussion of the parallel algorithm we need resolve the following issues:

- Partitioning problem: Partition the algorithm into tasks that can be executed concurrently by independent processing elements.
- Scheduling problem: Assign tasks to one or more processors for execution.
- Synchronization problem: Specify an order of execution that guarantees correct results.

We discuss the details of these issues as they relate to the relaxation algorithm.

Algorithm Partitioning

The natural partitioning of the relaxation algorithm for parallel computing is to assign the computations for every variable to different processors. In the case of the Connection Machine this partitioning is the most efficient since we may expect to have at least as many processing elements as we have variables.

In cases where the size of the problem exceeds the size of the machine a more ingenious partitioning scheme is possible. We observe that the algorithm iterates on the price of one node at a time given information from adjacent nodes. We can therefore process simultaneously nodes that are not directly connected to each other. Identifying sets of disconnected nodes is equivalent to coloring the underlying graph and operating in parallel on nodes with the same color. This observation was first made in Zenios and Mulvey[12] and the underlying graph-coloring problem can be solved efficiently using graph coloring heuristics.

Task Scheduling

In our implementation we use static task scheduling. The network topology is mapped to the configuration of the machine before execution of the algorithm begins. At subsequent iterations every processor will operate on the same data. We point out that a dynamic scheduling mechanism is also possible, whereby nodes are assigned to different processors at every iteration. This scheme makes the algorithm fault-tolerant against situations where some processing elements are inoperative or there is a communication failure. This scheme would be inefficient on a machine with fixed interconnection and limited local memory and is not given any further consideration in this study.

Task Synchronization

The Connection Machine is a Single Instruction Multiple Data (SIMD) computer and synchronization of the tasks does not pose any problems. Upon completion of an iteration by all processors the dual variables have to be updated. This is the synchronization phase when every processor broadcasts its dual values to processors that need this information. We point out that an asynchronous version of the algorithm has been proposed by Bertsekas and El Baz,[19] whereby processors communicate at arbitrary intervals with arbitrarily long communication delays. It is proved in the same reference that this chaotic algorithm converges (under some mild conditions) and the same follows directly for our synchronous implementation.

Implementation

The primary consideration in implementing an algorithm on a system like the CM-1 is to develop an efficient mapping of the problem data to the processing elements. The choice adopted in our implementation provides us with the ability to perform operations on all nodes of the problem simultane-

Processor	Address	Data Stored Locally
(P)	ad (P)	
PT_{13}	0000	$ad(PH_{13})$, d_1, p_1, p_3 and data for arc (1,3)
PH_{13}	0011	$ad(PT_{13})$, d_3, p_1, p_3 and data for arc (1,3)
PT_{23}	0001	$ad(PH_{23})$, d_2, p_2, p_3 and data for arc (2,3)
PH_{23}	0100	$ad(PT_{23})$, d_3, p_2, p_3 and data for arc (2,3)
PT_{25}	0010	$ad(PH_{25})$, d_2, p_2, p_5 and data for arc (2,5)
PH_{25}	1000	$ad(PT_{25})$, d_5, p_2, p_5 and data for arc (2,5)
PT_{34}	0101	$ad(PH_{34})$, d_3, p_3, p_4 and data for arc (3,4)
PH_{34}	0110	$ad(PT_{34})$, d_4, p_3, p_4 and data for arc (3,4)
PT_{45}	0111	$ad(PH_{45})$, d_4, p_4, p_5 and data for arc (4,5)
PH_{45}	1001	$ad(PT_{45})$, d_5, p_4, p_5 and data for arc (4,5)

$SEG_1 = PT_{13}$
$SEG_2 = PT_{23}, PT_{25}$
$SEG_3 = PH_{13}, PH_{23}, PT_{34}$
$SEG_4 = PH_{34}, PT_{45}$
$SEG_5 = PH_{25}, PH_{45}$

A

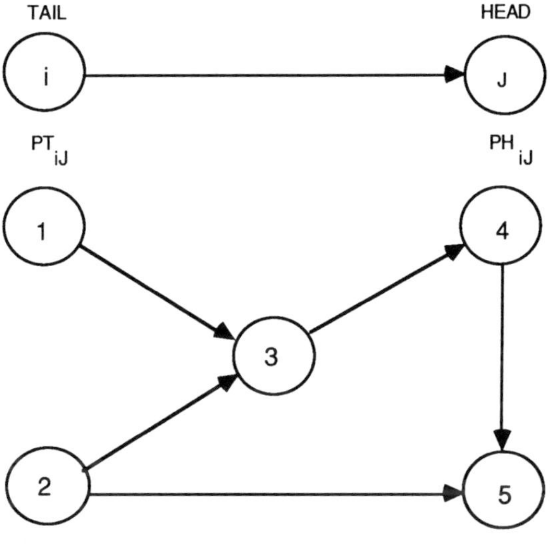

B

FIGURE 30. Mapping of a network model to CM-1 processors.

ously and minimizes data transfer through the communications network. Figure 30 shows the representation of a simple network using multiple processing elements and the data stored at each processor.

Every arc in the network is associated with two processing elements (PH_{ij} and PT_{ij}) where PH and PT indicate processors at the head and tail of an arc, respectively, and the subscript indicates arc (i,j) ε E. To allow use of segmented scan operations contiguous segments of processors are associated with node i. Processors corresponding to the same node are designated as a segment defined by $SEG_i = \{PT_{ij}, PH_{ki} \ \forall \ (i,j) \text{ and } (k,i) \ \varepsilon \ E\}$.

The data stored at each processor PH_{ij} and PT_{ij} include:

1. Arc data u_{ij}, l_{ij}, x_{ij} and the objective function $f_{ij}(x_{ij})$.
2. Dual variables of the nodes at both ends of the arc: p_i and p_j.
3. Deficit at the nodes associated with the processor; PH_{ij} stores d_j and PT_{ij} stores d_i.
4. The address of the processor at the other end of the arc; PH_{ij} stores the address PT_{ij} and vice versa.

With this representation of the network one iteration of the parallel algorithm is executed as follows:

The Parallel Relaxation Algorithm

Step 0: Set $k \leftarrow 0$. Choose starting prices $p^0 = \{p_1^0, p_2^0, ..., p_n^0\}$.

Step 1: Set $k \leftarrow k + 1$. Compute values of x_{ij} satisfying conditions (50) to (52) concurrently for all arcs (i,j)εE; PH_{ij} and PT_{ij} compute identical values for x_{ij}. Perform a Segmented-Plus-Can operation over the flows of all processors to compute the deficit at all nodes:

$$d_i = \sum_{PT_{ij} \varepsilon SEG_i} x_{ij} - \sum_{PH_{ki} \varepsilon SEG_i} x_{ki} \qquad (54)$$

If the maximum absolute deficit is less than a prespecified tolerance the algorithm terminates.

Step 2: The dual variable for all nodes is adjusted such that the flows corresponding to complementary slackness conditions result to zero deficit. This step is performed concurrently for all nodes by the *last* processor in every segment. (Recall that after the Segmented-Plus-Scan at Step 1 only the last processor in a segment has the correct value of the deficit.) A Segmented-Reverse-Copy-Scan operation is then used to copy the computed dual prices to the memory of all processors in a segment. Finally each processor broadcasts its dual values to the processor at the other end of the arc using a *PSet operation. For example, processor PH_{ij} will broadcast p_i to PT_{ij}. Return to Step 1.

This implementation may seem inefficient, especially with regard to storage requirement and processor utilization. Processors store duplicate information and we need twice as many processors as the number of arcs to solve the problem. Storage does not pose any serious problem; we store only a few bytes of information at every processor which is well below the 512 bytes limit.

TABLE 10
Test Problems

Problem	Size (nodes/arcs)	Objective value
Stick1	209 / 454	6.934392
Stick2	650 / 1412	3.124563
Stick3	782 / 1686	11.179780
Stick4	832 / 2264	1.566195
TRANST	10 / 20	1.956339
TRANS1	2500 / 7668	484.575671
TRANS2	5000 / 7834	1633.062210
TRANS3	7000 / 7960	2833.542504

Some operations are also performed redundantly: in the computation of flow variables the head and tail processors compute the same value. Furthermore, using two processors per arc allows data transfer without checking for memory access conflicts in the *PSet operations for dual updates. The inefficient utilization of processors comes as a tradeoff for the efficient utilization of the communications network and segmented scan operations. It reflects the basic design philosophy of the Connection Machine where the processors are not necessarily the scarce resource.

The current implementation of the algorithm is applicable only to problems with identical quadratic costs on all arcs. This assumption permits the analytic evaluation of dual variables in Step 2. There is no theoretical difficulty in relaxing this assumption. This would require, however, an iterative procedure within Step 2 in order to determine values of dual variables that reduce the deficit of all nodes to zero. This raises some implementation issues in the massively parallel environment. Several options are available for utilization of the multiple processors in this iterative step — i.e., implementing parallel linesearch algorithms — and those are the subject of further study.

Computational Experiments

The implementation of the relaxation algorithm was tested on a sequence of test problems arising from the study of percolation effects in insulating material and randomly generated quadratic transportation problems. The stick percolation problems are discussed in Dembo et al.[8] The transportation problems were generated using NETGEN[31] and a quadratic cost is assigned to every arc. Their characteristics are summarized in Table 10. The algorithm was implemented on a CM-1 with 16,384 processing elements, at the Perkin-Elmer Advanced Development Center. Only 8,192 elements were attached during our experiments with the stick problems and 16,384 were attached for the transportation problems. The CM-1 is connected to a Symbolics 3000 and the programming language is *LISP.

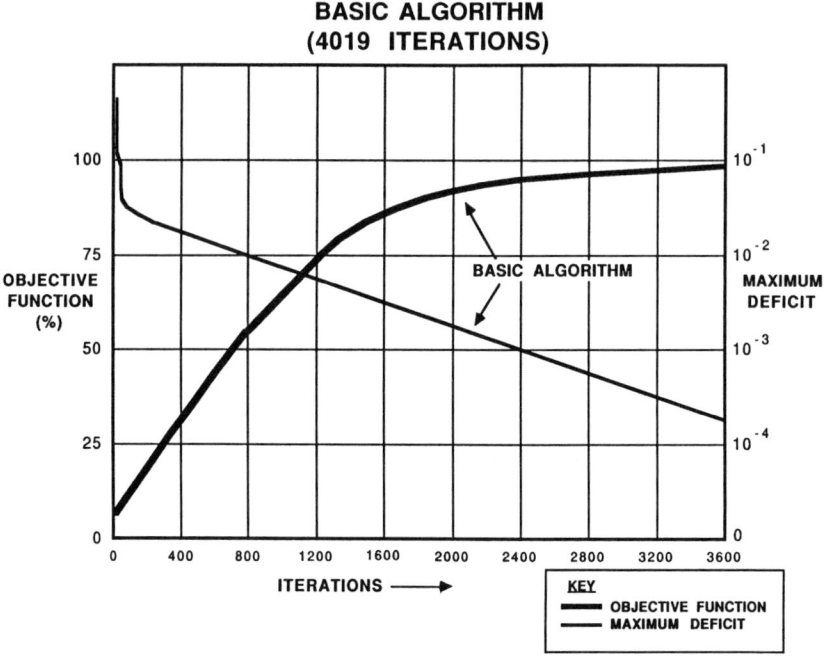

FIGURE 31. Performance of the relaxation algorithm on stick4.

Testing the Relaxation Algorithm on the CM-1

All test problems were solved with the same version of the algorithm under identical conditions. The performance of the algorithm is measured in terms of the CM-time required to solve a problem. CM-time is the elapsed wall clock time on the Connection Machine, always used in a dedicated environment. No assumptions are made on processor utilization and a CM-time of one unit indicates that the operation was performed in one unit of time no matter if it was performed by one processing element or all of them. The times reported in our results indicate active processor time and communication time. Detailed timing analysis of the algorithm indicates that about 65% of the CM-time is devoted to computations and 35% to communications.

Figure 31 depicts the performance of algorithms for a typical run (problem Stick4). It is interesting to observe that the experiences with sequential machines — i.e., that relaxation algorithm approaches optimality with large initial steps and then slow down — are true in the massively parallel environment as well. The fact that all variables are operated upon in parallel is what makes the algorithm attractive in this environment. In Table 11 we report the CM-time for all test problems when solved at different levels of accuracy. The three larger transportation problems (TRANS1-3) were solved using 32,768 virtual processors on a 16,384 PE system. The CM-time will decrease by a factor of two

TABLE 11
CM-time (seconds) for Test Problems using Relaxation

	Max. deficit				
Problem	10^{-2}	10^{-3}	10^{-4}	10^{-5}	10^{-6}
Stick1	3.32	5.54	8.08	10.55	15.27
Stick2	1.80	5.11	13.18	21.36	28.96
Stick3	6.85	17.45	31.84	45.19	61.88
Stick4	1.66	3.66	6.67	10.09	14.01
TRANST	0.14	0.18	0.24	0.31	0.36
TRANS1	0.37	0.52	0.64	0.79	0.91
TRANS2	1.77	3.47	5.13	6.86	8.79
TRANS3	3.82	10.07	17.83	25.63	34.62

TABLE 12
Objective Value for Test Problems using Relaxation

	Max. deficit				
Problem	10^{-2}	10^{-3}	10^{-4}	10^{-5}	10^{-6}
Stick1	6.891711	6.933923	6.934388	6.934392	6.934392
Stick2	2.984309	3.120907	3.124533	3.124563	3.124563
Stick3	10.514701	11.171151	11.179715	11.179774	11.179754
Stick4	1.566411	1.565182	1.566179	1.566194	1.566195
TRANST	1.941610	1.956750	1.956342	1.956338	1.956339
TRANS1	484.47250	484.56656	484.57452	484.57556	484.57565
TRANS2	1632.8004	1633.0485	1633.0656	1633.0673	1633.0674
TRANS3	2832.0590	2833.4802	2833.5393	2833.5422	2833.5425

when a full configuration CM-1 is used. The objective value obtained with these runs is reported in Table 12. When the algorithm executes until the maximum deficit is less than 10^{-6} the objective value is accurate to at least six decimal points.

The interesting observation from the results of Table 11 is that problem size is not the predominant factor affecting execution time. Solution time is constant for a given number of iterations and the number of iterations does not depend on the size of the problem. In fact, TRANS1, which is the largest and most dense problem, is solved in much less time than any of the smaller problems. This behavior violates previous computational experiences with network optimization, where the size of the problem is one of the predominant factors affecting solution time.

Alternative Parallel Implementation
We pointed out that there are two ways of partitioning the problem data for

TABLE 13
Comparing Alternative Parallel Implementations

Problem	With coloring		Without coloring	
	Iterations	CM-time	Iterations	CM-time
Stick1	3595	143.6	6532	62.6
Stick2	2170	86.1	3761	36.1
Stick3	17119	702.5	29954	287.1
Stick4	2245	101.1	3854	36.9

parallel execution. The results in the previous section were obtained with processors operating on the dual variables of all nodes simultaneously. Computations are performed locally based on information from adjacent nodes that is, in general, outdated by one iteration. This does not pose any problems to the global convergence properties of the algorithm — Bertsekas and El Baz[19] — but it degrades its performance. The amount of work performed by a processor based on outdated duals may be wasted if, in subsequent iterations, the variables change sign and/or magnitude. To avoid this inefficiency, Zenios and Mulvey[12] proposed a partitioning scheme based on graph coloring properties of the underlying network. Nodes are partitioned into sets of disconnected components and operations are performed in parallel on nodes that are not connected to each other. Partitioning the network into sets of disconnected nodes is equivalent to coloring the underlying graph and operating on nodes with the same color in parallel. This scheme was tested in our implementation of the CM-1. The graph coloring problem was solved using a graph coloring heuristic. Results with both parallel implementations (with and without coloring) are summarized in Table 13; the algorithm was run until the maximum deficit was reduced below 10^{-4}. We report both number of iterations and CM-time, where one iteration is completed when all nodes in the problem are operated upon exactly once. The coloring scheme consistently reduces the number of iterations. This is achieved at the expense of reduced parallelism and increased synchronization. Although the number of iterations is reduced by a factor of two the CM-time is increased by approximately the same factor. The explanation for this effect is simple: When using the coloring scheme not all operations are performed in parallel. All nodes with a given color are operated upon in parallel while processors with nodes of differing colors remain idle. Hence the overall solution time is increased. Use of coloring would improve efficiency when a problem exceeds in dimension the size of the machine and virtual processors would have to be used. Creating virtual processors based on the coloring scheme would then decrease the number of iterations without reducing the level of parallelism.

Global Flow Adjustments in the Relaxation Steps

We observe during the early stages of our experiment that it was often the

TABLE 14
Performance of Relaxation with Global Flow Adjustment

	With flow adjustment		Without flow adjustment	
Problem	Iterations	CM-time	Iterations	CM-time
Stick1	817	8.09	6532	62.6
Stick2	1375	13.18	3761	36.1
Stick3	3310	31.84	29954	287.1
Stick4	691	6.67	3854	36.9
TRANS3	748	13.27	1077	17.8

case of dual variables changing very slowly based on small changes on the adjacent nodes. This phenomenon was observed at the first few iterations. When, for example, the initial flows on all arcs are zero small changes in the dual variables will satisfy feasibility on a local basis. On recalculating the flows based on updated duals we detect large deficits and the process repeats. If long chains are present in the network, with supply and demand nodes at the two extremes, it is possible that duals at intermediate nodes in the chain will remain unchanged for several iterations.

To avoid this difficulty we devised a scheme whereby the overall state of the network and local trends of the duals are taken into consideration by individual processors when updating the prices. The dual value computed by each processor (Step 2 of the relaxation algorithm) is adjusted by a factor of $1 + \delta_{local} + \delta_{global}$, where δ_{local} is proportional to the change prices at the node during the last few iterations and δ_{global} depends on the difference between the highest and lowest dual prices and the change of total flow at the current iteration. As the algorithm approaches optimality the change in total flow and dual prices is small and the adjustment becomes less significant.

The use of δ_{local} reminds us of ideas from over-relaxation, but at present we cannot offer a theoretical justification why it works. Factor δ_{global} is introduced in order to take into account global changes in the state of the problem at the initial iterations. The results are quite impressive as shown in Table 14 — (termination criteria for maximum deficit less than 10^{-4}) — with savings in number of iterations and CM-time averaging a factor of seven. Although some fine-tuning was needed to determine the right setting of the adjustment factor, the resulting value was robust for all the stick problems. The transportation problems do not have long chains between supply and demand and the flow adjustment becomes less significant.

Comparing Parallel Relaxation with Sequential Algorithms

The ultimate objective of designs in parallel computing — and the related research in algorithm development — is to solve efficiently the large problems that arise from practical applications. To study the suitability of our algorithm and the CM-1 in achieving this goal we solved the test problems on three

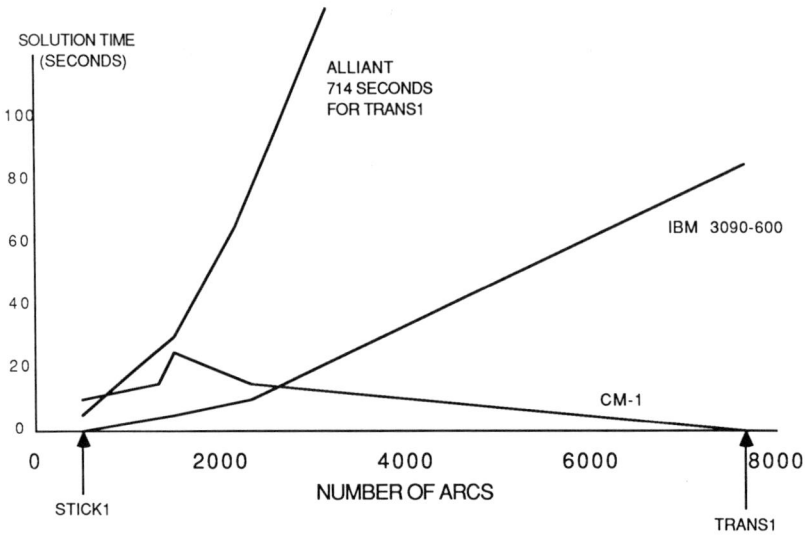

FIGURE 32. Network optimization on different computer architectures.

different computer systems with different algorithms.

The computers we use are the IBM 3081-D, IBM 3090-600 and the Alliant FX/8. IBM 3081-D represents the high end of large mainframes with a cycle time of 26 ns. The IBM 3090-600 belongs to the class of vector supercomputers with a clock cycle of 18.5 ns and capable of delivering operations at a peak rate of 116 MFLOPS. The Alliant FX/8 is classified as a mini-supercomputer capable of achieving performance of 94.4 MFLOPS. It is a shared memory vector multiprocessor. The particular configuration we used for our testing (the FX/8) is configured with eight processing elements sharing 80 Mbytes of global memory.

In our testing we use different algorithms for different machines. This may seem somewhat unorthodox by the standards of algorithm evaluation but it is the only sensible approach to take when attempting to evaluate computing technologies with so different architectures. The relaxation algorithm as implemented on the CM-1 is compared with the primal truncated Newton (PTN) algorithm on the large mainframe and vector multiprocessors. The later algorithm is described by Ahlfeld et al.[32] The software system — NLPNETG — has been modified both for the architectures of the IBM 3090-600 (vectorization) and for the Alliant FX/8 (vectorization and multitasking). Vectorization and multitasking of this (and another) algorithm are discussed in the papers by Zenios and Mulvey.[11,12] We point out that the NLPNETG software has been under development and refinement over a period of 5 years, whereby the relaxation algorithm was implemented on an experimental basis as a research

TABLE 15
Solution Time for Problems with Varying Densities

Problem	Density	IBM 3090-600	CM-1
TRANS1	6	92.72	0.64
TRANS2	3	108.83	5.13
TRANS3	2	83.18	17.83

code. NLPNETG is coded in standard FORTRAN 77. Figure 32 shows the solution time for the test problems with both algorithms on all four computer systems. The NLPNETG code was run until the reduced gradient was less than 10^{-2}; termination criteria for the relaxation algorithm was a maximum deficit below 10^{-4}. We observe that the parallel implementation of the relaxation algorithm outperforms the Alliant FX/8 and IBM 3090-600 for the larger problems. We also compare in Table 15 the performance of the relaxation algorithm on the CM-1 with PTN on the IBM 3090-600 for the transportation problems. These problems have approximately the same number of arcs but varying densities in the underlying network. The relaxation algorithm performs much better on dense problems.

Conclusions

We have reported here on the computational study of a relaxation algorithm on the Connection Machine. The implementation in itself has been quite instructive. Several of our computing preconceptions had to be drastically altered in order to achieve an efficient implementation. Data structure and software organization are not geared toward efficient use of the processors and minimization of the underlying computations. Instead, we aim at optimizing the communication delays and utilizing efficiently the large number of processors that are available on the machine. We have also shown that reducing the number of iterations (using the graph coloring scheme) is not a worthy goal since it does not imply a reduction in execution time. Actually the solution time increased. The number of iterations is reduced at the expense of increased time per iteration.

The computational results indicate that massively parallel computers, as represented here by the CM-1, hold promise for solving efficiently very large network optimization problems. While for small to medium size problems vector supercomputers and shared memory multiprocessors outperform the CM-1, this situation is reversed as the size of the problem increases. Additional efficiency is anticipated when using the CM-2. In a more general context we point out that the relaxation algorithm — shown elsewhere to be a poor match for commonly used nonlinear network programming algorithms — is making a strong comeback in the domain of parallel computing.

Massively Parallel Implementation of some Common 0/1 Knapsack Approximation Algorithms

Introduction

The 0/1 Knapsack problem is a well-known optimization problem appearing naturally and often in many common applications. It also arises in the search for sub-optimal solutions to larger and more complex problems. Its prototypical instance involves a knapsack of fixed weight capacity and N objects of differing weights and profits. The optimal solution is the subset of those N objects having maximum total profit and a collective weight not exceeding the carrying capacity of the knapsack. Fractional parts of items are not allowed. We will also refer to this maximum total profit as the solution of the problem instance.

It has been proven that this problem is intractable ("NP-hard" in the research literature), meaning that the expected effort needed to find the optimal solution is an exponential function of the size of the particular problem instance. Since so many interesting problem instances are of considerable size, this exponential time behavior generally precludes the possibility of finding the true optimal solution.

Because of this proven intractability, a great deal of effort has been expended toward finding approximate solutions to the 0/1 Knapsack problem. Of particular interest are so-called "approximation algorithms", a term reserved for heuristics of provably good accuracy. (References on intractable problems and on approximation algorithms include the texts of Horowitz and Sahni[33] and Gary and Johnson.[34] The requirement of provable accuracy is not of mere academic interest; since the optimal solution is normally unknown, and the success of a heuristic varies with the particular problem instance, reliance on unproven heuristics can lead to very poor or erratic approximations.

This section discusses the adaptation of two well-known approximation algorithms into a massively parallel, single-instruction multiple-data-set (SIMD) form. This work was performed on a Connection Machine with 16,384 processors.

Approximation Algorithms

For an intractable problem such as the 0/1 Knapsack, the only method of determining the true optimum solution is with an exhaustive search, or its equivalent. Because this search involves checking every possible subset of the N items, it is only practical for problems of size, say, N < 40. Since so many problems of interest are substantially larger than that, we are forced to lower our sights and aim for an approximation to the optimal solution. The goal then is to develop algorithms which approximate this unknown to some known degree in reasonable time.

A desirable goal would be to develop an approximation algorithm which produces a "solution" having total profit guaranteed to be within a fixed amount of the optimal solution. For the 0/1 Knapsack problem it has been shown that finding such an approximate solution with an absolute error guarantee is itself an NP-hard problem.

However, if we ask for approximate solutions with total profits known to differ from optimal by a fixed percent, these are available. Known as ε-approximations, where ε is the percentage deviation of approximate solution to optimal solution, we will discuss two, herein called "extended greedy" and "k+1" (see Horowitz and Sahni[33] and Sahni[35] for more detail).

If the N items are ordered in terms of nonincreasing profit-to-weight ratios, the simple greedy approach consists of adding items to the knapsack in that order, stopping when the next item's weight would cause the knapsack's capacity to be exceeded. The extended greedy performs this much, then continues along the list, skipping the heavier items and adding lighter ones as long as the capacity is not exceeded. The operation count for the serial version of extended greedy is $O(N \log N)$, due to the sorting of the data, and $\varepsilon = 1.0$. In practice the percentage error is quite a bit less than 100%.

The k+1 algorithm works with $\sum_{j=1}^{k} \binom{n}{j}$ replicates of the problem. Each begins with a different "seed" of j items $(j = 1,....,k)$ in the knapsack. The extended greedy algorithm is then performed for all of the replicates (skipping over the particular items already selected), and the one with best total profit is the method's approximate solution. For instance, with $k = 1$, N copies of the problem are set up, and a different item placed in each knapsack initially. The remaining items are ordered by profit/weight ratio, and the extended greedy algorithm performed.

This algorithm has operation count $O(kN^{k+1})$ in its serial form. If $p^\wedge$ is the $(k+1)^{st}$-largest profit, and P the total profit of the approximate solution, then the method's fractional accuracy ε is given by $\min\{1/(k+1), p^\wedge/P\}$. The form of this error bound allows us to tailor the algorithm to guarantee any desired ε by appropriate choice of k. For most problems it is too time-consuming to use k other than 1 or 2; however, this error bound is generally loose. Sahni[35] tested a variety of problem instances and found the k+1 algorithm, with $k = 1$, generally to be accurate to within 5%.

Parallelization

To program the extended greedy algorithm, we begin with one item represented per processor — i.e., profit and weight are stored locally. (For problems of size >16384 each processor may be made to represent 2, 4, 8, or more virtual processors. Assuming N is not a perfect power of 2, excess processors are turned off.) These items are sorted according to profit/weight ratio and shuffled into processors 0 through N-1 in just two commands. The remainder of the algorithms consists of choosing subsets of processors, corresponding to sections of the list, and "accumulating" them into the knapsack. The worst case operations count for this parallelization is $O(N)$, but in practice the behavior is usually better. In this worst-case estimate we have used the fact that the parallel sorting algorithm implemented on the CM is faster than $O(N)$. Also, we have assumed that the "plus-scan" operation (computing all k partial sums for a

TABLE 16
Knapsack Run Times

N	Extended greedy actual error	(k+1) predicted error bound	(k+1) actual error
15	12 instances solved exactly; 8 others have errors in range 0.67—4.6% (mean = 2.97%)	12.6—17.2%	16 instances solved exactly; 4 others have errors in range 0.67—2.9% (mean = 1.87%)
20	13 instances solved exactly; 7 others have errors in range 0.10—1.24% (mean = 1.24%	9.8—15.0%	16 instances solved exactly; 4 others have errors in range 0.23—3.3% (mean = 1.33%)
25	9 instances solved exactly; 11 others have errors in range 0.02—3.5% (mean = 1.09%)	7.9—10.3%	14 instances solved exactly; 6 others have errors in range 0.02—1.3% (mean = 0.56%)

sequence of length k spread across k processors) is a unit-time primitive operation. Thinking Machines Corp.[36] has found that the plus-scan operation is accomplished in time comparable to inter-processor memory references.

For the k+1 algorithm, the profit and weight data is not stored in the machine at all, but is broadcast sequentially. In this case each processor represents one of the replicates, and is initialized with its own seeding profit and weight value(s). The order list is then broadcast, each processor adding items to its particular collection, turning itself off if the next items would make it exceed knapsack capacity. When done, the processor with maximum total profit is readily found, and its initial choice of k items identified. The approximately optimal subset of items is either found during this process, or re-created with extended greedy once the best choice of initial items has been found. The operations count for the parallelized k+1 algorithm is O(N), assuming that the processors know their seed values ahead of time. This assumption is justifiable when k = 1, but for larger values, "seeding" the replicates may be a time-consuming part of the algorithm.

Because so many replicates can be handled simultaneously, this parallel algorithm is a natural for large problems (N > 1000) when k = 1 provides sufficient accuracy. In addition, it provides the flexibility of choosing larger values of k, guaranteeing better accuracy *a priori,* for smaller problems.

Results

Sahni[35] provides ample evidence of the accuracy of both the extended greedy and k+1 algorithms. In Table 16 we present analogous figures for a smaller sample of test cases. For each case integer profits and weights are chosen at random from the range [1,1000], and the knapsack capacity is limited to half the N-weight total. The same data set is used as input for extended greedy, k+1 (k = 1), and exhaustive search algorithms. The latter, of course,

TABLE 17
Mean Error Bound

N	100	500	1000	5000	10000
Mean error bound (5 trials)	2.53%	0.499%	0.243%	0.049%	0.025%

TABLE 18
Knapsack Run Times

	Run times		
N	Parallel extended greedy	Parallel k+1 approx.	Serial k+1 approx.
50	.115	.335	N.A.
100	.121	.502	.25
500	.140	1.92	3.5
1000	.132	3.71	14.6
5000	.141	17.9	350
10000	.135	35.7	N.A.
50000	.451	206	N.A.
100000	1.31	604	N.A.

provides the true optimum solution. The *a priori* error bounds are set as explained above; this bound is 100% for the extended greedy, and not listed in the table. Twenty trials were run for each of N = 15, 20, and 25. Note that the actual accuracy of these algorithms will vary significantly with the distribution of profits and weights.

For large problems we, of course, cannot find the true optimal solution. Table 17 lists the (k+1) algorithm error bounds for larger instances of the trials described above, for k = 1. The linearity of these results is attributable to the fact that profits and weights are uniformly distributed. The total profits calculated from these runs were very close whether found by the k+1 algorithm or the extended greedy: within 0.05% at worst. This indicates that extended greedy is very nearly as accurate as the k+1 algorithm for large problems, under the tested condition.

Run times are shown in Table 18, in seconds. Each of our entries represents the mean of three trials of the sort described above, except for the last (N = 100,000), which is the result of one run. Our run times for the k+1 approximation (k = 1) grow linearly with N until the Connection Machine's processors are "virtualized" fourfold at N = 50,000. The linear growth relation is empirically found to be t = 0.0036N + 0.148 for the data below. Also shown are comparable run times on an IBM 360/65, admittedly outdated, from Table IV of Sahni,[35] and our timings for the extended greedy algorithm. The latter times are constant (t = .13 s) regardless of N, for an unvirtualized machine.

Conclusions

Massively parallel versions of two common 0/1 Knapsack approximation algorithms have been demonstrated. These heuristics are readily adapted to the novel computer architecture of the Connection Machine, and the run times are very satisfactory for large N. In fact, it appears that problem instances of unprecedented size may now be handled with this implementation.

Run times for the parallelized extended greedy algorithm grow much more slowly than do those for the k+1 scheme. The latter has the advantage, however, of better *a priori* error bounds. For many instances of the 0/1 Knapsack problem, the two produce very similar approximations to the optimal solution. (Their relative performances are affected by the distribution of profits and weights, and extended greedy cannot be depended upon to produce a solution close to the theoretically better algorithm.) If a number of instances with profits and weights drawn from the same distributions are attacked, and the investigators can convince themselves that the two algorithms perform similarly, then two or three runs of k+1 can provide "typical" error bounds, and these can be applied to future runs of the more economical extended greedy algorithm.

7.3.3. Image Processing

One example is presented of an application to the field of image processing, automatic target detection.

AUTOMATIC TARGET DETECTION ON THE CONNECTION MACHINE

Introduction

A pattern recognition algorithm to detect vehicles in aerial black and white photographs against a variety of background types has been implemented on the massively parallel Connection Machine computer at the Perkin-Elmer Advanced Development Center.

The algorithm in its present form is primarily intended for target cueing. An image analyst can train the algorithm on a small subset of the total collection image set and then allow the Connection Machine to designate candidate targets for the whole set. In this scenario, the analyst need not examine frames that have no targets cued.

First Stage Processing

The first stage of the algorithm consists of three parts: (1) feature extraction, (2) training, and (3) pixel classification.

The feature extraction step generates multiple feature planes derived from the input image as shown in Figure 33. The set of features that can be employed is quite large and is only suggested by the list in Figure 33. In practice, the most commonly employed set consists of five features:

(a) MAX-MIN texture
(b) Sobel

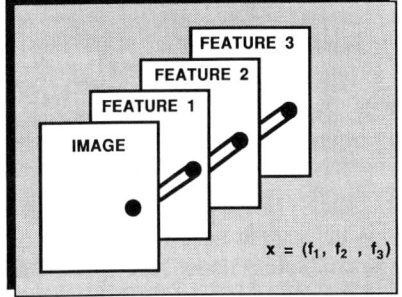

FIGURE 33. Automatic target detection algorithm summary.

(c) CFAR annulus sum
(d) Local area sum
(e) Raw gray level

At each pixel, these five feature values constitute a feature vector which can be represented as a point in five-dimensional feature space. The algorithm is trained by designating a few interesting and uninteresting, or target and background, points using a mouse. All of the pixels not so designated can then be classified as interesting or uninteresting based on whether their feature vectors are closer in the Euclidean sense to an interesting or uninteresting trained point in the five-dimensional feature space.

The five features are computed as follows.

MAX-MIN texture feature

Mitchell et al.[37] developed a technique for image texture analysis which they call MAX-MIN. The algorithm used is computationally simple and can be implemented in hardware for real-time analysis. Their algorithm was based on an intuitive determination that important texture information is contained in the relative frequency of local extremes in the intensity.

Mitchell et al. characterize the texture of the image at a given point by counting the number of extrema in a window centered at that point. Through the use of a thresholding parameter, they have introduced a hystersis-like smoothing to ignore reversals of amplitude smaller than the threshold.

A number of parallel scans are made through the window, and the number of extrema counted (according to a given criterion) are summed together for all scan lines to give an overall measure of the texture at that point.

The variables involved in the processes are:

1. The coordinates of the point in question;
2. The size of the window area to be examined;
3. The direction of the scans made through the image (0 and 90);
4. The threshold value that determines whether an extremum is large enough to be counted.

In Figure 34, the pixel values along the scan are shown in the heavy lines, while the lighter lines that bound the pixel values denote the upper and lower bounds of the hystersis limits that provides a criterion for the MAX-MIN counting process. The arrows at the top indicate instances where the MAX-MIN threshold has reversed direction.

The hystersis limits are set as follows. The upper limit is initially given the value of the starting pixel. The lower limit is simply equal to the upper limit minus the threshold. Now, going along the scan, the upper limit will be increased (pulling the lower limit up with it) as required to contain the pixel information. When the pixel value drops below the lower limit, the lower limit will be reduced (pulling the upper limit down with it) as required to contain the pixel information. At each reversal of the direction of "drift" of the interval, one threshold crossing count will be registered. The process continues counting alternate decreases and increases of the thresholds until the ends of the scan line is reached. Note that a count is not recorded if the pixel value merely equals the limit; a pixel must exceed the limit to register a count.

Sobel

Pratt[38] has suggested a 3 × 3 nonlinear edge enhancement operator described by the pixel numbering convention of Figure 35.

The edge enhancement plane is defined as

$$G(J,K) = \sqrt{X^2 + Y^2} \qquad (55)$$

where

$$X = (A_2 + 2A_3 + A_4) - (A_0 + 2A_7 + A_6)$$
$$Y = (A_0 + 2A_1 + A_2) - (A_6 + 2A_5 + A_4) \qquad (56)$$

In this context, G(j,k) is termed the Sobel feature.

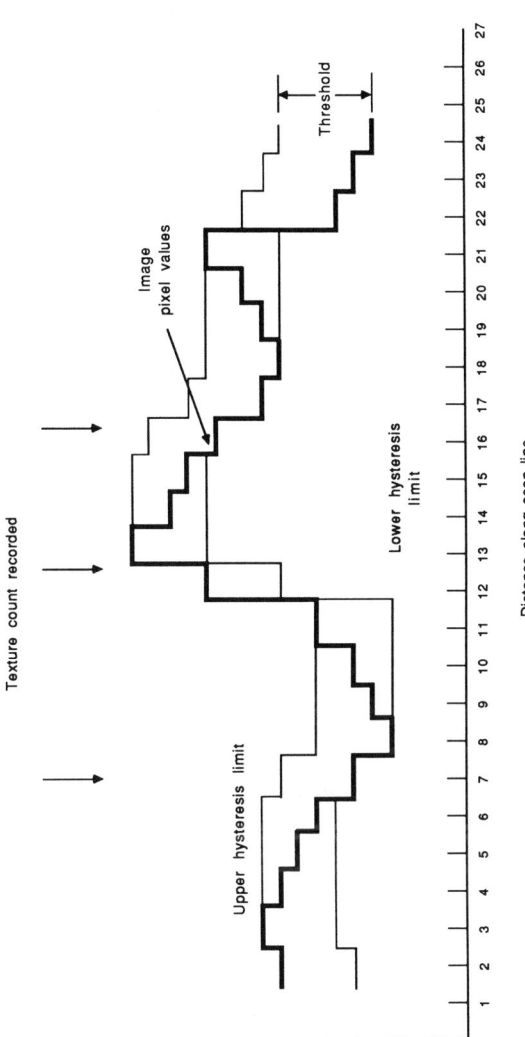

FIGURE 34. Operation of the MAX-MIN algorithm.

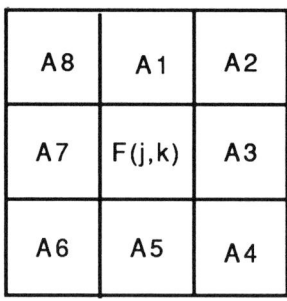

FIGURE 35. Numbering for the 3 × 3 Sobel edge detection operator.

CFAR annulus sum

The CFAR (constant false alarm rate) feature at pixel (i,j) consists of the sum of the gray levels of the pixels in a square window centered on pixel (i,j) minus the sum of the gray levels in a smaller window also centered at (i,j). Let the larger window have 2k+1 pixels on the edge and let the smaller window have 2m+1 pixels on an edge. k and m are parameters that may be chosen at run time. Figure 36 shows the annular region shaded for k = 3 and m = 2.

Local area sum

The local area sum feature at pixel (i,j) consists of the gray level values of all the pixels in a 2n+1 by 2n+1 square window centered on the location (i,j) added together. n can be selected arbitrarily.

The **gray level** at (i,j) is the fifth and final feature commonly used. Many other features can be employed.

The training consists of identifying at least one interesting (target) and one uninteresting (background) point using the mouse. Any number of each category can be designated.

Pixel classification is performed by assigning each pixel in the image to the category of the nearest training point in feature space. The distance metric in the five-dimensional feature space is either Euclidean or Manhattan, i.e., the sum of the absolute values of the differences in the coordinates of the end points. The latter is faster because it avoids squaring and square roots.

Second and Third Stage Processing

The second stage of processing consists in aggregating interesting (target) pixels into groups or "blobs" based on adjacency. This operation is generally termed connected components. All blobs are then subjected to a shrink and expand operation in which all perimeter pixels are deleted and then all new perimeter pixels grow outward by one pixel. This serves to clean up the classification results.

The final stage is mensuration and size screening. Since each trained inter-

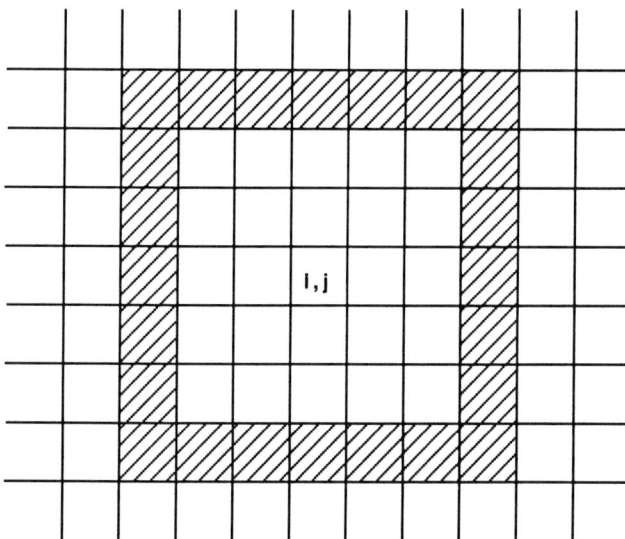

FIGURE 36. CFAR region.

esting (target) point is almost always itself a member of a blob, a maximum trained blob size and a minimum trained blob size can be determined. This allows the specification of an acceptable blob range. For example, blobs smaller than say one half of the smallest trained blob or larger than twice as large as the largest trained blob would be eliminated as objects not in the same size category as the trained targets.

The automatic target detection algorithm applied to a 256×256 image on 8K processors (virtualized by a factor of 8) runs in about 2 s of computation time on a first generation Connection Machine. A similar algorithm (not exact) runs for 20 to 30 min on a VAX 11/780. It is expected that an additional speedup factor of 160 would be realized on a full sized second generation Connection Machine.

Conclusion

The algorithm is not considered to be fully developed. Optimal feature choices and parameter settings have not been determined. Currently, principal components analysis is being conducted to ascertain the best discriminants for each target/background combination. These settings will be stored and invoked as classification rules for future target detection applications.

7.3.4. Miscellaneous

Three applications of the Connection Machine are grouped here. The first, Free-Text Search, discusses the dialogue one may have with a massive text data base through matching surrogate bit strings and rapid sort operations. The second describes the implementation of neural networks on the CM hypercube. The last deals with a benchmark comparison of the CM with CRAY and IBM machines for orbital collision prediction of eighteen space objects over a 1 week period.

Parallel Free-Text Search

Locating information in natural-language documents contained in large databases is an important problem. Such databases may contain articles from wire services and newspapers, abstracts and full-text articles from journals and encyclopedias, as well as bibliographies and judicial rulings. Search facilities for such databases are provided by vendors like Dialog, Lexis, Nexis, Westlaw, Medlars, and VU-Text. In many communities, particularly legal and biomedical research, access to such databases has become critical. The size of such databases can be virtually unlimited; the Lexis database, for example, reportedly contains a total of 125 Gbytes of information.

The two general methods of accessing such databases are free-text search and keyword search. Free-text search systems allow the user to search the text of the document for arbitrary combinations of words, whereas keyword systems require that indexers read all the documents in a database and assign them keywords from a controlled vocabulary. In the latter system, the user searches for information or documents using some combination of keywords from this controlled vocabulary. The relative merits of these systems continue to be debated. Advocates of manual indexing systems argue that free-text search systems suffer from poor recall (they fail to retrieve all relevant documents) and poor precision (they deliver too many irrelevant documents). Proponents of free-text search counter that techniques are available to alleviate these two problems and that manual indexing systems have problems of their own: specifically, that the manual indexing process is unreliable, based on the observation that different indexers will often assign radically different keywords to the same article.

Although techniques are available to improve the recall and precision of searches in free-text database systems, they are in some cases not widely used because they make computational demands beyond the practical limits of conventional computers.

The usual method of organizing a free-text database to construct an inverted index of all words it contains. To locate documents containing a given word, the index for that word is located (via hashing, for example) and retrieved. Documents containing combinations of words are found by computing the intersection and union of indexes. However, inverting a database in this way often increases the storage utilization 200% over the raw text.

This article will take no stand on the issue of the relative merits of free-text versus controlled-vocabulary searching. Rather, it will present a new implementation of free-text searching that takes advantage of the possibilities offered by a massively parallel computer, the Connection Machine (CM), with up to 65,536 processing elements. In this implementation, the representation of the documents in memory permits a very fast search for the presence of a word in a document. With the CM, each processing element stores between one and three documents; queries are then broadcast to the entire machine and the results collected. Because of the massive parallelism, the resulting system is fast enough to permit the application of exhaustive methods not previously feasible for large databases.

Two applications of this technology are discussed. The first is a benchmark of the CM as a query evaluator. In this application, a Boolean query is evaluated against a sample database. Using this measurement and taking into account the approximate performance of the I/O system, we estimate that the CM can evaluate a 20,000-term query against a 15 Gbyte database in 3 min, of which 2 min are devoted to I/O and 1 min to computation. The second application, a prototype of an interactive document-retrieval system, uses a technique called relevance feedback to produce an interface that is fast and easy to use, and produces a high-quality search.

Data Structure

The starting point of the information-retrieval system is the representation of documents in the CM's memory. The basic requirement is to be able to very quickly test the machine's memory for the presence of a word. In addition, the representation must be reasonably compact. A representation was selected that was originally developed for spelling-correction dictionaries and is known as surrogate coding. The advantage of surrogate coding is that probing a surrogate table for the presence of a word requires only AND-ing together a small number of bits; the disadvantages are that word order and proximity information are lost, and probes sometimes return false positives.

A surrogate table is a binary array k bits long. In the example that follows, we will use $k = 30$, although in actual implementations we have used values of 512 and 1024. The table is initialized in all zeros.

To insert a document into the table, i independent hash codes between 0 and $k - 1$ are generated. Here, we will use $i = 4$, although, in practice, i is usually between 10 and 30. If we insert the word "chemistry" into an empty table, and hash (chemistry) = (1, 6, 29, 20), setting these four bits to 1 produces the following table:

010000100000000000001000000001

If we then insert a second word "biology" into the table, with hash (biology) = (16, 22, 29, 3), setting these four bits in the table produces the following layout. Note that one of these bits (29) was already set.

000100000000000010000010000001

To probe for a word, we generate its hash codes and then AND together the corresponding bits. To probe for "chemistry", we AND together bits 1, 6, 29, and 20. Since all four are 1, the probe returns PRESENT.

Suppose we probe for a second word, "physics", which was not inserted in the table. Assuming that hash (physics) = (3, 7, 16, 26), we check the table and find that bits 7 and 26 are 0. Thus, the probe returns ABSENT.

It is possible for a probe to return a false positive. In the case of hash (mathematics) = (3, 6, 13, 20), checking those four bits will show that all are 1. Thus, the probe returns PRESENT, even though "mathematics" was never inserted in the table.

The probability of a false positive depends on the size of the table (k), the number of words encoded in the table (w), and the number of bits that are set for each word (i). Decreasing w reduces the false-hit rate, but also increases storage needs. Increasing i, up to a point, also reduces the false-hit rate, but increases processing time. An analysis that will provide some guidance in making the most appropriate trade-off is presented next.

Analysis of the Data Structure

The probability of a false positive may be determined by finding the probability that a random bit is 1 as a function of k,w, and i, and then using this result to calculate the probability of i random bits all being 1. Given w words, and i bits set for each word, a total of w x i bits will be set. We will now calculate $P_1(w,i,k)$, the probability that a random bit in the table will be set. With a k-bit table T and a sequence S of w x i random integers uniformly distributed between 0 and k - 1, an element T_j of T is 1 if and only if j occurs in S; otherwise, it is 0. Therefore,

$$P_1(w,i,k) = 1 - P_0(w,i,k) \tag{57}$$

where $P_0(w,i,k)$ is the probability that a random number j, uniformly distributed between 1 and k - 1, does not occur in the sequence S. This gives us the following two formulations:

$$P_0(w,i,k) = \left(\frac{k-1}{k}\right)^{wi} \tag{58}$$

$$P_1(w,i,k) = 1 - \left(\frac{k-1}{k}\right)^{wi} \tag{59}$$

We now need to determine the probability that i random bits are all 1: We will call this $P_{false}(w,i,k)$. This is the probability of a false hit: That is, the probability that a word not present in the document used to create the table will return PRESENT when it is probed for.

TABLE 19
Probability of a False Hit

w = Words per table
i = Bits set for each word, fixed at 10
k = Bits per table, fixed at 512
P_{false} = Probability of a false positive being returned when probing one table for one word

w	P_{false}
5	4.91×10^{-11}
10	3.12×10^{-8}
15	1.13×10^{-6}
20	1.26×10^{-5}
25	7.46×10^{-5}
30	2.96×10^{-4}

$$P_{false} = (P_1)^i = \left(1 - \left(\frac{k-1}{k}\right)^{wi}\right)^i \tag{60}$$

Table 19 shows some values of P_{false}, assuming k = 512 and i = 10. This important fact to remember about Table 19 is that a small change in w causes a large change in P_{false}; that is, the false-hit rate may be dramatically reduced by a small decrease in the number of words in each table. Depending on the false-hit rate that is considered acceptable in any particular application, the values of w, i, and k can be tuned to yield optimal system performance.

Application

In practice, the application of this algorithm is slightly more complex than suggested above. First, each processor in the CM contains 4096 bits of memory, some of which are used as scratch memory. This allows six tables of 512 bits, or three tables of 1024 bits. Moreover, as the above discussion indicates, it is necessary to limit the number of words in each table. For a 512-bit table, limit of 15 to 30 words is reasonable. If a document contains more than the allotted number of words, several tables must be used for each document. Thus, if 30 words are being put into each table and a 90-word document is encountered, it is necessary to use three tables, which are generally put in three different processors. Such a set of tables is called a chain, where each chain reports to a designated head processor.

Probes for words are never used in isolation, but rather as components of Boolean queries (where they are combined via Boolean connectives), or as components of simple queries (where each word is given a numerical weight).

In simple queries, point values are assigned to a set of words, and each document in the database is then scored by totaling the scores of the words it contains. This is done by adding up the score for the words in each processor and then totaling the scores of the processors in each chain.

The documents in the database that are assigned non-zero scores are retrieved best-to-worst. The global-maximum operation is used to find the processor in the CM with the highest total score; that processor is masked out, and the process repeated to extract the second highest score, and so forth. The process can be halted when a specified number of documents has been retrieved, or when only documents with a score below a specified threshold remain in the machine.

A Benchmark

In September 1985, a benchmark of the Boolean query algorithm was performed on a 16,384-processor prototype of the CM. The test database consisted of 31,994 documents, totaling 18 Mbytes. A document consisted of several fixed-format numerical fields, totaling 100 bits, and two variable-length text fields. Each virtual processor in the CM had two tables of 512 bits each, and the packing density was limited to 35 words per table. Several trial queries, containing between 25 and 20,000 atomic terms, were evaluated. The times required to execute the queries varied between 0.004 s for a 25-term query and 0.295 s for a 20,000-term query. For a full-sized CM with 65,536 processors, execution times would remain the same, while the database could grow to 128,000 documents totaling 71 Mbytes.

Databases larger than 71 Mbytes will require the use of the swapping disk to page additional segments of the database into the CM's memory. Swapping in the data should require approximately 0.6 s per database segment. From then on, evaluating a query again takes between 0.004 s (25 terms) and 0.295 s (20,000 terms) per database segment. Using these figures, the time required to search a 15-Gbyte database should vary between 2 min for a 25-term query and 3 min for a 20,000-term query — a substantial improvement over existing search systems. The results of this benchmark are summarized in Table 20.

Relevance feedback constructs simple queries from the texts of documents that have already been judged relevant. After some search method has been used to locate a small set of possibly relevant documents, the user scans these documents and marks any documents that are obviously relevant as good, and any that are obviously irrelevant as bad. The text of the marked documents is then scanned for words. The more good documents a word occurs in, the higher its score.

Since simple queries built using this technique may contain hundreds of terms, its practical application on large databases will require a very powerful machine. Applying the CM to this task, since it has both sufficient processing power and sufficient I/O bandwidth, results in an excellent search in terms of quality, ease of use, and performance.

TABLE 20
Results of the Boolean Query Benchmark

Measured benchmark and projected times

	Query size				
	25 terms	100 terms	1,000 terms	10,000 terms	20,000 terms
Load, one batch(s)	0.567	0.567	0.567	0.567	0.567
Execute query, one batch(s)	0.004	0.005	0.022	0.149	0.295
Total time per batch(s)	0.571	0.572	0.589	0.716	0.862
Hits per batch	22	22	25	50	68
Total time for 15 Gbytes (min.)	2.0	2.1	2.1	2.5	3.0

	Benchmark batch size (16K processors)	Full batch size (64K processors)	Full database size (211 batches)
Raw data	18 Mbytes	71 Mbytes	15 Gbytes
Compressed data	8 Mbytes	32 Mbytes	7 Gbytes
Documents	31,994	128,000	27,000,000

Conclusion

Using the CM, it is possible to perform a fast parallel search of a database. For a database of 18 Mbytes (31,994 documents) on a 16,384 element machine, the measured time to execute a query varies from 0.004 s for a Boolean query with 25 terms (compute time only) to 0.295 s for 20,000 terms. For a 15-Gbyte database, the estimated time to execute a query varies between 2 min for a Boolean query with 25 terms to 3 min for a Boolean query of 20,000 terms (compute plus I/O time). It is also possible to search a database using simple queries and achieve similarly high search speeds: 60 ms for a 200-term query on a 112-Mbyte database.

Using these facilities, it is possible to implement database search techniques that are not feasible given sequential machines and inverted databases. A demonstration system utilizing relevance feedback has been written and tested. It combines high precision and recall with ease of use and fast response, and represents an advance over existing free-text database search technology.

NEURAL NETWORK IMPLEMENTATION APPROACHES FOR THE CONNECTION MACHINE

Introduction

Simulations of neural network models on digital computers perform various computations by applying linear or nonlinear functions, defined in a program, to weighted sums of integer or real numbers retrieved and stored by array references. The numerical values are model-dependent parameters like time averaged spiking frequency (activation), synaptic efficacy (weight), the error in error back propagation models, and computational temperature in thermodynamic models. The interconnect structure of a particular model is implied by indexing relationships between arrays defined in a program. On the Connection Machine (CM), these relationships are expressed in hardware processors interconnected by a 16-dimensional hypercube communication network. Mappings are constructed to define higher dimensional interconnectivity between processors on top of the fundamental geometry of the communication network. Parallel transfers are defined over these mappings. These mappings may be dynamic. CM parallel operations transform array indexing from a temporal succession of references to memory to a single temporal reference to spatially distributed processors.

Two alternative approaches to implementing neural network simulations on the CM are described. Both approaches use "data parallelism"[39] provided by the *Lisp virtual machine. Data and control structures associated for each approach and performance data for a Hopfield model implemented with each approach are presented.

Data Structures

The functional components of a neural network model implemented in *Lisp are stored in a uniform parallel variable (pvar) data structure on the CM. The data structure may be viewed as columns of pvars. Columns are given to all CM virtual processors. Each CM physical processor may support 16 virtual processors. In the first approach described, CM processors are used to represent the edge set of a model's graph structure. In the second approach described, each processor can represent a unit, an outgoing link, or an incoming link in a model's structure. Movement of activation (or error) through a model's interconnect structure is simulated by moving numeric values over the CM's hypercube. Many such movements can result from the execution of a single CM macroinstruction. The CM transparently handles message buffering and collision resolution. However, some care is required on the part of the user to ensure that message traffic is distributed over enough processors so that messages don't stack up at certain processors, forcing the CM to sequentially handle large numbers of buffered messages. Each approach requires serial transfers of model parameters and states over the communication channel between the host and the CM at certain times in a simulation.

The first approach, "the edge list approach", distributes the edge list of a network graph to the CM, one edge per CM processor. Interconnect weights for each edge are stored in the memory of the processors. An array on the host machine stores the current activation for all units. This approach may be considered to represent abstract synapses on the CM. The interconnect structure of a model is described by product sets on an ordered pair of identification (id) numbers, rid and sid. The rid is the id of units receiving activation and sid the id of units sending activation. Each id is a unique integer. In a hierarchical network, the ids of input units are never in the set of rids and the ids of output

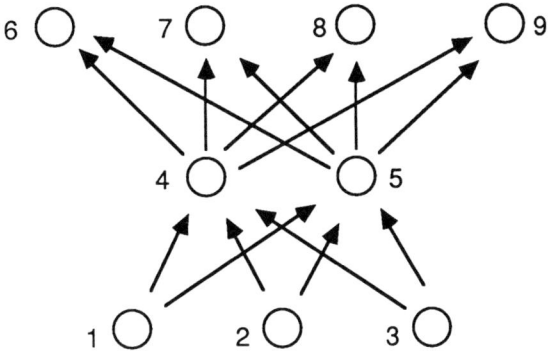

FIGURE 37. Edge list representation of a R3→ R2→ R4 interconnect structure.

units are never in the set of sids. Various set relations (e.g., inverse, reflexive, symmetric, etc.) defined over id ranges can be used as a high level representation of a network's interconnect structure. These relations can be translated into pvar columns. The limits to the interconnect complexity of a simulated model are the virtual processor memory limits of the CM configuration used and the stack space required by functions used to compute the weighted sums of activation. Figure 37 shows a $\mathbf{R}^3 \to \mathbf{R}^2 \to \mathbf{R}^4$ interconnect structure and its edge list representation and the CM.

This representation can use as few as six pvars for a model with Hebbian adaptation: rid (i), sid (j), interconnect weight (w_{ij}), ract (a_i), sact (a_j), and learn rate (η). Error back propagation requires the addition of: error (e_i), old interconnect weight ($w_{ij}(t-1)$), and the momentum term (α). The receiver and sender unit identification pvars are described above. The interconnect weight pvar stores the weight for the interconnect. The activation pvar, sact, stores the current activation, a_j, transferred to the unit specified by rid from the unit specified by sid. The activation pvar, ract, stores the current weighted activation $a_j w_{ij}$. The error pvar stores the error for the unit specified by the sid. A variety of proclaims (e.g., integer, floating point, boolean, and field) exist in *Lisp to define the type and size of pvars. Proclaims conserve memory and speed up execution. Using a small number of pvars limits the amount of memory used in each CM processor so that maximum virtualization of the hardware processors can be realized. Any neural model can be specified in this fashion. Sigma-pi models require multiple input activation pvars be specified. Some edges may have a different number of input activation pvars than others. To maintain the uniform data structure of this approach a tag pvar has to be used to determine which input activation pvars are in use on a particular edge.

The edge list approach allows the structure of a simulated model to "physically" change because edges may be added (up to the virtual processor limit), or deleted at any time without affecting the operation of the control structure. Edges may also be placed in any processor because the subselection (on rid or sid) operation performed before a particular update operation ensures that all processors (edges) with the desired units are selected for the update.

The second simulation approach, "the composite approach", uses a more complicated data structure where units, incoming links, and outgoing links are represented. Update routines for this approach use parallel segmented scans to form the weighted sum of input activation. Parallel segmented scans allow a MIMD-like computation of the weighted sums for many units at once. Pvar columns have unique values for unit, incoming link, and outgoing link representations. The data structures for input units, hidden units, and output units are composed of sets of the three pvar column types. Figure 38 shows the representation for the same model as in Figure 37 implemented with the composite approach.

In Figure 38, CM processors acting as units, outgoing links, and incoming links are represented respectively by circles, triangles, and squares. CM cube address pointers used to direct the parallel transfer of activation are shown by arrows below the structure. These points define the model interconnect mapping. Multiple sets are represented by operation-arrow icons above the structure. A basic composite approach pvar set for a model with Hebbian adaptation is: forward B, forward A, forward transfer address, interconnect weight (w_{ij}), act-1 (a_i), act-2 (a_j), threshold, learn rate (η), current unit id (i), attached unit id(j), level, and column type. Back propagation of error requires the addition of: backward B, backward A, backward transfer address, error (e_j), previous interconnect weight ($w_{ij}(t-1)$), and the momentum term (α). The forward and backward Boolean pvars control the segmented scanning operations over unit constructs. Pvar A of each type controls the plus scanning and pvar B of each type controls the copy scanning. The forward transfer pvar stores cube addresses for forward (ascending cube address) parallel transfer of activation. The backward transfer pvar stores cube addresses for backward (descending cube address) parallel transfer of error. The interconnect weight, activation, and error pvars have the same functions as in the edge list approach. The current unit id stores the current unit's id number. The attached unit id stores the id number of an attached unit. This is the edge list of the network's graph. The contents of these pvars only have meaning in link pvar columns. The level pvar stores the level of a unit in a hierarchical network. The type pvar stores a unique arbitrary tag for the pvar column type. These last three pvars are used to subselect processor ranges to reduce the number of processors involved in an operation.

Again, edges and units can be added or deleted. Processor memories for deleted units are zeroed out. A new structure can be placed in any unused processors. The level, column type, current unit id, and attached unit id values must be consistent with the desired model interconnectivity.

The number of CM virtual processors required to present a given model on

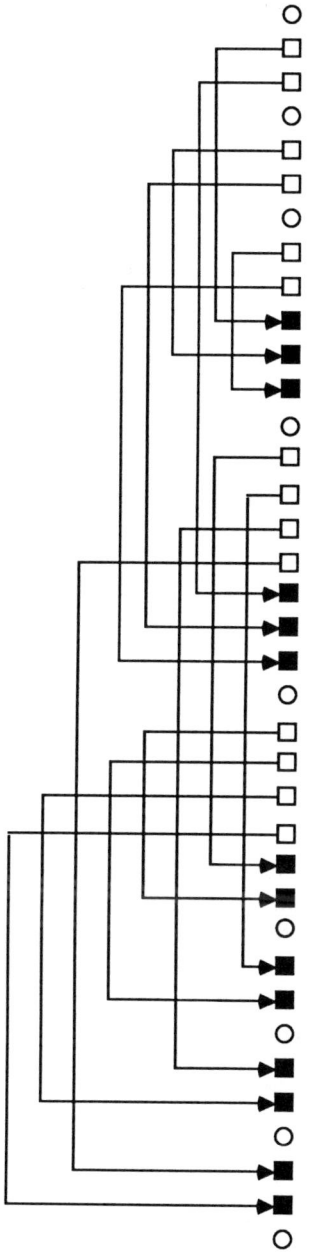

FIGURE 38. Composite representation of a R3→ R2→ R4 interconnect structure.

TABLE 21
Model Sizes and CM Processors Required

Run no.	Grid size	Number of units	Edge List N(N–1)	Quart CM virt. procs.	Virt. level
1	8^2	64	4032	8192	0
2	9^2	81	6480	8192	0
3	11^2	121	14520	16384	0
4	13^2	169	28392	32768	2
5	16^2	256	65280	65536	4
6	19^2	361	129960	131072	8

Run no.	Grid size	Number of units	Composite N+2N(N–1)	Quart CM virt. procs.	Virt. level
7	8^2	64	8128	8192	0
8	9^2	81	13041	16384	0
9	11^2	121	29161	32768	2
10	13^2	169	56953	65536	4
11	16^2	256	130816	131072	8
12	19^2	361	260281	262144	16

the CM differs for each approach. Given N units and N(N–1) non-zero interconnects (e.g., a symmetric model), the edge list approach simply distributes N(N–1) edges to N(N–1) CM virtual processors. The composite approach requires two virtual processes for each interconnect and one virtual processor for each unit or N + 2N(N – 1) CM virtual processors total. The difference between the number of processors required by the two approaches is N^2. Table 21 shows the processor and CM virtualization requirements for each approach over a range of model sizes.

Control Structures

The control code for neural network simulations (in *Lisp or C*) is stored and executed sequentially on a host computer (e.g., Symbolics 36xx and VAX 86xx) connected to the CM by a high speed communication line. Neural network simulations executed in *Lisp use a small subset of the total instruction set: processor selection reset (*all), processor selection (*when), parallel content assignment (*set), global summation (*sum), parallel multiplication (*!!), parallel summation (+!!), parallel exponentiation (exp!!), the parallel global memory references (*pset) and (pref!!), and the parallel segmented scans (copy!! and +!!). Selecting CM processors puts them in a "list of active processors" (loop) where their contents may be arithmetically manipulated in parallel. Copies of the list of active processors may be made and used at any time. A subset of the processors in the loop may be "subselected" at any time, reducing the loop contents. The processor selection reset clears the current selected set by setting all processors as selected. Parallel content assignment allows pvars in the currently selected processor set to be assigned allowed values in one step. Global summation executes a tree reduction sum across the CM processors by grid or cube address for particular pvars. Parallel multiplications and additions multiply and add pvars for all selected CM processors in one step. The parallel exponential applies the function, e^x, to the contents of a specified pvar, x, over all selected processors. Parallel segmented scans apply two functions, copy!! and +!!, to subsets of CM processors by scanning across grid or cube addresses. Scanning may be forward or backward (i.e., by ascending or descending cube address order, respectively).

Programs A and B show the edge list approach kernels required for Hebbian learning for a $\mathbf{R}^2 \to \mathbf{R}^2$ model. The loop construct in Program A drives the activation update

$$a_i(t+1) = F[\Sigma \, w_{ij}(t+1) a_j(t)] \tag{61}$$

operation. The usual loop to compute each weighted sum for a particular unit has been replaced by four parallel operations: a selection reset (*all), a subselection of all the processors for which the particular unit is a receiver of activation (*when (=!! rid (!! (1+u)))), a parallel multiplication (*!! weight sact), and a tree reduction sum (*sum...). Activation is spread for a particular unit, to all others it is connected to, by storing the newly computed activation in an array on the host, then subselecting the processors where the particular unit is a sender of activation (*when (=!! sid (!! (1+u)))), and broadcasting the array value on the host to those processors.

```
(dotimes (u 4)
  (*all (*when (=!! rid (!! (1+u)))
    (setf (aref activation u)
      (some-nonlinearity (*sum (*!! weight sact))))
    (*set ract (!! (aref activation u))))
  (*all (*when (=!! sid (!! (1+u)))
    (*set sact (!! (aref activation u))))))
```

PROGRAM A. Activation Update Kernel for the Edge List Approach.

$$w_{ij}(t+1) = \eta a_i(t+1) a_j(t+1) \tag{62}$$

```
(*all
  (*set weight
    (*!! learn-rate ract sact)))
```

PROGRAM B. Hebbian Weight Modification Kernel for the Edge List Approach.

The edge list activation update kernel is essentially serial because the steps involved can only be applied to one unit at a time. The weight modification is parallel. For error back propagation a separate loop for computing the errors for the units on each layer of a model is required. Activation update and error back propagation also require transfers to and from arrays on the host on every iteration step incurring a concomitant overhead.

Other common computations used for neural networks can be computed in parallel using the edge list approach. Program C shows the code kernel for parallel computation of Lyapunov energy equations

$$E = -\frac{1}{2}\sum_{\substack{i \neq j}}^{N} w_{ij} a_i a_j + \sum_{i=j}^{N} I_i a_i \qquad (63)$$

where i = 1 to number of units (N).

```
(+ (* -.5 (*sum (*!! weight ract sact))) (*sum
        (*!! input sact)))
```

PROGRAM C. Kernel for Computation of the Lyapunov Energy Equation.

Although an input pvar, input, is defined for all edges, it is only non-zero for those edges associated with input units. Program D shows the pvar structure for parallel computation of a Hopfield weight prescription, with segmented scanning to produce the weights in one step,

$$w_{ij} = \sum_{r=1}^{S} (2a_i^r - 1)(2a_j^r - 1) \qquad (64)$$

where $w_{ii} = 0$, $w_{ij} = w_{ji}$, and r = 1 to the number of patterns, S, to be stored.

seg	t	n		n	t	n		n
ract	v_1^1	v_1^2	...	v_1^S	v_1^1	v_1^2	...	v_1^S ...
sact	v_2^1	v_2^2	...	v_2^S	v_3^1	v_3^2	...	v_3^S ...
weight				w_{12}				w_{13}

PROGRAM D. Pvar Structure for Parallel Computation of Hopfield Weight Prescription.

Program E shows the *Lisp kernel used on the pvar in Program D.

```
(set weight
    (scan '+!! (*!! (-!! (*!! ract (!! 2)) (!! 1)) (-
!! (*!! sact (!! 2)) (!! 1)))) :segment-pvar seg
:include-self t)
```

PROGRAM E. Parallel Computation of Hopfield Weight Prescription.

The inefficiencies of the edge list activation update are solved by the updating method in the composite approach. Program F shows the *Lisp kernel for activation update using the composite approach. Program G shows the *Lisp kernel for the Hebbian learning operation in the composite approach.

```
(*all
  (*when (=!! level (!! 1))
    (*set act (scan!! act-1 'copy!! :segment-pvar
      forwardb
      :include-self t))
    (*set act (*!! act-1 weight))
    (*when (=!! type (!! 2)) (*pset :overwrite act-1
      act-1
      ftransfer)))
  (*when (=!! level (!! 2))
    (*set act (scan!! act-1 '+!! :segment-pvar
      forwarda
      :include-self t))
  (*when (=!! type (!! 1)) (some-nonlinearity!!
    act-1))))
```

PROGRAM F. Activation Update for the Composite Approach.

```
(*all
  (*set act-1 (scan!!act-1 ¢copy!! :segment-pvar forwardb
                                   :include-self t))
  (*when (=!! type (!! 2))
    (*set act-2 (pref!! act-1 btransfer)))
      (*set weight
        (+!! weight
          (*!! learn-rate act-1 act-2)))))
```

PROGRAM G. Hebbian Weight Update Kernel for the Composite Approach.

It is immediately obvious that no looping is involved. Any number of interconnects may be updated by the proper subselection. However, the more subselection is used the less efficient the computation becomes because less processors are involved.

Complexity Analysis

The performance results presented in the next section can be largely anticipated from an analysis of the space and time requirements of the CM implementation approaches. For simplicity a $\mathbf{R}^n \to \mathbf{R}^n$ model with Hebbian adaptation is used. The order of magnitude requirements for activation and weight updating are compared for both CM implementation approaches and a basic serial matrix arithmetic approach.

For the given model, the space requirements on a conventional serial machine are $2n + n^2$ locations or $O(n^2)$. The growth of the space requirement is dominated by the $n \times n$ weight matrix defining the system interconnect structure. The edge list approach uses six pvars for each processor and uses $n \times n$ processors for the mapping, or $6n^2$ locations or $O(n^2)$. The composite approach uses 11 pvars. There are $2n$ processors for units and $2n^2$ processors for interconnects in the given model. The composite approach uses $11(2n+2n^2)$ locations or $O(n^2)$. The CM implementations take up roughly the same space as the serial implementation, but the space for the serial implementation is composed of passive memory whereas the space for the CM implementation is composed of interconnected processors with memory.

The time analysis for the approaches compares the time order of magnitudes to compute the activation update and the Hebbian weight update. On a serial machine, the n weighted sums computed for the activation update require n^2 multiplications and $n(n-1)$ additions. There are $2n^2-n$ operations or time order of magnitude $O(n^2)$. The time order of magnitude for the weight matrix update is $O(n^2)$ since there are n^2 weight matrix elements.

The edge list approach forms n weighted sums by performing a parallel product of all of the weights and activations in the model, (*!! weight sact), and then a tree reduction sum, (*sum ...), of the products for the n units (see Program B). There are $1+n(n\log_2 n)$ operations or time order of magnitude $O(n^2)$. This is the same order of magnitude as obtained on a serial machine. Further, the performance of the activation update is a function of the number of interconnects to be processed.

The composite approach forms n weighted sums in nine steps (see Program F): five selection operations; the segmented copy scan before the parallel multiplication; the parallel multiplication; the parallel transfer of the products; and the segmented plus scan, which forms the n sums in one step. This gives the composite activation update a time order of magnitude $O(1)$. Performance is independent of the number of interconnects processed. The discussion below shows that this is not quite true.

The n^2 weights in the model can be updated in three parallel steps using the edge list approach (see Program B). The n^2 weights in the model can be updated in eight parallel steps using the composite approach (see Program G). In either case, the weight update operation has a time order of magnitude $O(1)$.

The time complexity results obtained for the composite approach apply to computation of the Lyaponov energy equation[40] and the Hopfield weighting prescription,[41] given that pvar structures, which can be scanned (see Program D) are used. The same operations performed serially are time order of magnitude $O(n^2)$.

The above operations all incur a one time overhead cost for generating the addresses in the pointer pvars, used for parallel transfers, and arranging the values in segments for scanning. What the above analysis shows is that time complexity is traded for space complexity. The goal of CM programming is to use as many processes as possible at every step.

Performance Comparison

Simulations of a Hopfield spin-glass model[40] were run for six different model sizes over the same number (16,384) of physical CM processors to provide a performance comparison between implementation approaches. The Hopfield network was chosen for the performance comparison because of its simple and well-known convergence dynamics and because it uses a small set of pvars which allows a wide range of network sizes (degrees of virtualization) to be run. Twelve treatments are run. Six with the edge list approach and six with the composite approach. Table 21 shows the model sizes run for each treatment. Each treatment was run at the virtualization level just necessary to accommodate the number of processors required for each simulation.

Two exemplar patterns are stored. Five test patterns are matched against the two exemplars. Two test patterns have their centers removed, two have a row and column removed, and one is a random pattern. Each exemplar was hand picked and tested to ensure that it did not produce cross-talk. The number of rows and columns in the exemplars and patterns increase as the size of the networks for the treatment increases. Since the performance of the CM is at issue, rather than the performance of the network model used, a simple model and a simple pattern set were chosen to minimize consideration of the influence of model dynamics on performance.

Performance is presented by plotting execution speed versus model size. Size is measured by the number of interconnects in a model. The execution speed metric is interconnects updated per second, $N*(N-1)/t$, where N is the number of units in a model and t is the time used to update the activations for all of the units in a model. All of the units were updated three times for each pattern. Convergence was determined by the output activation remaining stable over the final two updates. The value of t for a treatment is the average of 15 samples of t. Figure 39 shows the activation update cycle time for both approaches. Figure 40 shows the interconnect update speed plots for both approaches. The edge list approach is plotted in black. The composite approach is plotted in white. The performance shown excludes overhead for interpretation of the *Lisp instructions. The model size categories for each plot correspond to the model sizes and levels of CM virtualization shown in Table 21.

Figure 40 shows an order of magnitude performance difference between the approaches and a roll off in performance for each approach as a function of the number of virtual processors supported by each physical processor. The performance turn-around is at 4x virtualization for the edge list approach and 2x virtualization for the composite approach.

Conclusions

Representing the interconnect structure of neural network models with mappings defined over the set of fine grain processors provided by the CM architecture provides good performance for a modest programming effort utilizing only a small subset of the instructions provided by *Lisp. Further, the

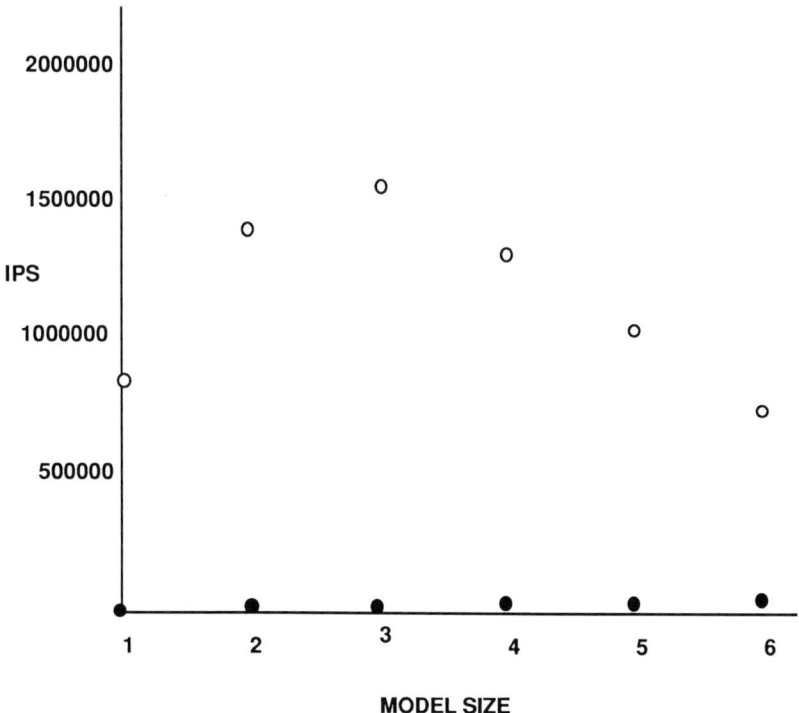

FIGURE 39. Activation update cycle times.

performance will continue to scale up linearly as long as not more than 2x virtualization is required. While the complexity analysis of the composite activation updates suggests that its performance should be independent of the number of interconnects to be processed, the performance results show that the performance is indirectly dependent on the number of interconnects to be processed because the level of virtualization required (after the physical processors are exhausted) is dependent on the number of interconnects to be processed and virtualization decreases performance linearly. The complexity analysis of the edge list activation update shows that its performance should be roughly the same as serial implementation on comparable machines. The results suggest that the composite approach is to be preferred over the edge list approach but not to be used at a virtualization level higher than 2x.

The mechanism of the composite activation update suggest that hierarchical networks simulated in this fashion will compare in performance to single layer networks because the parallel transfers provide a type of pipeline for activation for synchronously updated hierarchical networks while providing simultaneous activation transfers for asynchronously updated single layer networks. Researchers at Thinking Machines Corporation and the M.I.T. AI

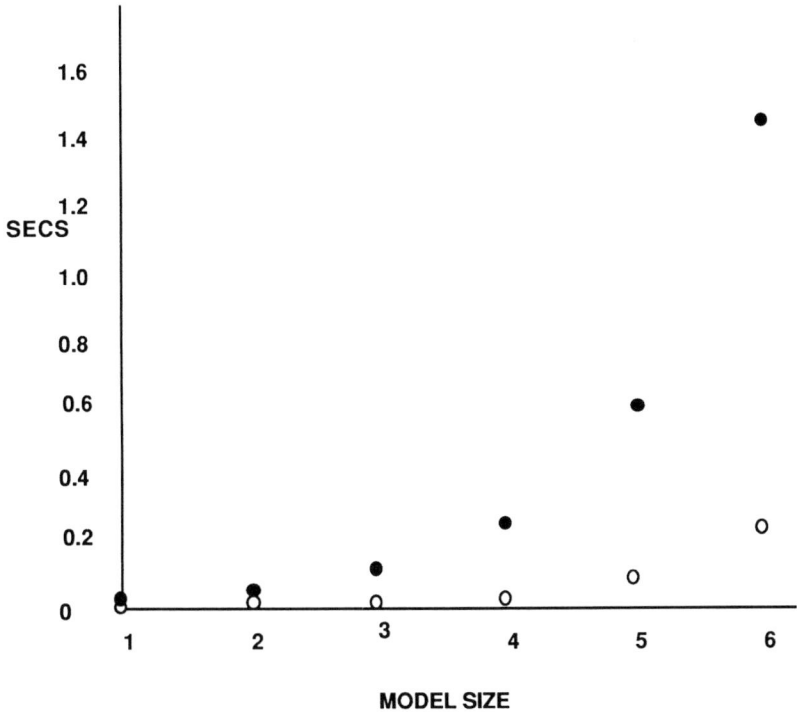

FIGURE 40. Edge list interconnect update speeds.

Laboratory in Cambridge, Massachusetts use a similar approach for an implementation of NETtalk. Their approach overlaps the weights of connected units and simultaneously pipelines activation forward and error backward.[41]

Performance better than that presented can be gained by translation of the control code from interpreted *Lisp to Paris and use of the CM2. In addition to not being interpreted, Paris allows explicit control over important registers that aren't accessible through *Lisp. The CM2 offers a number of features which enhance performance of neural network simulations: a *Lisp compiler, larger processor memory (64K), and floating-point processors. The compiler and floating-point processors increase execution speeds while the larger processor memories provide a larger number of virtual processors at the performance turnaround points, allowing higher performance through higher CM utilization.

ORBIT COLLISION PROBLEM BENCHMARK STUDY

Introduction
The continuing onslaught of new high performance computer models and

TABLE 22
Conventional Supercomputer Comparison Parameters

Computer model	Cycle time (ns)	No. processors	Theor. peak MIPS	Theor. peak MFLOPS	LINPAC MFLOPS	Machine cost ($M)
IBM 3090/X00	18.5	1	28	N/A	6.8	2.20
CRAY X-MP/1	9.5	1	105	210	39.0	5.75
CRAY X-MP/4	9.5	4	420	840	713.0	11.75
CRAY Y-MP/8	6	8	1300	2600	N/A	20.00
CM-2/full	149.0	64K	4500	10000	N/A	4.70
CM-2/half	149.0	32K	2250	5000	N/A	2.60
CM-2/quarter	149.0	16K	1125	2500	N/A	1.60

Notes: Cycle times for IBM 3090 and CRAY X-MP correspond to the MIPS, MFLOPS, and LINPAC MFLOPS shown; current versions of these have cycle times of 17.6 and 8.5 ns, respectively. LINPAC MFLOPS for the IBM and CRAY X-MP/1 are from Table 1 of References 42 and 43. The LINPAC MFLOPS for the CRAY X-MP/4 are from Table 6 of Reference 43. See Table 23 for other CM-2 performance data. Cost data for the CRAY models were provided by Cray Research, Inc. All CM-2 models costed had 32-bit floating-point chips and a $0.2M high performance host. CM-2 cost data were provided by The Thinking Machines Corporation (TMC). The IBM cost shown is that of an IBM 3090/180E from Reference 44.

architectures is both gratifying and confusing to those with high performance computer applications. On one hand it appears that the applications can be performed faster and less expensively on these machines. Performance of the new computers is much better than the old ones in terms of MIPS, theoretical and LINPAC MFLOPS, machine cycle times, Whetstones, Dryestones, simple examples, and relative costs. Table 22 presents such comparisons for the computers addressed in this benchmark. On the other hand, new applications software and new algorithms are usually needed to achieve the inherent "compute power" of the new hardware. This is especially true for parallel computers. In most cases no simple model exists that can establish an accurate prediction of what the speed-up would be for any specific application when it is moved from one machine to another.

The most accurate way to determine the relative performance of different computers for a given application is to benchmark the application, or an analogous one, on the candidate machines of interest. Even then the scope of the benchmark needs to be adequate to ensure that reasonable comparisons are developed.

The primary objective of the orbit collision problem benchmark study is to establish "apples-to-apples" run time comparisons of several high performance computers using a relatively easy to understand technical problem. A secondary objective is to provide insight into the problem of benchmarking high performance computers having significantly different architectures.

The computers benchmarked include a single processor IBM-3090/200,

TABLE 23
CM-2 Specifications (Provided by TMC)

Typical application performance*		**Single precision floating point**	
(Fixed point)		Average (4K × 4K matrix	
General computing	2500 MIPS	multiply)	3500 MFLOPS
Terrain mapping	1000 MIPS	Dot product	10,000 MFLOPS
Document search	6000 MIPS		
		General specifications	
Interprocessor communication		Processors	65,536
(32-bit messages)		Memory	512 Mbytes
Regular pattern	250 million per sec	Memory bandwidth	300 Gbits/s
Random pattern	80 million per sec		
Sort 65,536 32-bit Keys	30 MS	**Input/output channels**	
		Number of channels	8
Variable precision fixed point		Capacity per channel	40 Mbytes/s
64-bit integer add	1500 MIPS	Maximum transfer rate	320 Mbytes/s
32-bit integer add	2500 MIPS		
16-bit integer add	3300 MIPS	**Physical dimensions**	
8-bit integer add	4000 MIPS	Size	56" × 56" × 62"
64-bit move	2000 MIPS	Weight	2,600 lbs
32-bit move	3000 MIPS		
16-bit move	3800 MIPS	**Environmental requirements**	
8-bit move	4500 MIPS	(does not include host)	
		Power dissipation	28 kW
Double precision floating point		Power input	Four 30-amp 3-phase
Average (4K × 4K matrix			110/208 V
multiply)	2500 MFLOPS	Operating	
Dot product	5000 MFLOPS	temperature	70 F 5 F
		Operating relative	
		humidity	50% 10%

* Thinking Machines Corporation believes all specifications are accurate as of the date of publication. Thinking Machines Corporation cannot, however, be responsible for inadvertent errors. Product specifications are subject to change without notice. For further detail, see the Product Specification Sheet. MIPS = millions of instructions per second; MFLOPS = millions of floating point operations per second.

CRAY X-MP and Y-MP computers with one and multiple processors and a Connection Machine 2 (CM-2) with various numbers of processors.

The benchmark technical problem is to determine the times, if any, at which certain satellites have near collisions with other satellites over a 7-day period. This problem is of more than academic interest in that there are approximately 6,000 orbiting objects larger than 10 cm across.

A number of ground rules were established to ensure that the solutions developed for the different computers are equivalent. Two general cases were benchmarked. In case 1 each computer solves the problem in a brute force manner such that the number of computations needed for the solution is approximately equal from computer to computer. In case 2 algorithms are used that attempt to fully exploit the architecture and other features of each machine

TABLE 24
Summary Results of Orbital Benchmark

Model	No. CPU	Cost $M	Case 1 Best time (sec)	Case 1 M or S	Case 1 C×P ratio[a]	Case 2 Best time (sec)	Case 2 M or S	Case 2 C×P ratio[a]	Case 2 Speed-up[b]
IBM 3090/200	1	2.20	1147	M	5.87	140	M	0.72	8
CRAY X-MP	1	5.75	68	M	1	0.67	M	0.01	100
CRAY X-MP	4	11.75	20	M	0.58	0.20	S	0.0058	100
CRAY Y-MP	1	N/A	49	M	N/A	0.50	M	N/A	100
CRAY Y-MP	8	20.00	7	M	0.36	0.07	S	0.0036	100
CM-2	8K	N/A	30	S	N/A	1.27	S	N/A	24
CM-2	16K	1.60	13	M	0.05	0.62	S	0.0025	21
CM-2	32K	2.80	8	M	0.06	0.32	M	0.0023	25
CM-2	64K	4.70	6	S	0.07	0.17	S	0.002	35

Notes: M or S—measured or scaled.

[a] C×P ratio = Cost × Performance (relative to case 1 CRAY X-MP/1 C×P)
 = (machine cost × run times) / 391, where 391 = 5.75×68).
[b] Speedup is case 2 time relative to case 1 time for same computer.

benchmarked. Both cases are required to achieve the same answers.

This report presents the study findings and conclusions and then describes the benchmark ground rules and conditions, solution methods, and quantitative results.

Findings and Conclusions

(1) For this benchmark the fastest CRAY and CM-2 run times are comparable to each other and are much faster than those achieved on the IBM.

Case 1: IBM 1,147 s vs. CRAY Y-MP/8 7 s vs. CM-2 6 s.
Case 2: IBM 140 s vs. CM-2 .17 s vs. CRAY Y-MP/8 .07 s.

See Table 24 and Figures 41 and 42 for other data.

(2) For this benchmark the CM-2 provides its high performance in a more economical manner than do any of the CRAY models. See Table 24 and Figures 41 and 42 for specific data.

(3) Programmers who are experienced on the machine of interest are required to get the best performance out of the CRAY and CM-2. Even these experts may need several iterations to achieve good results. The first attempts of novice programmers were factors of 7 and 12 slower than the fastest results

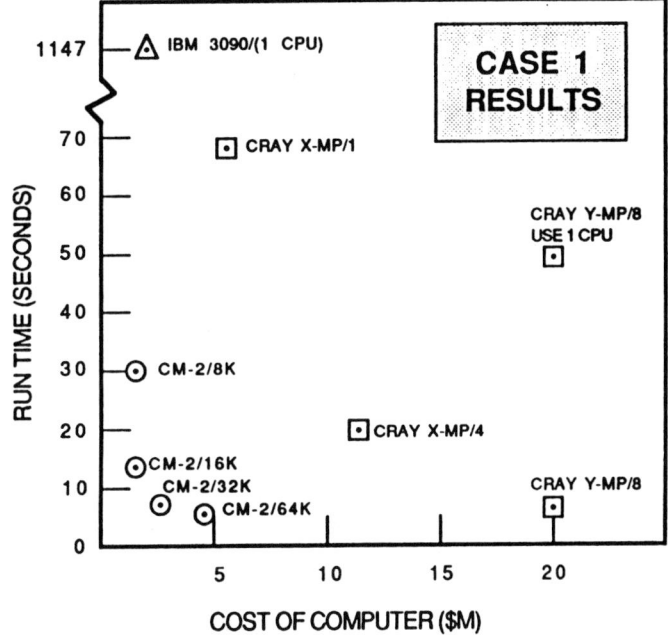

FIGURE 41. Orbit collision benchmark case 1 results.

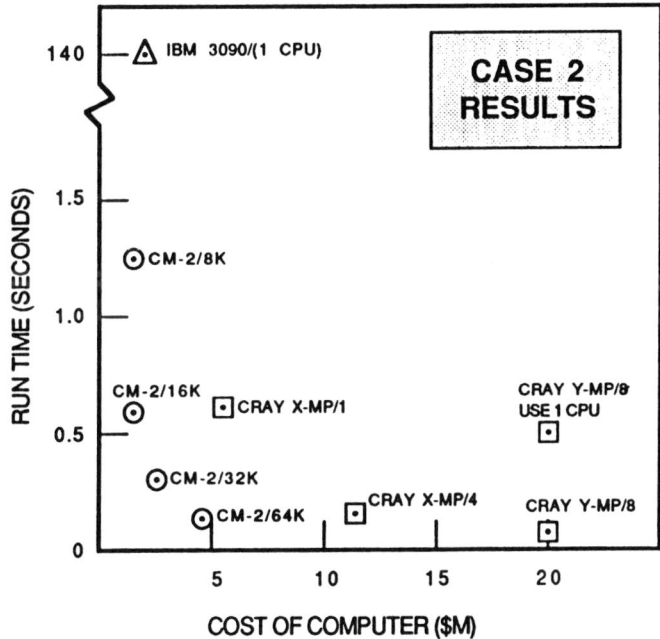

FIGURE 42. Orbit collision benchmark case 2 results.

achieved on the CM and the CRAY, respectively.

(4) Algorithms that exploit the nuances of the problem together with the architecture of the computer provided very substantial reductions in run times on all three types of computers.

> The IBM 3090 was 8 times faster for Case 2.
> All CRAY models were about 200 times faster for Case 2.
> The CM-2 models were 21 to 35 times faster for Case 2.

(5) Run time is not always inversely proportional to the number of processors applied for either the CRAY or CM-2. Benchmark runs should be used if accurate answers are needed.

Ground Rules and Conditions

General Problem — The benchmark technical problem is to determine the times during a 7-day interval at which the distance between any of 3 specified satellite orbits and 15 other satellite orbits is less than a specified magnitude. The times of closest approach (minima) when the satellites are within the specified distance are also to be determined. All times are to be determined to within 1 s. Examples of how the distance between two satellites varies with time are presented in Figure 43.

General Conditions — The solutions for both benchmark cases are constrained to the conditions listed below.

1. All orbit computations must use the NORAD SGP4 equations to single-precision (32-bit) accuracy.
2. All benchmark runs must use the same input orbit parameters, threshold separation distance (1,000 km), start date/time, end date/time, etc. Note: a 1,000 km threshold separation distance was used for the benchmark runs to ensure that several "near collisions" would occur.
3. All benchmarks must produce the same quantitative results: (1) Time at which the satellites first get within 1,000 km of each other; (2) Times of each distance minima less than 1,000 km; (3) Times at which the satellites first exceed 1,000 km separation after being that close to each other.
4. The measured run time for each benchmark is determined as wall time with outputs suppressed on a dedicated computer.

Conditions for Case 1 — The solutions for benchmark case 1 are required to compute the positions of each of the 18 satellites and the distances between each of the 45 satellite combinations of interest at least once each second over the 7-day period. This requires approximately 10.9 million satellite XYZ position computations and approximately 27.2 million separation distance computations to be made in each Case 1 run.

Conditions for Case 2 — The solutions for benchmark Case 2 are to be

PROBLEM

- FIND "NEAR COLLISIONS", IF ANY, OF 3 SPECIFIED ORBITS
  WITH 15 OTHER SPECIFIED ORBITS OVER A 7 DAY PERIOD

- DETERMINE THE FOLLOWING TO 1 SECOND ACCURACY:
  -- DISTANCE THRESHOLD (1,000 km) CROSSING TIMES
  -- DISTANCE MINIMA WITHIN THE THRESHOLD DISTANCE

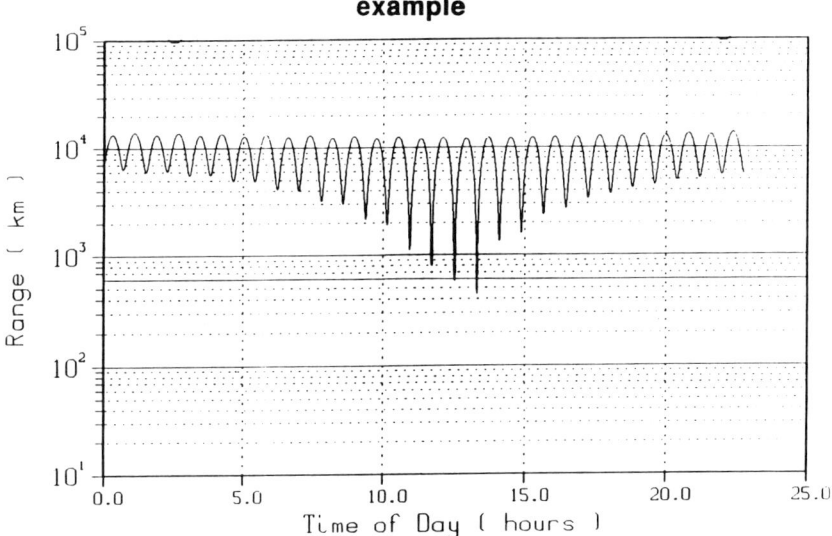

FIGURE 43. Examples of Satellite separation range vs. time of day.

based on algorithms that exploit the architecture and features of the mahcine being benchmarked.

Solution Methods

Benchmark solutions were established for an IBM 3090/200 using one processor without a vector facility; CRAY X-MP/Y-MP series computers using single and multiple processors with vector processing; and Connection Machine 2 computers having 32-bit floating-point chips using various numbers of active processors.

IBM 3090 Solution Methods — The IBM 3090 Case 1 solution consisted of a series of loops which calculated the 18 satellite XYZ positions, using an off-the-shelf FORTRAN '77 SGP4 routine, and 45 distances of interest at 1-s intervals for a week. Points of interest were determined by simple tests. The IBM Case 2 solution calculated the same values at approximately 3 min intervals. Parabolas were fit to each set of three sequential solutions and tested for occurrences of events of interest. When an event of interest was identified, parabolic interpretation and more fine-grained SGP4 calculations were used to establish the solutions meeting the required conditions.

CRAY X-MP/Y-MP Solution Methods — The FORTRAN '77 code developed for the IBM Case 1 and 2 solutions were provided to Cray Research, Inc. Cray experts used basically the same methods but refined the code for the CRAY computers, added features to split the work between multiple processors (for case 1 only), and optimized the code to take advantage of the CRAY vector/pipeline processing features.[45]

CM-2 Solutions — For Case 1 each Connection Machine processor represented 1, 2, 4, or 8 points in time, depending on the "virtualization level" selected, and computations for each satellite were performed sequentially. Eleven Case 1 runs were made over a range of virtualization levels and numbers of processors.

A two-step approach, similar to those used for the IBM and CRAY, was developed for Case 2. In Step 1, position and distance calculations were made at 3.5-min intervals. Parabolas were then fit to each successive set of three points to determine an approximate solution. The processors were then reallocated to simultaneously perform position and distance calculations at 1-s steps for appropriate satellites and intervals as established in Step 1.

REFERENCES

1. **Rao, S. M., Wilton, D. R., and Glisson, A. W.**, Electromagnetic scattering by surfaces of arbitrary shape, *IEEE Trans. Antennas Propag.*, AP-30, 409, 1982.
2. **Umashankar, K. R., Taflove, A., and Rao, S. M.**, Electromagnetic scattering by arbitrary shaped three dimensional homogeneous lossy dielectric objects, *IEEE Trans. Antennas Propag.*, AP-34, 758, 1986.
3. **Richtmyer, R. D. and Morton, K. W.**, *Difference Methods for Initial Value Problems*, 2nd ed., Interscience, New York, 1967.
4. **Peaceman, D. W. and Rachford, H. H.**, *J. Soc. Ind. Appl. Math.*, 3, 28, 1955.
5. **Heller, D.**, *SIAM Rev.*, 20, 740, 1978.
6. **Thinking Machines Corp.**, *The Essential *Lisp Manual*, 1986.
7. **Kraay, T.**, private communication.
8. **Dembo, R. S., Mulvey, J. M., and Zenios, S. A.**, Large Scale Nonlinear Network Models and their Application, Rep. EES-86-18, Civil Engineering and Operations Research, Princeton University, 1986.
9. **Chang, M. D., Engquist, M., Finkel, R., and Meyer, R. R.**, A Parallel Algorithm for Generalized Networks, Tech. Rep. No. 642, Computer Science Department, University of Wisconsin, Madison, 1987.
10. **Chen, R. J. and Meyer, R. R.**, Parallel Optimization for Traffic Assignment, manuscript, Computer Science Department, University of Wisconsin, Madison, 1987.
11. **Zenios, S. A. and Mulvey, J. M.**, Nonlinear network programming on vector supercomputers: a study on the CRAY X-MP, *Oper. Res.*, 34(5), 1986.

12. **Zenios, S. A. and Mulvey, J. M.,** Vectorization and Multitasking of Nonlinear Network Programming Algorithms, Rep. 87-03-03, Decision Sciences Department, The Wharton School, University of Pennsylvania, Philadelphia, March 1987.
13. **Phillips, A. T. and Rosen, J. B.,** A Parallel Algorithm for Solving the Linear Complementarity Problem, Computer Science Techn. Rep., University of Minnesota, Minneapolis, 1987.
14. **Mangasarian, O. L. and De Leone, R.,** Parallel Successive Overrelaxation Methods for Symmetric Linear Complementarity and Linear Programs, Mathematics Research Center Rep. 2947, University of Wisconsin, Madison, 1986.
15. **Kindervater, G. A. P. and Lenstra, J. K.,** Parallel Computing in Combinatorial Optimization, Rep. OS-R8614, Center for Mathematics and Computer Science, Amsterdam, The Netherlands, 1986.
16. **Meyer, R. R. and Zenios, S. A., Eds.,** Parallel optimization on novel computer architectures, *Annals of Operations Research,* A. C. Baltzer Scientific Publishing Co., Switzerland, 1988 (to appear).
17. **Mangasarian, O. L. and Meyers, R. R., Eds.,** Parallel Optimization, to appear as a Mathematical Programming Study, North-Holland, 1988.
18. **Chazan, D. and Miranker, W.,** Chaotic relaxation, *Linear Algebra and its Applications,* 199, 1969.
19. **Bertsekas, D. P. and El Baz, D.,** Distributed asynchronous relaxation methods for convex network flow problems, *SIAM J. Control Optimization,* 25(1), 74, 1987.
20. **Zenios, S. A. and Mulvey, J. M.,** A distributed algorithm for convex network optimization problems, *Parallel Computing,* 1987 (to appear).
21. **Bertsekas, D. P. and Eckstein, J.,** Distributed asynchronous relaxation methods for linear network flow problems, in *Proc. IFAC '87,* Munich, Germany, Pergamon Press, Oxford, England, 1987.
22. **Bertsekas, D. P.,** A unified framework for primal-dual methods in minimum cost network flow problems, *Math. Programming,* 32(2), 125, 1985.
23. **Bertsekas, D. P., Tsitsiklis, J. N., and Athans, M.,** Convergence theories of asynchronous computation: a survey, in *Stochastic Programming,* Archetti, F., Di Pilo, G., and Lucertini, M., Eds., Springer-Verlag, New York, 1986, 107.
24. **Goldberg, A. V.,** Efficient Graph Algorithms for Sequential and Parallel Computers, Ph.D. thesis, Electrical Engineering and Computer Science, Massachusetts Institute of Technology, Cambridge, MA, February 1987.
25. **Bertsekas, D. P.,** A distributed asynchronous relaxation algorithm for the assignment problem, in *24th IEEE Conf. on Decision and Control,* Ft. Lauderdale, FL, December, 1985.
26. **Bertsekas, D. P. and Tseng, P.,** Relaxation Methods of Minimum Cost Network Flow Problems, MIT Rep. LIDS-P-1339, October 1983.
27. **Stern, T. E.,** A class of decentralized routing algorithms using relaxation, *IEEE Trans. Commun.,* COM-25(10), October 1977.
28. **Ohuchi, A. and Kaji, I.,** Lagrangian dual coordinatewise maximization algorithm for network transportation problems with quadratic costs, *Networks,* 14, 515, 1984.
29. **Zenios, S. A. and Mulvey, J. M.,** Relaxation techniques for strictly convex network problems, in *Annals of Operations Research,* Vol. 5 (special volume on algorithms and software for optimization), Monma, C. L., Ed., A. C. Baltzer Scientific Publishing Co., Switzerland, 1986.
30. **Bertsekas, D. P., Hossein, P., and Tseng, P.,** Relaxation methods for network flow problems with convex arc costs, *SIAM J. Control Optimization,* 1987 (to appear).
31. **Klingman, D., Napier, A., and Stutz, J.,** NETGEN — a program for generating large-scale (un)capacitated assignments, transportation, and minimum cost flow network problems, *Manage. Sci.,* 20, 814, 1974.

32. **Ahlfeld, D. P., Dembo, R. S., Mulvey, J. M., and Zenios, S. A.,** Nonlinear programming on generalized networks, ACM *Trans. Math. Software,* December 1987 (to appear).
33. **Horowitz, E. and Sahni, S.,** *Fundamentals of Computer Algorithms,* Computer Science Press, Potomac, MD, 1978.
34. **Gary, M. and Johnson, D.,** *Computers and Intractability: A Guide to the Theory of NP-Completeness,* W. H. Freeman, San Francisco, 1979.
35. **Sahni, S.,** Approximate algorithms for the 0/1 Knapsack problem, *J. ACM,* 22(1), 115, 1975.
36. **Blelloch, G.,** Scans as primitive parallel operations, in *Proc. 1987 Int. Conf. Parallel Processing,* August 1987, 355.
37. **Mitchell, O. R., Myers, C. R., and Boyne, W.,** A MAX-MIN measure for image texture analysis, *IEEE Trans. Comp.,* C-26, 407, 1977.
38. **Pratt, W. K.,** *Digital Image Processing,* John Wiley & Sons, New York, 1978, 487.
39. **Thinking Machines Corp.,** Introduction to Data Level Parallelism, Tech. Rep. 86.14, April 1986.
40. **Hopfield, J. J.,** Neural networks and physical systems with emergent collective computational abilities, *Proc. Natl. Acad. Sci. U.S.A.,* 79, 2554, 1982.
41. **Blelloch, G. and Rosenberg, C.,** Network Learning on the Connection Machine, MIT Tech. Rep., 1987.
42. **Dongarra, J. J.,** Performance of Various Computers Using Standard Linear Equations Software in a FORTRAN Environment, Argonne National Laboratory, July 26, 1986.
43. **Dongarra, J. J.,** Performance of Various Computers Using Standard Linear Equations Software in a FORTRAN Environment, Argonne National Laboratory, November 29, 1987.
44. *Datapro 70, the EDP Buyer's Guide,* Vol. 1 — Computers, McGraw-Hill, New York, 1988.
45. **Williams, M. G.,** Orbit Benchmark, Cray Research, Inc., July 26, 1988.

8 RESEARCH SIMD COMPUTERS

THE GAM II AND THE CLIP4 RESEARCH SIMD COMPUTERS ARE PRESENTED AS EXAMPLES OF UNIVERSITY SIMD DEVELOPMENT EFFORTS

8.1 THE GAM II PYRAMID

The GAM II Pyramid is a hierarchical structure with a total of 1365 processing elements. The sequencer designed for controlling this structure is capable of executing pyramid oriented primitives similar to an add and subtract operations, as well as independent procedures similar to convolution. The pyramid contains three control buses embedded in the hardware in anticipation of a future control system that will contain three independent control units.

8.1.1. Introduction

The GAM II Pyramid is a six level pyramid, which is a one level expansion of the previous GAM I Pyramid, used for image processing applications. Both systems were developed in George Mason University's Advanced Computer Architecture Laboratory. The GAM II Pyramid, Figure 44, is a hierarchical Single Instruction Multiple Data (SIMD) system.

8.1.2. The Pyramid Structure

The GAM II Pyramid contains 1365 processing elements, made up of 172 custom microcircuits designed for the Massively Parallel Processor. The pyramid structure is broken down into the following sections: the basic processing element organization, the daughter cards that contain the processing elements, the back plane that contains the daughter cards and the processing element adder network.

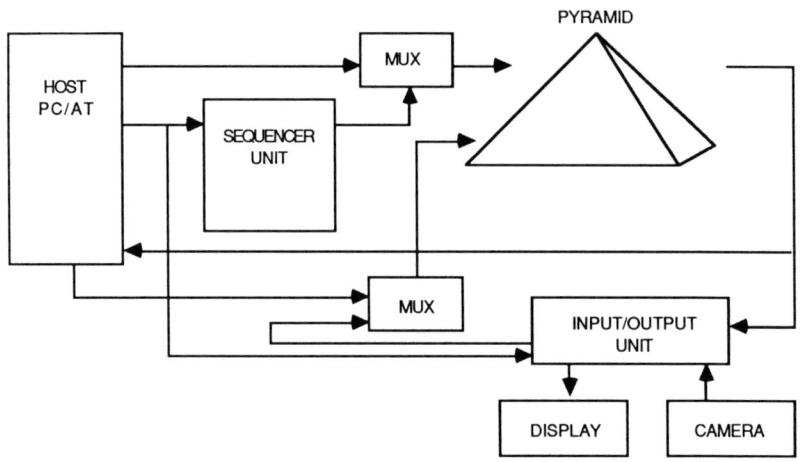

FIGURE 44. The GAM II Pyramid System.

8.1.3. The Processing Element

The GAM II Pyramid's processing elements are identical to the Massively Parallel Processor's, with an expanded inter level communication network. Each processing element is connected in a quad tree architecture that is six levels deep. A level is an $N \times N$ square mesh whose edges on the bottom three levels are connected in a toroidal topology and on the top three levels connected to a logical zero. A processing element can communicate to four siblings (those to the North, South, East and West), four children (V, X, Y and Z) and one parent. A Sum-OR circuit on each level signals if any processing element has a value "one".

8.1.4. The Daughter Card

The six levels are built by using 45 identical daughter boards developed at the Advanced Computer Architecture Laboratory. Each GAM II Daughter board has four MPP microchips, and thus has 32 processing elements configured in an 8×4 array slice. There are four Static Random Access memory (SRAM) microchips configured in an $8K \times 8$. Eight three-state switch microchips are used to perform level transfers, two per MPP microchip. The three-state switches are attached to every processing element bus, and are enabled whenever a level transfer cycle is performed. Once the switches are enabled, the sending level performs a memory write to the three-state switches and the receiving level performs a memory read from the three-state switches.

8.1.5. The Back Plane

The back plane of the GAM II Pyramid contains three identical control buses and a communications network that is unique to each card connector.

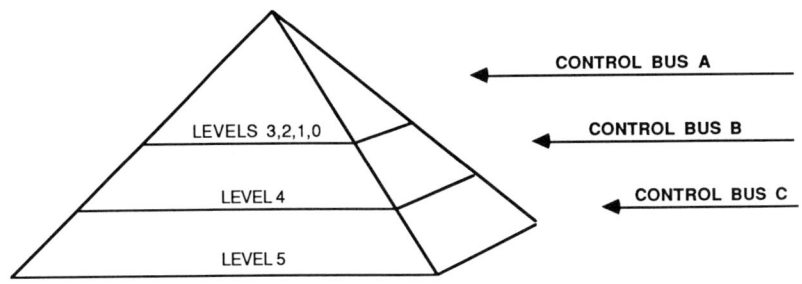

FIGURE 45. Control buses vs. pyramid levels in GAM II.

The inter-daughter card communication determines the logical location of that particular set of processing elements. The back plane is an active back plane that performs signal buffering and some logical operations on the control signals.

One of the three control buses controls the base level, another controls the level above the base and the last controls the top four levels, Figure 45. Though currently there is only one controller, the pyramid can handle up to three controllers with each issuing a unique instruction at the particular levels. Each control bus is buffered through a set of latches that could be configured to be transparent for debugging purposes or as part of an instruction pipeline to facilitate higher clock speeds. Discrete logic, on the back plane, has been added that allows for level disabling when not in use and generates a dummy memory read instruction whenever a level transfer operation is in progress.

8.1.6. The Adder Network

The Adder Network is a collection of high density 64K × 8 EPROM and binary adders. It is used to produce a sum of 256 1-bit inputs. The EPROM is programmed to produce the 5-bit sum of the 16-input address bits where each can have a value "one". There are 16 EPROMs that each sum 16 bits for a total of 256 input bits and produce 16 partial sums. The 16 partial sums are then added together in pairs by 8 binary adders to produce 8 partial sums. The process is repeated until all the partial sums are totaled and one number remains. The addition is carried out in parallel and the total delay is 400 ns which is well within one cycle of the 500 ns targeted cycle time of the GAM II Pyramid clock.

8.1.7. The Sequencer

The sequencer is the lowest instruction level interface to the GAM II Pyramid arrays, Figure 46. Remote calls are issued to the sequencer to execute a primitive on the pyramid structure similar to an addition or multiplication, or to execute a full procedure that manipulates scalar data as well as pyramid data. The Sequencer Unit is capable of issuing instructions to the pyramid at a rate

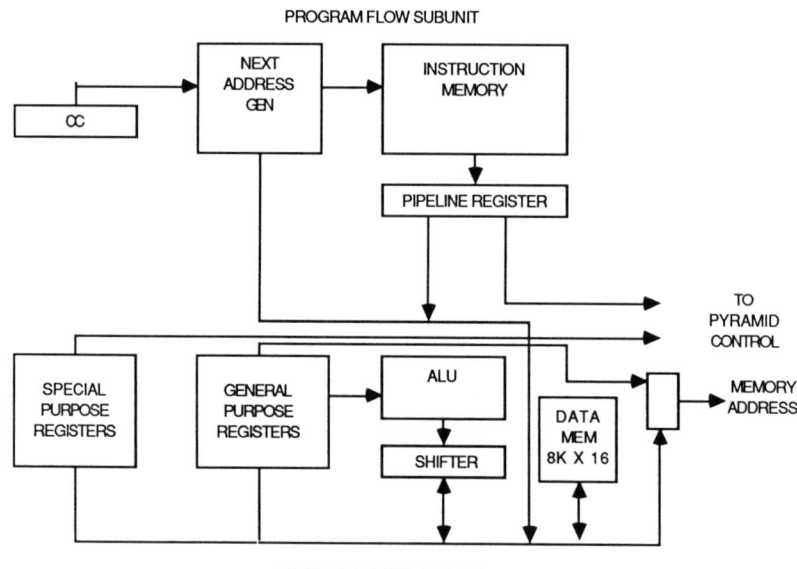

FIGURE 46. Sequencer unit.

of 2 MHz. The function of the sequencer is to handle program flow control, global scalar calculation and pyramid instruction generation. The sequencer hardware is partitioned into two portions: the Program Flow Sub-Unit and the Data Execution Sub-Unit. The Program Flow Sub-Unit handles the sequencing of program instructions. The Data Execution Sub-Unit handles scalar data computation and storage. Pyramid instructions are generated in coalition with the two Sub-Units.

8.1.8. The Program Flow Sub-Unit

This portion of the sequencer contains the next address generator, the Micro Memory, the Pipeline Register and the Condition Code register.

8.1.9. The Next Address Generator

The Next Address Generator is based on Advanced Micro Device's 2930, which is a bit-sliced program flow control microchip. The Next Address Generator is composed of four such microchips connected to form a 16-bit address bus of which only 13 bits are used. There is an adder module as well as four input sources: an instruction pointer, an auxiliary register, an external data bus and a 17 register deep stack. The microchip can perform its full 32 instructions on the four input sources to produce a straight-through address or an offset branch address.

8.1.10. Micro Memory

The Micro Memory is composed of a configurable Static Random Access Memory. The memory is built using 8K by 8-bit memory modules with a 120 ns access and constructed to be addressed as an 8K by 96-bit memory or as a 48K by 16-bit memory. The 96-bit wide memory is used when executing primitives or procedures, during which the memory is in a read only mode. However, during host transfer operations, for example program load time or when the system debugger is operational, the memory is configured as a 16-bit data bus with read and write capabilities.

8.1.11. Pipeline Register

The Pipeline Register is a 96-bit register that holds a sequencer instruction. An instruction has four major fields. The first is a K constant which is used as a branch address or as a constant scalar for the Data Execution Sub-Unit. The second field of the Pipeline Register is used to control the Data Execution Sub-Unit. The third field is used for pyramid control.

8.1.12. Condition Code Register

The Condition Code (CC) register is a set of 16 flags that are used by the sequencer to perform branching functions. The CC register contains the basic scalar flags generated by the Data Executive Sub-Unit as well as flags that are used to handle communications between the host and the sequencer. The Sum-OR values from all the levels of the pyramid are also latched into the CC register.

8.1.13. The Data Executive Sub-Unit

The Data Executive Sub-Unit is composed of three basic parts. The General Purpose Registers and their accumulator, the Special Purpose Registers and the Data Memory.

8.1.14. The General Purpose Registers

There are sixteen 16-bit registers that are used for general purpose data calculation. These registers are attached to an Arithmetic Logic Unit and a Shifter. They can be used to store results of addition, subtraction and logical operations on scalars. An extremely flexible function of these registers is the generation of pyramid addresses. The registers can contain the address along with offsets that could be loaded from the K constant or the host, an addition or multiplication can take place and a new address will be generated. This same address can be used in a post-increment or a pre-decrement mode to support sequential address traversing for multiple-bit pyramid data.

8.1.15. Special Purpose Registers

Sixteen 8-bit registers are used for level and child select generation. These registers are triple ported registers that can output the level mask pattern and the child select values simultaneously. Any of these register values can be enabled to the Sequencer Data Bus.

8.1.16. Data Memory

This is a bank of 8K by 16-bit of Static Random Access Memory connected to the sequencer's data bus. The address of the memory is latched in a Data Memory Address Register (DMAR) from the data bus. Data is then read or written to the memory also from the data bus. This memory is used whenever a procedure runs out of register memory. This allows a compiled procedure, from a high level language, to contain large data structures.

8.1.17. Array Instruction Generation

Array instructions are generated by the different portions of the sequencer. The basic processing element operations are issued from the pipeline register. Array memory source and destination addresses are stored in the Data Memory and the General Purpose Registers where their ALU is used to compute relative offsets, increment and decrement operations. Level masking information as well as an alternative set of child enable signals are maintained in the Special Purpose Registers.

8.1.18. The Input and Output Unit

This unit is used to digitize analog input signals from a video camera and to generate analog video signals to display images from the pyramid, Figure 47. An image has a 128 × 128 pixel resolution and 6 bits of gray shades. The unit has 128 kbytes of image memory that is capable of storing up to 8 images. An image is stored in sixteen 32 by 32 pixel frames that are shifted in and out of the pyramid by using the S-Registers on the base level of the pyramid (Level 5). The image memory stores data in bit planes and can communicate to the host, camera and the display device through a corner turning interface of shift registers. If a byte is needed from the image memory, a block of 8 bytes is written to the corner turning block that in turn will be decoded for the appropriate byte.

8.1.19. The Host System

The Host System is an IBM AT compatible system that operates under DOS 3.3. The Host has a 50 Mbyte hard disk. Text is displayed on a Monochrome display terminal and pyramid array graphics are displayed with 256 colors on a Video Gate Array (VGA) display. The Host interfaces to the pyramid by using a set of 32 8-bit registers. The Host can control the Pyramid Arrays, the Sequencer Unit and the Input and Output Unit.

Since an AT clone is used as a front end to the pyramid, PC software is available for program system development. Pyramid system software was developed using C and PASCAL, some of which include: an interactive micro assembly language with a simulated micro sequencer (PYRASM), an interpre-

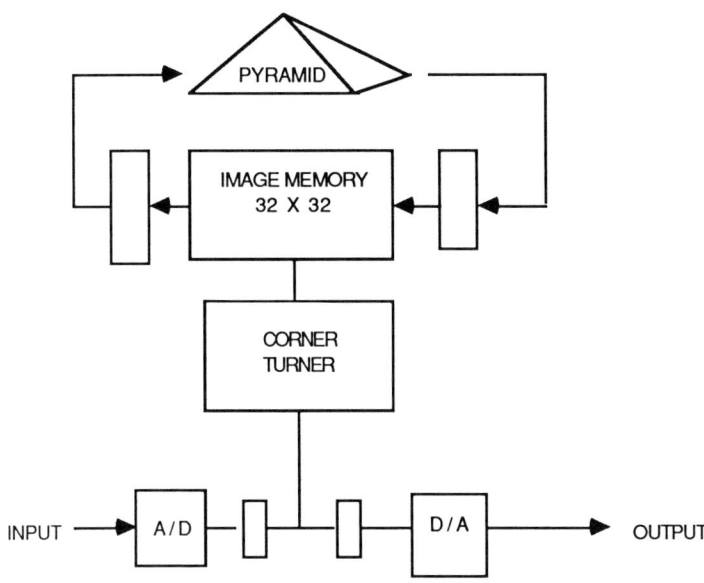

FIGURE 47. Input and output unit.

tive high level language called Function IV and a compiler that generates microcode from a high level language similar to C.

The Host was also extremely valuable in debugging the pyramid, since through software it can single step any micro cycle. A state can be stopped and restarted as though it were running continuously. Special software has also been developed; S_Bug for example, is a debugger for the Sequencer Unit.

8.1.20. Future Control System Expansion

Connecting up to three sequencer type controllers is being investigated for future expansion of the GAM II Pyramid. If a single controller is used, it will be able to clock only one level at a time, the remaining unclocked levels will be idle. In such a situation, for every apex clock there are 1364 processing elements that are idle. A system being investigated contains multiple controllers that can be dynamically attached to any of the three control buses, Figure 48. All array operations are memory to memory reference operations. If a controller is allowed to attach to a level and not be interrupted until it is done with that level, no context switching is needed. The data memory locations are pre-allocated at compile time, a controller will attach to a level and read its own data memory area, update the data with the desired operations and restore the updated data back to memory. Another controller can attach itself to that level at that time.

The controllers will be capable of running complete procedures and thus

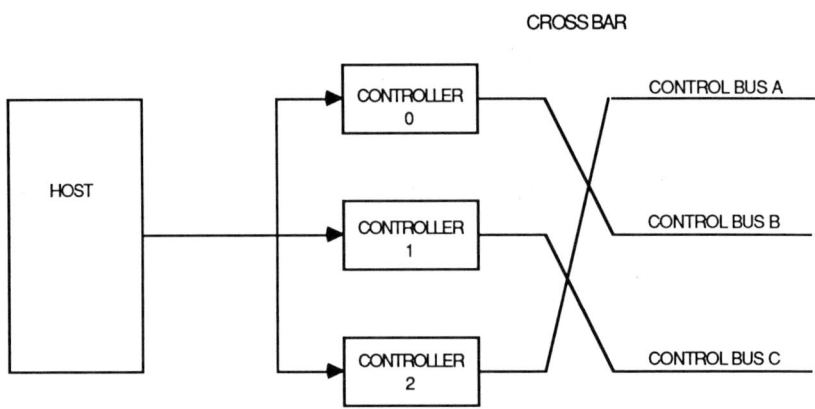

FIGURE 48. Multiple SIMD controller configuration.

enabling the parallelization to occur at the high level language level. Most of the program will be executing on the Host system. Parallelism is achieved by the use of Fork and Join operators that control the asynchronous execution of the procedures.

A controller can produce multiple level requests per procedure. This feature is useful to facilitate data transfers between levels and alleviate the problem of deadlock. Any controller that wishes to perform a level transfer will request two levels only when all its other computations are complete and it has already released its current level. The next step is to request the two levels the controller needs. Should there be a conflict that results in deadlock, the controller attached to the level with the most processing elements gets its request fulfilled while the other controller waits until the two are available.

8.2 CLIP4

8.2.1. Introduction

Pattern processing is a comparatively new science but an extremely old practice. It is probably true to say that most living creatures owe their survival very largely to their ability to process patterns, particularly patterns revealing danger, food, or a potential mate.

Today, the requirements for pattern processing are less basic but no less challenging. It is now machines, as well as animals, that need a pattern processing capability. In this section, the philosophy behind the CLIP development program is discussed and a technical description of the latest array in the CLIP series, CLIP4, will be presented. The CLIP program has been concerned primarily with image processing and pattern recognition in images, i.e., pictures. There would appear to be no general agreement as to what any of these terms mean precisely; however, it can be assumed that the purpose of the CLIP series of machines has been either to extract information from pictures or to create new pictures from the original inputs. Although in some cases a CLIP array might prove effective in processing other types of patterns (speech, seismograph records, railway timetables, "structures"), the principal aim has been to optimize the array architecture for picture processing. This account will therefore treat CLIP4 as a pattern processor specialized for image processing and the term "pattern" when used should be taken to imply "picture".

Finally, it should be clearly understood that an attempt has been made in the CLIP program to design a fast image processor which could be constructed cheaply within the financial range of, say, a university or hospital research laboratory. The "scale" of the system is intended to be comparable with that of the larger minicomputers. Inevitably, this has led to some degree of compromise. Notwithstanding this, it has been suggested that the concessions made to cheapness have not caused an unacceptable degradation of the performance of the CLIP machines, nor have they led to any particularly annoying design features. One can look to the ever-decreasing cost of semiconductor device manufacture to recover in the foreseeable future some of the speed which has been sacrificed in the fight to keep down costs.

8.2.2. Array Processing

CLIP is a Cellular Logic Image Processor. The principle of array processing of images is as follows: the image is assumed to be adequately reproduced by means of an n by n array of square picture elements (pixels) whose intensities are chosen from a range of L discrete values which are uniformly spaced over a selected brightness range. Obviously, both n and L must be large enough to give acceptable spatial and gray-tone resolutions for the task to be performed on the image. A crude test which can be applied to determine whether the reproduction is "adequate" is to construct the digitized image, using a suitable display device, and then to see whether the required pattern recognition task can be performed by eye. Alternatively, if the required processing is some form of image enhancement, then the digitized image should not appear inferior in quality to the original image. Neither test is very satisfactory; in particular, the superior pattern recognition ability of the eye permits quite severe degradation of an image before recognition is actually prevented, whereas a computer-based system might be caused to fail completely. It would be useful to be able to apply strictly quantitative tests for digitizing sufficiency but such tests have not yet been devised.

An array of processors is assembled in which there is a one-to-one correspondence between pixels and processors. The intensity in each pixel in input into the corresponding processor, each pixel being represented by $g = \log_2 L$ binary bits. Storage for pixels and for other data, including output data, is

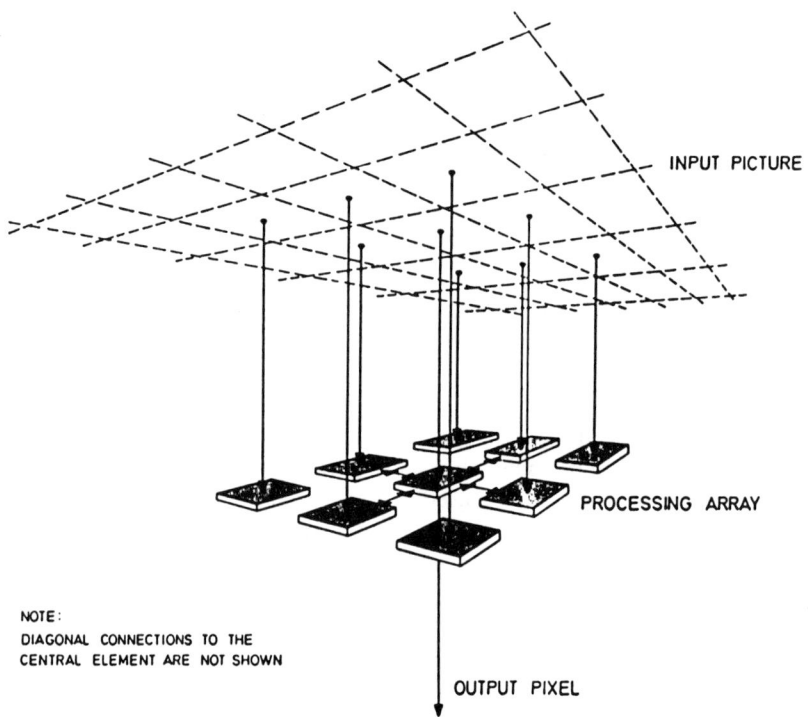

FIGURE 49. Connections to one processor in a cellular logic array.

provided within the array. Corresponding bits in each processor are given the same address in the array. Thus, a 6-bit image might be stored in an array in addresses D_1 to D_6 so that the least significant bit in every pixel is to be found in address D_1 in every processor. Each bit at a given address in any one processor can be regarded as an element of a bit plane extending over the whole array; it is the bit plane to which an address such as D_1 in the above example applies.

Pattern processing is concerned with the structure of images so that it is clearly insufficient for the array operations to be confined to point functions in which the output from any processor is a function only of the pixel intensity input to the processor. There must be some mechanism by which each processor can inspect pixels in other parts of the array. The simplest way to achieve this is to form connections between every processor and its immediate eight neighboring processors. This scheme is illustrated in Figure 49. Unfortunately, the cost of building an array will depend critically on the complexity of the interconnection structure so that it must be reduced to the absolute minimum consistent with performance requirements. In particular, from the point of view of wiring costs, it is extremely relevant to determine whether it is necessary to provide eight incoming connections as well as eight outgoing connections for every processor. On the one hand, an equally important decision, which will strongly influence the processor design, concerns the desirability or otherwise of propagating multiple bit (as opposed to single bit) information between neighboring processors.

In arrays of the CLIP type, an array instruction is broadcast to every processor in the array and is obeyed simultaneously by every processor, operating on data loaded from its internal storage into suitable input buffers. Thus, if two arrays of numbers A_Q and A_R are stored in addresses S_Q and S_R (each representing a stack of bit planes), then a typical array instruction might be such as to perform the calculation

$$A_Z = A_Q \cdot A_R \tag{65}$$

and to store the resulting array of numbers in the bit-plane stack address S_Z. The multiplication would probably be performed element by element so that

$$A_Z(x,y) = A_Q(x,y) \cdot A_R(x,y) \tag{66}$$

for all $1 \leq x \leq n$, $1 \leq y \leq n$.

If this calculation is only one example of many that the array must be able to perform, indeed, if the system is expected to be able to behave as an array of general purpose digital computers, then two points are immediately apparent: (1) each processor will be expensive, and (2) control of the processors will involve very sophisticated control and data buses. In practice, despite the low price of currently available microprocessors, the cost of constructing this sort of array would be prohibitive. It is also far from clear that an array with this architecture would be particularly efficient. A further point to be taken into consideration is that much useful pattern processing is carried out using binary images, i.e., black and white images representable as a single bit plane containing only 1s and 0s. With images of this type, most of the computational power of the array processors would be unused. But complex arithmetic operations can be synthesized as strings of single bit Boolean functions. It therefore is worthwhile to explore what might be achieved by reducing the processor to the extremely simple form of a single-bit Boolean processor.

8.2.3. Processor Design: General Considerations

Even with an array of simple Boolean processors, it is desirable to restrict severely the complexity of the processor design. Suppose an array is composed of nine-input, nine-output Boolean processors, as shown in Figure 50. If every output is an independent Boolean function of the nine inputs, then each processor must be able to perform some 10^{1387} functions. Even to select any one of these functions would require more than 4600 control lines to each processor. Generally, a processor with D input lines and E output lines could, in principle, perform a maximum number of transformations given by

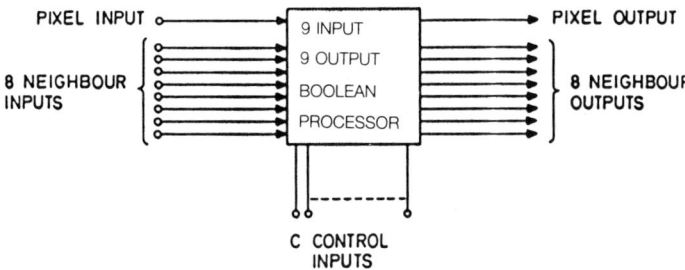

FIGURE 50. Specification for a general Boolean processor in an array.

$$T = (2^E)^{(2^D)} \qquad (67)$$

and requires $C = \log_2 T$ control inputs. Clearly, it is very important to limit drastically the number of input states (2^D) to the processor in order to keep T, the number of possible transformations, manageably low. In fact, since the internal logic circuits of the processor are also very complex when T is large, it is found that a two-input, two-output processor ($D = E = 2$, $T = 256$, $C = 8$) is getting near to the limit of what is practical for arrays of useful dimensions for image processing. The problem is then to decide how to make best use of the available inputs and outputs. If one input and one output are used for the pixel data input and output, respectively, then the two remaining connections must be responsible for all transmission of data between neighboring processors. Assuming that the incoming data must be relatable to the location of the neighbor outputting the data, some form of gating must be provided in the interconnection lines. In CLIP4, the gates are associated with the inputs and the processor specification is as shown in Figure 51. A represents a single-bit input buffer and D is the single-bit output which can be directed to a selected address of the in-array storage (referred to as the "D-memories"). Having gated each interconnection input, the inputs are fed into an OR-gate, thereby making immediately available OR-ing operations on subsets of the neighbor outputs.

Although the original concept of a nine-input, nine-output processor has been vastly reduced in the CLIP4 design, it is nevertheless still possible to construct any of the 10^{1387} functions by a sequence of operations on interconnection inputs gated one by one from each neighbor in turn, making use of the array D-memories to store intermediate results. In practice, the majority of the possible functions are of little, if any, value. The more valuable functions are often achievable by quite short sequences of operations, even single operations.

8.2.4. The CLIP4 Processor

The research and development program leading to the design of CLIP4 has been stimulated by work in many other laboratories, some of which is reported

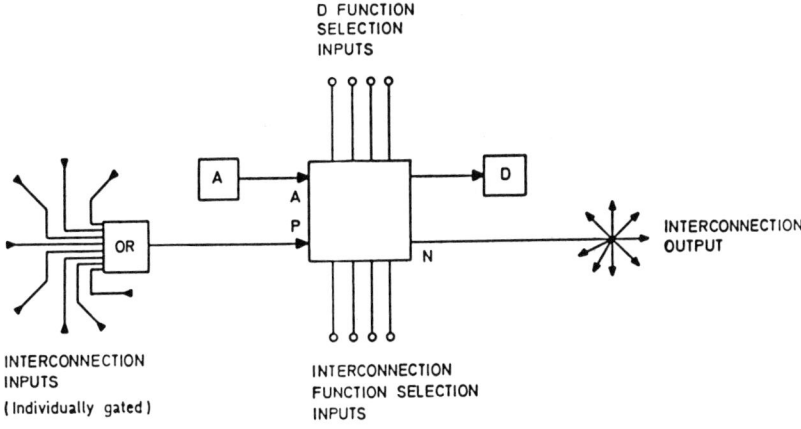

FIGURE 51. Interconnections for the CLIP4 processor.

TABLE 25
University College London Arrays Leading to the CLIP4

Array	Completion date	Array size	Technology
UCPR1	1967	20 × 20	Transistors, resistors
DIODE array	1969	5 × 5	Diodes, resistors, mechanical switches, etc.
CLIP1	1971	10 × 10	Small scale IC (TTL)
CLIP2	1972	16 × 12	Small scale IC (TTL)
CLIP3	1973	16 × 12	Small scale IC (TTL)
CLIP3 (scanned)	1974	96 × 96	Small & medium scale IC (MOS and TTL)
CLIP4	1980	96 × 96	LSIC (NMOS — Custom designed)

elsewhere. At University College London, the program has been based around a sequence of design studies, each of which resulted in the construction of a small pilot array. Table 25 summarizes some features of the arrays.

The broad principles governing array design have already been discussed. In this section, the actual design of CLIP4 will be examined in detail. A convenient way of approaching this topic is to take commonly occurring pattern processing tasks and to show how the various parts of the logic circuitry in the CLIP4 processor handle each task.

8.2.4.1. Boolean Functions of Two Binary Images

If there are two binary (black and white) images I_1 and I_2 stored in two bit planes in the D-memories of the array, then it is often required to produce and store a third binary image I_3 which is a cell-by-cell Boolean function of the two

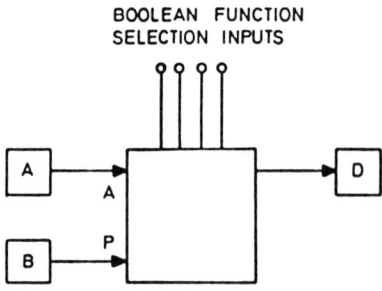

FIGURE 52. Circuit for Boolean functions of two binary images.

input images. Using symbols (., +, @) to represent (AND, OR, EXCLUSIVE OR), we may wish, for example, to compute:

$$I_3 = I_1 @ I_2 \qquad (68)$$

where $I_3(x,y) = I_1(x,y) @ I_2(x,y)$ for all $1 \leq x \leq n$, $1 \leq y \leq n$. Figure 52 shows the relevant part of the processor for this and all similar operations. The image I_1 is loaded into an input buffer A by an array instruction LDA 1 (assuming that I_1 is stored in D_1, I_2 in D_2 and I_3 will be stored in D_3). Similarly, the instruction LDB 2 loads B with I_2. Note that the output from the B buffer is labeled P in the figure. In this simplified section of the circuit P and B are identical; when more of the circuit is involved, the output from B is combined with other variables so that B and P are no longer the same quantity.

The required Boolean function is selected by putting an appropriate code on the four control inputs. As only 16 Boolean functions of 2 variables exist, 4 control inputs provide the necessary control capability. The instruction determining the internal logic function of the processor is SET P @ A. Once the input binary patterns have been loaded and the function selected, the final "process and store" instruction initiates the function and, after the correct interval, loads the output pattern D into the D-memory address specified in the instruction. The complete sequence, usually described as "CLIP operation" is therefore:

```
SET P @ A
LDA 1
LDB 2
PST 3
```

For subsequent operations, it is not necessary to repeat unchanged instructions. For example, if the next operation is to perform the same Boolean operation on I_3 and a new pattern I_4 (in D_4), again storing the results in D_3, then the next instruction sequence would be:

```
            LDA  3
            LDB  4
            PST  3
```

But if I_3 and I_4 (I_3 being the result of the first exclusive OR operation) are then to be recombined to give an ANDed image in D_5, the following sequence would be used:

```
            SET  P.A
            PST  5
```

The programs are written in a language devised for CLIP4, called CAP4.

8.2.4.2. Shifting Binary Images

Binary images can very simply be shifted by one or more array positions in any of the eight array directions. In fact, the array has been constructed so as to permit reconfiguration as a hexagonal array. The inclusion of the symbol H at the end of any SET instruction effects this change. The possibility, therefore, of shifting the image in four more directions also exists. Shift directions are specified by enabling appropriate gates in the interconnection inputs. Figure 53 shows how the directions are labeled. It should be particularly noted that the specified directions point into the cell, not away from them. Figure 51 gives the relevant part of the CLIP4 logic circuit both for shifting binary images and for the local neighborhood operations described in the next section. Since shifting only involves one binary image, the B buffer is no longer connected into the circuit. The P input to the processor is formed by the output of the OR-gate summing the enabled interconnection inputs.

Suppose it is required to shift the image upwards by one array row. The direction is 6 and this is specified in the SET instruction. Conventionally, white cells in the picture are represented as 1's, black as 0's. If the binary image is loaded into A from its storage location D_1, then a Boolean function is required which will produce an interconnection output N which is 1 whenever A is 1, i.e., N = A. By selecting a second function D = P, the output pattern will be composed of 1-cells wherever a cell in the row beneath in the input image is a 1. Thus the image is shifted vertically. The instruction sequence for this operation is

```
            SET  P, [6] A
            LDA  1
            PST  1
```

An option exists for specifying what happens at the edges of the array. Since one row will shift in from below the array, this row can be chosen to be all 1's or all 0's by writing or not writing, respectively, an E next to the direction label.

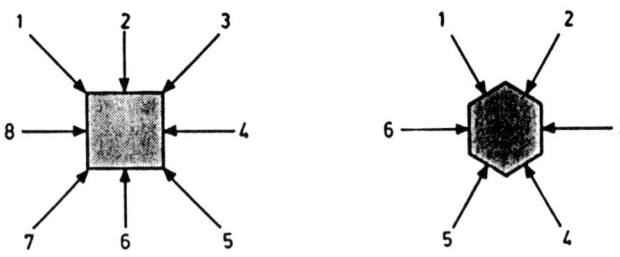

FIGURE 53. Interconnection directions in CLIP4 arrays.

The exact effect of inclusion of an E in the instruction is to connect to 1 all interconnection inputs which "overflow" the array.

Similar to shifting is the process of projecting, involving propagation. If the Boolean function for N is replaced by (P + A), then an interconnection signal is not only generated by 1-cells, but is also passed on to all cells lying in the same direction. In the above example, every cell vertically above (i.e., displaced in direction 6) a 1-cell receives a propagation signal from its neighbor below. It is as though the 1-cells cast a shadow for illumination in direction 6. Note in passing that a delay is required after the initiation of the PST instruction to allow time for propagation to cease before clocking the D output into store.

8.2.4.3. Local Neighborhood Operations

These are central to the effective use of parallel processing images. Propagation is always involved but may be confined to the immediate 3 by 3 neighborhood or may proceed further afield. The principle for immediate neighborhood functions is as follows: consider a particular element in a bit-plane having neighbors:

$$(a_i...a_j,...a_k,...a_8)$$

The processor output N will be determined by a selected Boolean function to be either A or -A (using the minus symbol to indicate NOT A). Certain interconnection gates will be enabled so that:

$$P = OR(a_j,...a_k)$$
$$\text{or else } P = OR(-a_j,...-a_k) \tag{69}$$

The second expression can be rewritten as

$$P = -AND(a_j,...a_k) \tag{70}$$

Next, the second Boolean function combines A and P to compute D, so that:

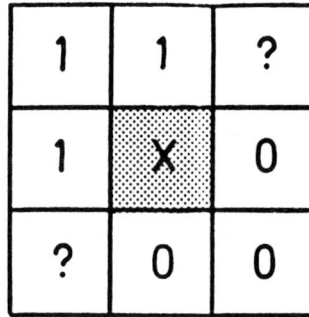

FIGURE 54. A defined local neighborhood.

$$D = -F_D[OR(a_j,....a_k), A] \qquad (71)$$

or

$$D = -F_D[AND(a_j,....a_k), A] \qquad (72)$$

noting that the inversion before the AND in the second expression can be omitted as F_D is a perfectly general Boolean function of the two variables P and A.

Two extremely important local neighborhood operations are EXPAND and SHRINK, the terms usually being applied to white objects on a black background. EXPAND surrounds every 1-cell by a 3 by 3 neighborhood of 1-cells; SHRINK surrounds every 0-cell by a 3 by 3 neighborhood of 0-cells. After loading A with the input binary image, the two SET instructions for these operations are

```
EXPAND:  SET  P + A, [1-8]  A
SHRINK:  SET   -P.A, [1-8] -A
```

A short sequence of three CLIP operations can be used to find cells with any specified 3 by 3 surround, such as that shown in Figure 54. If the input binary pattern is D_1 and the cells meeting the surround specification are to be represented as 1-cells in D_2, the sequence is

```
SET -P.A, [1 2 8 E] -A
LDA 1
PST 2
SET -P.A, [4 5 6 E]  A
PST 3
SET  P.A.
LDA 2
```

LDB 3
PST 2

The first operation loads D_2 with 1-cells which do not have 0-cell neighbors in the top left-hand group; the second operation loads D_3 with 1-cells which do not have 1-cell neighbors in the bottom right-hand group; the third operation finds cells satisfying both conditions. The Es are included to eliminate cells on the borders of the array since these presumably cannot be regarded as having the defined neighborhood.

Propagation which carries information beyond the 3 by 3 neighborhoods can be particularly useful as a means of "sorting out" binary images containing more than one connected object. Black objects touching the array edges can be removed by propagating from the E connections through 0-cells and outputting as 1-cells in the original image and also all 0-cells on the propagation path. The instruction used is

$$SET\ P + a, [1-8\ E]\ P. - A$$

8.2.4.4. Labeled Propagation Operations

As an extension of the ideas discussed in the last paragraph, a facility has been included in the CLIP4 processor to enable propagation to be initiated at any selected element, or group of elements, in the array. Figure 55 shows how the B buffer can be connected into the propagation path by means of an additional gated input to the interconnection OR-gate. This input is enabled by writing a B in the square brackets with the direction list, etc.; the enabling is implied when pure Boolean operations are executed.

8.2.4.5. Arithmetic Operations

Much of pattern processing is connected with arithmetic operations on pixel values in various neighborhoods surrounding each cell. Such operations are often referred to as "gray-level" processes as opposed to "binary" processes. The earlier CLIP arrays were seen as processors for binary images and were effective over a range of tasks which did not involve any numerical calculation. Even so, the 1-bit Boolean processor lends itself well to arithmetic operations and a few simple additions to the circuit can greatly improve its efficiency by providing a fast automatic carry facility.

Before considering the details of the arithmetic operations in the CLIP4 cell, it would be helpful to examine the organization of the data storage in the n by n cell array (see Figure 56). Each address in the D memories can be visualized as a plane of single bit stores, d. If the location of each cell is given by its (x,y) coordinate in the array, then a bit-plane will be the set of bit stores defined by:

$$D_j = [d_{jxy} : x, y = 1, 2, \ldots n]$$

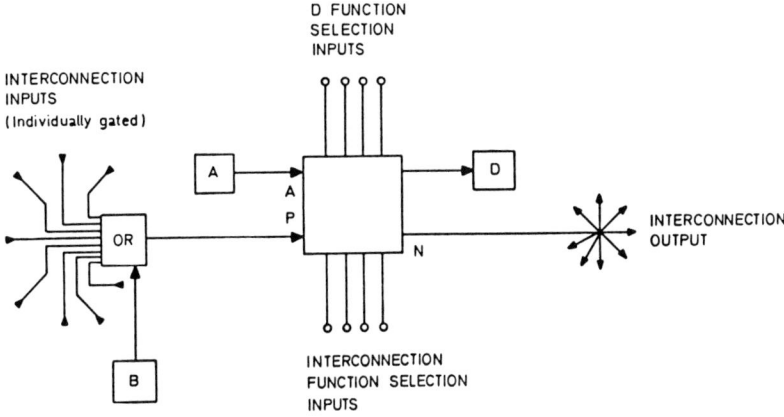

FIGURE 55. Logic circuit for "labeled propagation".

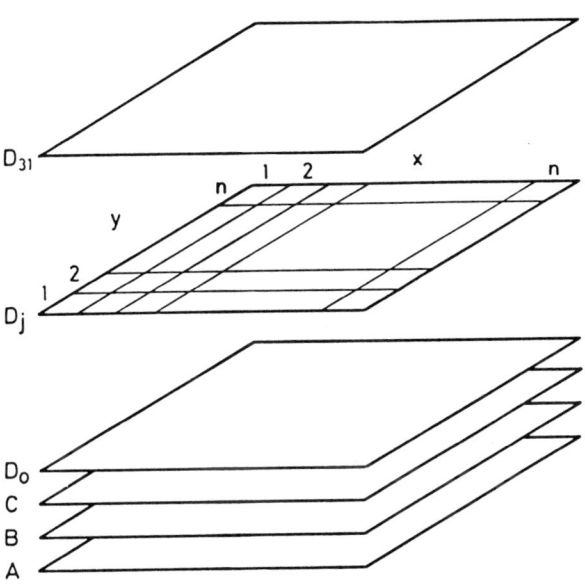

FIGURE 56. Data storage in the CLIP4 array.

A pixel will be formed as a column of single bit stores passed through g bit planes. Thus the pixel at location (x,y) is defined by:

$$P_{xy} = [d_{jxy} : j = k, \ k+1, \ldots k+g-1]$$

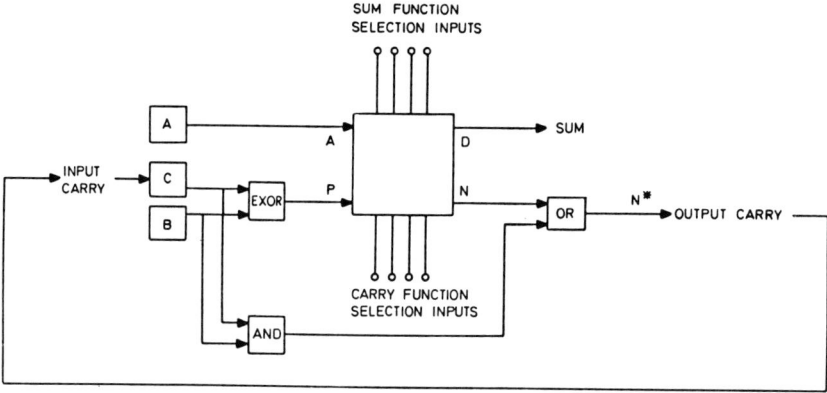

FIGURE 57. Organization of the CLIP4 circuit for bit-plane arithmetic.

D_k will contain all the least significant bits in the binary numbers representing the gray-levels in each pixel; D_{k+g-1} will contain the most significant bits. Note again that

$$g = \log_2 L$$

implying g binary bit planes are needed to represent L gray-levels. The ordering and addressing of the bit planes in the 32 available addresses in the D memories is arbitrary and may be varied to suit the needs of the program.

When numerical calculations are to be performed on data in which there is no longer an exact correspondence between the n^2 pixels and the n^2 processors, it is sometimes convenient to represent data in the binary column mode, in which n binary numbers are stored in the n columns of a single bit plane. A binary number will be represented by the column:

$$[d_{jxy} : y = 1, 2, \ldots, n]$$

Normally, the precision provided by n bits will be unnecessarily high for typical values of n. Methods have been devised for using the available storage more effectively.

In order to implement the procedures for bit-plane arithmetic, the CLIP4 circuit is reorganized as shown in Figure 57. The buffer C is provided as an automatic carry store. As an example of the use of the circuit in this form, consider two images stored as bit plane stacks (D_1, D_2, D_3) and (D_4, D_5, D_6), in which D_1 and D_4 contain the least significant bit planes in each image. The aim of the program is to produce a new image in (D_7, D_8, D_9) in which each pixel has an intensity which is the average of the corresponding pixels in the two input images. The first step loads the carry plane C:

```
              SET A, [B] P.A, R
              LDA 1
              LDB 4
              PST 1
```

Three points to note here are the use of the code R which enables the carry circuits in the processor, and the 'dummy' (SET A, LDA 1, PST 1) which allows the C plane to be loaded with the carry from the addition of the two least significant bit planes but does not enter new data into the D-memories. Instead, the D_1 plane is merely rewritten into store. This is a technical requirement stemming from the way in which control signals are associated within the integrated circuits used in CLIP4. The third point to note is that the sum plane has been discarded since this will be the least significant bit plane of an image whose intensities must all be halved in order to produce the required averages.

The next two bit planes are added thus:

```
              SET P @ A, [BC] P.A, R
              LDA 2
              LDB 5
              PST 7
              LDA 3
              LDB 6
              PST 8
```

and, finally, the last carry is transferred into D_9 by the instructions:

```
              SET P, [C] 0
              PST 9
```

Similar programs can be written for subtraction (involving sign changes on some of the Boolean variables) and the principle can be extended to include multiplication, division, and, indeed, any other mathematical operation. However, it can often be simpler and quicker to adopt table look-up methods, storing the relevant tables in the array.

Binary column arithmetic, used when calculations are to be performed on arrays of numbers of the order n rather than n^2, or when high precision is needed, is achieved using a slightly different arrangement of the processor circuits (see Figure 58). Since the numbers to be added are now stored as two rows of columns in two D-addresses (say D_1 and D_2), carry information will be propagated from processor to processor up the array columns. D_1 will be loaded into the A buffers and D_2 into the B buffers. Data in corresponding columns in each plane are added and the results output as a new plane containing the sums (D_3, say). The way in which the data is organized is shown

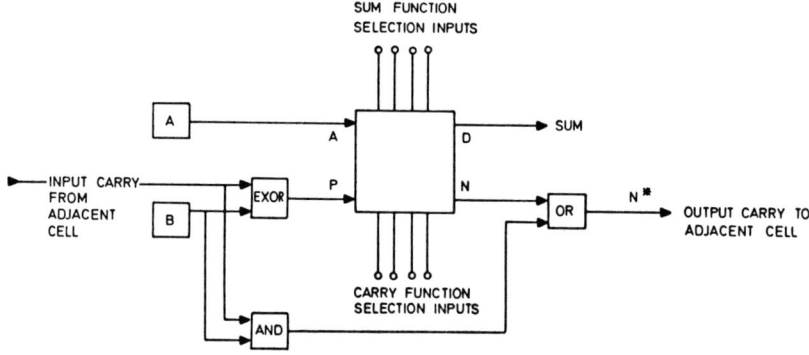

FIGURE 58. Organization of the CLIP4 circuit for binary column arithmetic.

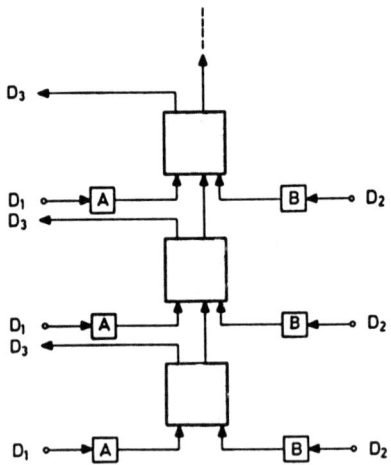

FIGURE 59. Data organization of processors for binary column arithmetic.

in Figure 59. An addition is accomplished by the single operation:

```
SET  P @ A,  [6] P.A, R
LDA  1
LDB  2
PST  3
```

Subtraction and other arithmetic operations are compounded in a similar manner to that for bit-plane arithmetic. For both methods, the most significant digit can be interpreted as a sign bit and twos-complement arithmetic used to allow negative integers to be handled. Floating-point arithmetic is a possibility

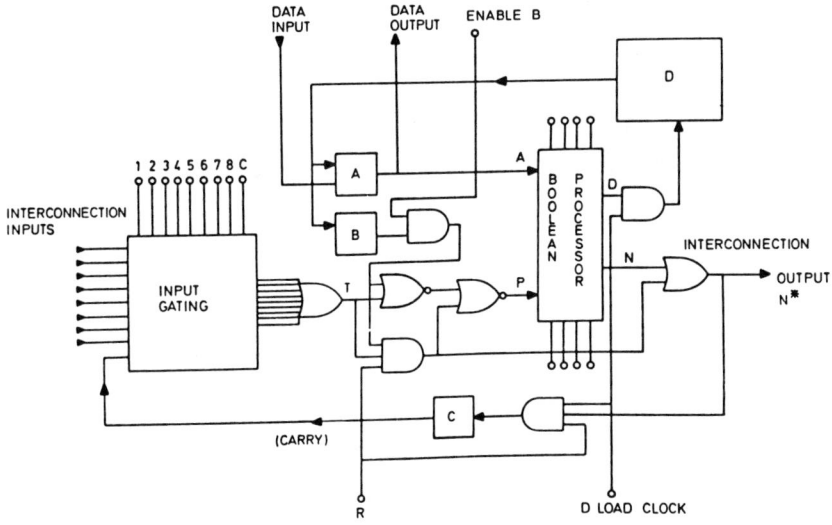

FIGURE 60. The complete logic circuit for CLIP4.

but has still to be studied in depth.

The complete logic for CLIP4 comprises all the subsections which have been considered above and is shown in Figure 60. As each SET instruction is interpreted, the relevant parts of the logic circuit are enabled by appropriate control signals.

8.2.5. Further CLIP4 Functions

In addition to the processor functions described above, some further functions have been built into the CLIP4 system. These will now be briefly discussed.

8.2.5.1. Counting

Various algorithms have been developed to enable an array such as CLIP4 to count 1-cells in a bit plane. This requirement frequently occurs where an array is used for "visual" inspection and where quantitative measurements are to be made on an image. Despite the ingenuity shown by the algorithm designers, it can be argued that counting does not lend itself well to parallel processing and results have been disappointing (in terms of execution time).

In order to provide a better alternative method, the CLIP4 array is arranged so that the A buffers are connected together to form row shift registers. These are primarily intended to handle the input/output functions for the array, but a subsidiary purpose is to permit rapid bit counting. The ends of the shift registers are connected to a "tree" of parallel adding circuits. A few clock pulses after the shift registers have clocked their data out of the array, the required sum appears in register 14 of the array controller. The time taken by

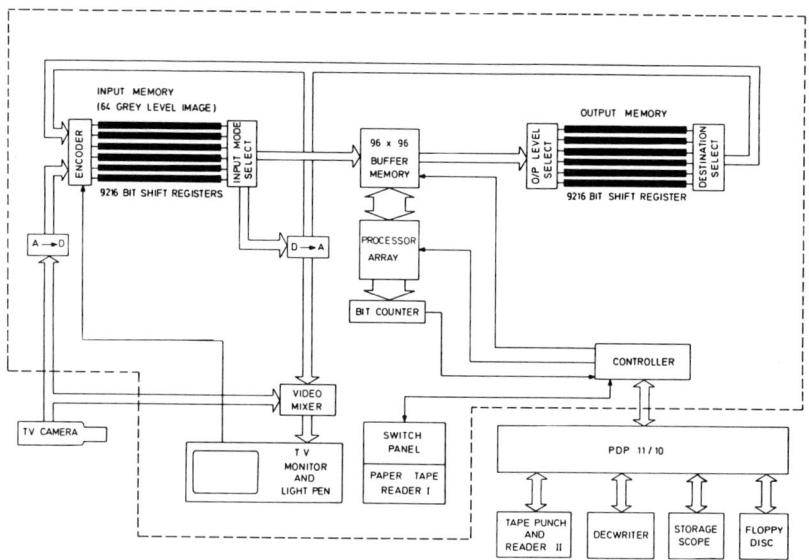

FIGURE 61. The CLIP4 system configuration.

this function is roughly equivalent to five CLIP operations. To count the 1-bits in D_1, the instruction sequence is simply:

```
LDA 1
COUNT
```

8.2.5.2. The INSERT Instruction

Data is usually entered into the array as an "image" but it is convenient to be able to load the array with the contents of an array controller register (again, register 14). An instruction INSERT is provided which has the effect of connecting the interconnection input at the array edges to an E value which is 1 or 0 depending on the value of the most significant bit in register 14. As the INSERT instruction is executed, it enters this value as a row (or column) in the array and then shifts both the register and array by one step.

8.2.5.3. Input and Output of Images

CLIP4 (see Figure 61) is still to be regarded more as a design study pilot model rather than as a production image processor. The input-output structure is intended to provide only adequate facilities during an initial assessment period. Within these terms of reference, the system is quite satisfactory. Both input and output memories are provided for a 96 by 96, 6-bit pixel image derived from the central one ninth of the area of a television frame. Analogue to digital conversion is provided at the input and digital to analogue conversion at the output. The input circuits also incorporate a four mode interface between the input memory and the array so that the program can demand upper or lower thresholding, a gray-level "slice" or a bit plane in the instruction IPA which transfers a binary image from the input memory into the A buffers of the array. New data is input from the television camera by the instruction SCAN; the instruction SCANS empties the output memory back into the input memory. The 6 bit-planes of the output memory are loaded one at a time from the array and display as a gray-level image on a standard television monitor. A light pen is available and acts on the most significant bit plane of the input memory.

8.2.5.4. Serial Computer Functions and Register Operations

The controller for CLIP4 is responsible for extracting instructions from the instruction memory and applying appropriate waveforms to the array and input-output circuits. The controller includes 15 general purpose 16-bit registers and provides the programmer with a conventional range of register functions, including register arithmetic. Provision is also made for interfacing to a PDP 11 computer, thereby simplifying the peripheral interfacing problems for the user and permitting easy access to assemblers, editors, etc.

8.2.5.5. Array Tests

Two array tests are included in the system which are invaluable. These are

1. **Empty array test** — An OR-gate has inputs from every D output in the array. The controller can instruct a conditional branch based on the output of this OR-gate. Thus a branch can be taken when a bit plane is all zeros, or not all zeros, as the case may be.

2. **Propagation test** — As some propagation paths in complex patterns can be long, it is inefficient to delay all CLIP operations sufficiently to allow maximum possible propagation paths to be completed (before clocking D-data into memory). The same OR-gate used for the previous test is therefore adapted to look at the interconnect outputs and the processor "busy" signal is not released until activity on the OR-gate output has ceased.

8.2.6. The Integrated Circuit

The delay between the commissioning of CLIP3 and the eventual completion of CLIP4 has been entirely due to the difficulties experienced in producing a large-scale integrated circuit embodying the CLIP4 processor. First approaches to a potential designer were made in April 1972; an organization proving willing to accept a design contract entered into discussion with University College London in October 1974. Two further companies worked on the project in subsequent years and the first correctly working prototype circuits were delivered in May 1979. No substantial changes in the logic specified for CLIP4 were made over the whole period.

Although this excessively long development period has been a source of almost unbearable frustration and disappointment to the author and his colleagues, it is pleasing to be able to report that the complete array has now been assembled and all other aspects of the hardware constructed so that a working system has been in full operation since the spring of 1980.

The integrated circuits are metal gate NMOS devices in 40-pin packages. Each circuit comprises eight processors, complete with 32 bits of RAM for each processor. The 4-phase clock controlling the chip runs at 1 MHz. 1152 circuits are needed to build the 96 by 96 processor array and these are mounted 12 to a board on 96 circuit boards. The whole system excluding the I/O facilities) is housed in one 7-ft cabinet.

The processors are connected within the chip to form a small block of array comprising two rows of four cells. Interconnections between these cells are built into the chip and output pins provide for interconnections with neighboring circuits. In principle, arrays of any size could be produced by assembling the requisite number of CLIP4 microcircuits. In practice, difficulties might be encountered in distributing control signals over very large arrays; experience to be gained with the present system should shed light on this particular problem, if it turns out to be significant.

8.2.7. Operating Speeds

The integrated circuit technology used for the CLIP4 chip is by no means the fastest available. This particular aspect of the design optimization was governed by the need for a fairly low cost system. Undoubtedly, at least another order of magnitude increase in speed could have been obtained by choosing a faster process. Nevertheless, very fast times are available.

Attempts have been made to estimate the theoretical and practical obtainable speed increases over conventional machines by using array processors. Unfortunately, the speed increases range from about 10 to over 10^5, depending on the task being performed. There are also additional virtually incalculable increases due to "housekeeping" advantages. It would obviously be difficult to suggest a meaningful average speed improvement to be expected with CLIP4. But it is perfectly possible to quote times for CLIP operations in the system now in use.

In summary, CLIP operations which are pure Boolean functions or which involve only a 3 by 3 neighborhood, require approximately 30 clock cycles, i.e., 30 µs. Propagation beyond the 3 by 3 neighborhood adds 3 µs to this time for each propagation step through a cell. The COUNT instruction takes approximately 100 µs to execute. Simple branches take 4 µs.

Applying these times to the image averaging program, we see that this would run for about 100 µs. In fact, slightly shorter times would be achieved as some redundant instructions are omitted from the normal 4-instruction CLIP operation sequences.

Input of binary data from the television camera takes less than 5 ms, consequently some 15 ms remain in each 20 ms television cycle which are available for processing. When all 6 bits of gray level information have to be input to the array, a similar 15 ms remains during every double cycle (40 ms).

These times indicate that the system can confidently be predicted to be of definite use as a real-time image processor in a variety of diverse applications.

9 COMMENTARY

Having visited a variety of machines, languages, and applications for SIMD computation in prior chapters, we are now in a position to examine some general issues for the technology as a whole. These include ease of programming, breadth of applicability, algorithmic richness, performance and future trends.

Fine grain, SIMD, parallel processing is a distinct and growing category of parallel processing characterized by large numbers of simple processors applied simultaneously to a given computation. Memory is disaggregated and associated with each of the processors. As the breed has evolved, a growing emphasis has been placed on the ease and speed of communications among any two of the individual processors in parallel.

These processors adopt an algorithmic philosophy termed data level parallelism, in contrast to the control level parallelism associated with coarse grain MIMD parallel processors that are characterized by fewer, individually more powerful asynchronous processors. In control level parallelism, one common approach is to break up existing algorithms and existing code into chunks with each assigned to run on its own processor. The programming effort is at the control level to keep all processors occupied by vectorizing and overlapping memory fetches for load balancing.

In contrast, data level SIMD parallelism assigns a processor to each piece of data. Here, the code flows through all the processors simultaneously. The effect of this contrast in parallelization philosophies is that the SIMD programmer spends proportionately more of his effort on the restructuring of algorithms per se while the MIMD programmer concentrates on the management and coordination of interdependent algorithm constituents, where the constituents usually remain sequential in character. This emphasis in SIMD on restructuring algorithms has had important payoffs in throughput by reduction of the order of the algorithm.

The order of an algorithm is a statement of how the number of operations required to execute the algorithm varies with the number of data items input. For example, an algorithm that adds 2 to each input data item is said to have order n, abbreviated 0[n], because the number of required operations is directly proportional to n, the number of input data items. An algorithm that produced the product of each input value with each of the other input values and itself would be order n^2, i.e. $O[n^2]$ since, as the number of inputs doubles, the number of operations increases by a factor of 4.

Computer technology is evolving. Next year computer clocks will be faster, memories will be larger and languages will be more powerful. But these are incremental improvements. To achieve tenfold, or greater increases in throughput over the short term one must generally find a way to reduce the order of the algorithm. If a new order n algorithm can be identified to obtain a result that previously required an $O[n^2]$ algorithm then as n increases, the savings in operations needed can be dramatic.

An example of such an improvement is the MAX function, i.e., given a list of n numbers determine the largest member of the list. On a sequential computer, this requires n − 1 compares and hence is considered an O[n] algorithm.

On 1-bit SIMD processors, for n less than or equal to twice the number of PEs, it has for some time been recognized that the MAX function can be performed in k $\log_2$ n steps, where k is the bit length of the numbers' binary representation. The method is illustrated on the left side of Figure 62. The algorithm pairs the various list entries and performs n/2 compares simultaneously, retaining the larger of the two values at each comparison. Hence this one operation discards half of the MAX candidates. The remaining candidates are paired and compared again at step 2, and again at step 3 and so forth until only one value is left; this last value is the maximum entry in the list, found after $\log_2$ n compares of k bit integers. When k = 32 and n = 64K, the order of the algorithm is reduced from 64K operations to 512 operations, a speedup of 128 times.

A more recently recognized MAX procedure is illustrated on the right side of Figure 62. One can see in hindsight that the k $\log_2$ n algorithm described above is not particularly efficient on an SIMD machine since after the first step most of the processors are idle. The newer method examines the high order bit of all the list members simultaneously and discards any list member that, on the basis of this first bit only, is no longer a candidate for being the largest list member. Next the second highest order bit is assessed in all of the remaining candidates and again some are discarded as no longer viable for being the maximum. This process continues for at most k steps; it terminates as soon as the maximum list entry is determined which may be after the first step. This is an O[1] algorithm since it is independent of n. For 32 bit integers, the number of operations is at most 32, a 16 fold improvement over the $O[\log_2 n]$ method and 2048 times better than the sequential algorithm, for n = 64K.

Another example is maze solving. The solving of a maze is one type of tree

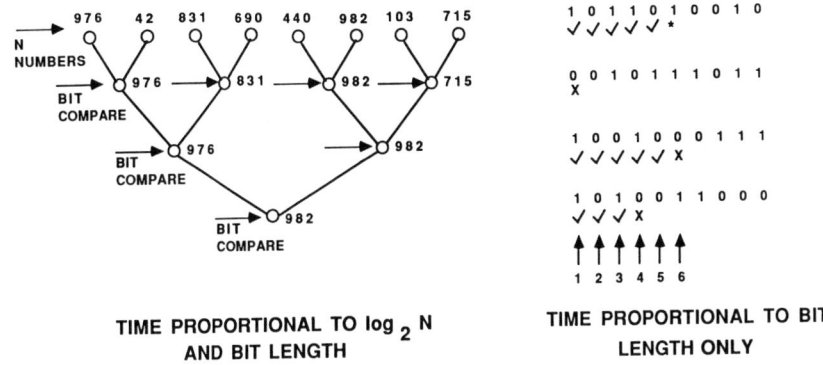

FIGURE 62. Parallel advantage in MAX function.

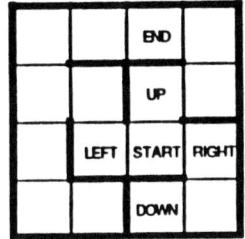

FIGURE 63. Maze.

search problem. As shown in Figure 63, four choices are available to move from the starting cell of the maze (up, down, left and right), which correspond to branches at a node of a logic tree. After the initial move, only three choices are available because the vacated cell is excluded. Maze walls correspond to cutting off branches from the tree because they reduce the choices available at certain cells of the maze.

On conventional computers there are two ways to search a tree: depth first or breadth first. To do depth first search of the tree in Figure 64, a conventional computer would examine one path from the start, always turning left, all the way to the dead end before it would begin to explore alternative path choices. In a breadth first search, the computer first would identify all of the choices available at the start cell. Then it would identify all of the choices produced by the first set of choices. Either type of search is terminated when the end cell is located or when no more choices are available. If the search is terminated when the end cell is first located, the path may not be optimal.

The SIMD technique for maze solving explores all paths simultaneously — each maze cell is assigned to its own processor. First, the start cell processor (numbered 1 in Figure 65) sends a message to its adjacent cells (which are

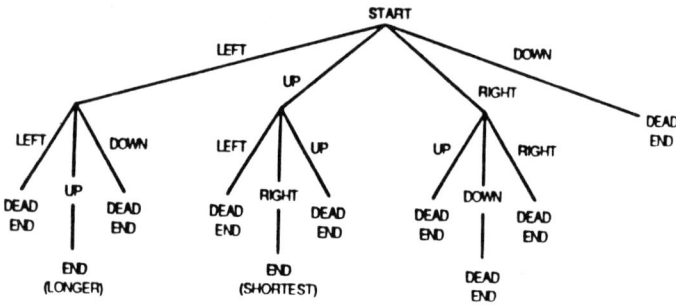

FIGURE 64. Maze logic tree.

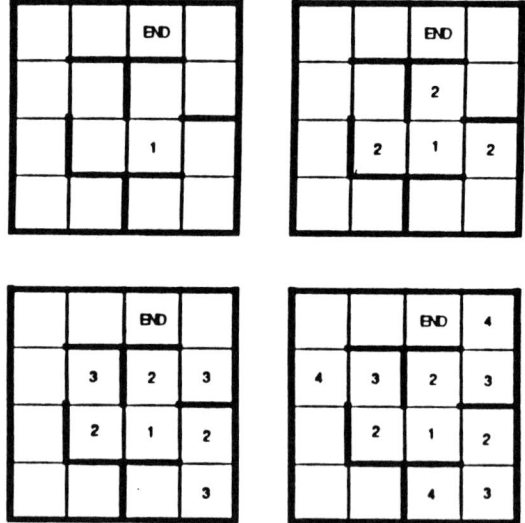

FIGURE 65. Maze solving steps.

numbered 2 in Figure 65). Next, each of these cells forward the message to all of their possible next cells.

Each cell remembers the origin of the message it received. When a message reaches the end cell, the optimal path is determined by backtracking. In Figure 65, the end cell turned out be number 5. Therefore, the optimal path is 5, 4, 3, 2 and 1. Other paths from the start cell to the end cell exist, but they are longer. This parallel maze solution guarantees the optimal path.

Moreover, the optimal solution has been determined in a number of steps proportional to the length of the shortest path rather than the longest path or the full tree search. Here again the order of the algorithm has been reduced.

A third example is sorting. Figure 66 illustrates the process. Each stage has

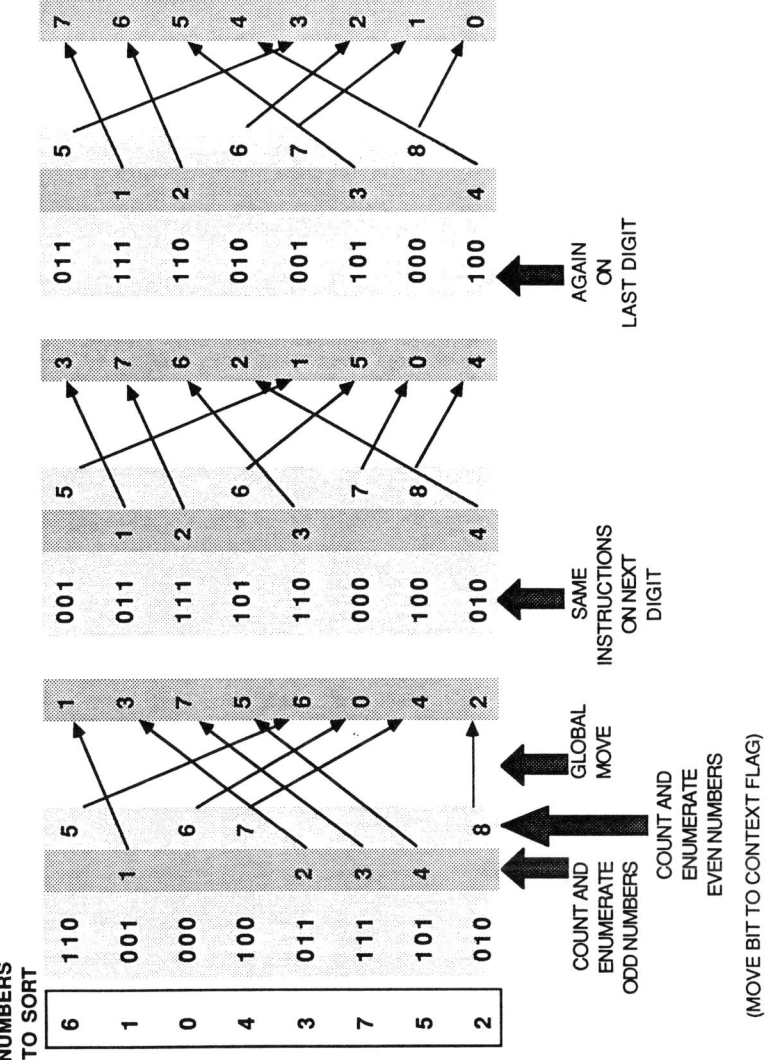

FIGURE 66. Sorting example.

three steps: count and enumerate odd numbers of the list, count and enumerate the even numbers, and perform a global move. The first stage uses the low order bit, the second stage uses the second to the lowest order bit, the third stage uses the third to the lowest order bit, and so forth. Note that the algorithm is proportional in operation count to the bit length of the numbers, not on the number of list entries.

The observation is that SIMD fosters better algorithms where better is measured in reduced operation counts. This is due in part to the SIMD programmer spending most of his effort restructuring the algorithm and hence having opportunity to see better alternatives if they are there to be seen. The better algorithms are also in part due to the algorithmic richness of the architecture so that those better alternatives are in fact there to be seen when the programmer looks for them.

There are two corollaries of the propositions discussed above. The first is that SIMD machines favor larger problems. The better algorithms generally are better because they reduce the order of the algorithms, from $O[n^2]$ to $O[n]$, from $O[n]$ to $O[\log_2 n]$ or the like. The reward for this change of order is greater for large n. If n is 8 then changing from $O[n]$ to $O[\log_2 n]$ reduces the operation count by only a factor of 2.67; if n is 16384 then the improvement factor is 1170. The most obvious example of this is when $O[n]$ reduces to $O[1]$ as when one wishes to add a constant to each of the pixels in an image.

The second corollary is that there is a reward for programmer skill in recognizing the improved algorithmic opportunities. This is not to say that programming SIMD is hard. Generally SIMD codes have fewer lines than equivalent sequential codes and the control commands are simpler so that the programs are more readable, and quicker to write once the algorithm is established. But recognizing the better algorithm is an acquired skill that seems only to come with experience. Many codes are rewritten over time as programmers realize that there are better and better ways to exploit parallelism for a given program. One example is the Fourier transform, an important widely used procedure.

The Giant-Fourier-Transform

The early radix-2 Cooley-Tukey FFT algorithm allowed fast and efficient computation of small transforms in special purpose hardware and on the general purpose computers of the late 1960's. Of course, more efficient algorithms were sought to provide larger transforms and faster answers within the limitations of technology. Higher radix and split radix algorithms traded increased complexity for a reduction in arithmetic operation. Not-in-place algorithms traded additional storage for overall speed. The need for ever larger and faster FFTs has continued as technology and algorithms for more efficient FFTs have continued to evolve.

Today, there are a growing number of applications for Giant-Fourier-Transforms, complex FFTs of size one million (i.e., 2^{20} or 1,048,576) and greater. VLSI circuitry and vertical integration have led to the development of multigigaflop top-of-the-line supercomputers. Unfortunately, the cost of these supercomputers is prohibitive to most users, driving the search to find more efficient algorithms and to implement them on less costly machines. To this end, MRJ has developed GFTs up to 64 million points (i.e., 2^{26} or 67,108,864) in core on the Connection Machine Model CM-2.

In past studies, using FFTs to size 16384, a VAX 11/780 with an FPS-164 attached processor was used; but, for new applications requiring GFTs, these facilities are inadequate. These applications required a million point GFT to run in a fraction of a second. In addition, error buildup can become extreme in large transforms, so a GFT was required that would work with a complete table of twiddle factors; and, because users would do the same size GFT many times, the table would be precomputed and its computation not counted toward executed time.

To size this problem, consider that the number of arithmetic operations for the radix-2 Stockham formulation of the complex FFT is

$$N(n) = 5n \cdot \log_2 n$$

For $n = 20$, the million point GFT, $N(n)$ is 104,857,600 so subsecond times imply a rate of over a hundred MFLOPS. Also considerable real memory is required. For 32-bit single precision, the complex input and the twiddle factor table each requires 8 Mbytes. Furthermore, this is true only for an inplace transform, which produces its output in shuffled (also called bit reversed) order. A final shuffle adds considerably to the execution time so that many modern algorithms trade additional memory to be able to directly compute the naturally ordered transform. If GFTs to size 64 million are required, the smallest usable memory size is one Gbyte. And for those supercomputers that use 64-bit precision only, that memory requirement doubles. Also, while a 2 Gbyte virtual address space is common for newer computers, insufficient real memory leads to "thrashing". Even for computers with sufficient real memory, this memory is usually arrayed in banks of very dense but low speed (compared to the CPU) memory chips. This reduces cost, size, and power consumption; and the memory behaves as if it were high speed as long as the banks are accessed sequentially. However, for the power of two memory strides of the FFT, memory bank conflicts greatly reduce execution speed.

The peak performance of a model CM-2, with 65,536 1-bit processors, 2048 32-bit Weitek floating point coprocessors, and 6.7 MHz clock, is 2.8 GFLOPS. It was introduced with 1/2 Gbyte of distributed memory, but availability of denser memory chips now allows an upgrade to 2 Gbytes.

The first thing one must realize about the CM-2 is that it is not a conventional supercomputer nor is it a pure SIMD hypercube computer. The 65,536 1-bit processors are not connected in a complete 16-dimensional hypercube. A 16-D full hypercube would require a very large number of interconnections:

$$I(n) = n \times 2^{n-1} = 16 \times 2^{15} = 524,288$$

Sixteen 1-bit processors on a single chip are connected in the CM-2 into a 4-D hypercube and 4096 of these chips are connected into a 12-D hypercube. The number of interconnections on-chip, in silicon, is

$$I_{oc} = 4096 \times 4 \times 2^3 = 131,072$$

The number of interconnections between the chip is only

$$I_{xc} = 12 \times 12^{11} = 24,576$$

Thus, this CM-2 implementation realizes a better than 3:1 reduction in interconnections and, moreover, a 20:1 reduction in actual wiring compared to a full discrete hypercube.

Interconnection with the Weitek 32-bit floating-point accelerators (FPAs) must be considered. There is one FPA chip per two processor chips. Interconnection between processor chips and FPAs is via a corner turning memory. Thirty-two 32-bit words are loaded in bit-order from the processors and read in word-order to the FPAs. Next, consider that the interchip message passing regime is store-and-forward of limited length serial packets. The lock step simultaneity of SIMD demands that every data pass between butterfly operations allows for the longest message transit time before the next computation.

So while radix-2 FFT algorithms can be neatly mapped onto a hypercube of FPAs, the above factors render the FFT on the CM-2 communications bound. The tricks (higher radix or mixed radix) used in traditional FORTRAN FFT codes are designed to reduce the number of arithmetic operations on compute bound hosts. Therefore, when these codes are translated into *LISP, a primary high level language for the CM-2, or with somewhat more effort, into PARIS assembly language, and run on the CM-2, the performance is naturally low. A GFT algorithm was sought which matched the CM-2's architecture rather than one which conformed to any existing FFT computation strategy.

The GFT posed three major problems:

- Computation and compaction of the twiddle factors.
- The butterfly shuffle.
- Communication efficiency during the stagings of the GFT.

Classical recursive computation of twiddle factors worked well for small FFTs. However, 26 applications of this technique for a 2^{26} transform caused unacceptable precision errors. Therefore, approximations were designed specifically for the CM that avoided error propagation. Also, the mathematical symmetry of the twiddle factors was exploited to reduce their storage in

TABLE 26
CM-2 FGT Performance

	FFT size						
	1 M	2 M	4 M	8 M	16 M	32 M	64 M
Preshuffle (sec)	.097	0.22	0.50	1.13	2.6	5.8	13
GFT time (sec)	.095	0.20	0.42	0.87	1.8	3.8	8
Total time (sec)	.192	0.42	0.92	2.00	4.4	9.6	21
Effective power (MFlops)	547	525	503	481	450	438	416

memory by almost a factor of four. The two latter problems were, by necessity, solved simultaneously.

Data had to be postured very precisely to maximally use the communication wires of the CM during the GFT stagings. The best arrangement was first identified and, subsequently, algorithms were explored to effect rearrangement from the fastest input form to the desired output form. In a 2^n size GFT, the 65,536 CPUs must behave as 2^n virtual CPUs. Principles of diclique formation led us to an innovative in place "butterfly shuffle". Only a 128-bit buffer per physical CPU was required to support the algorithm. Normally, for the GFT, the number of general communications calls is on the order of the square of the virtual to physical processor ratio (V/P ratio). The algorithm reduced this to the order of the V/P ratio, drastically reducing the time of the bit-reversal process.

Finally, stagings of the GFT fell into three classes:

- Communication between chips
- Communication between processors on the same chip
- Communication-less staging (communication between virtual processors which are actually the same physical processor)

The temporal sequence of stagings is as reflected above. First, the shuffle postures the data for optimal off chip communications. Then an interim general communication (on the same order of run times as the shuffle) arranges data for optimal on chip communications. The final calculations produce data in place on each CPU resulting in the desired, perfectly ordered GFT.

The GFT described above was developed in the PARIS language on a quarter-machine and then was ported to a full CM-2 for benchmarking. To exploit the speed of the Weitek FPAs, 32-bit precision was used. Timing is presented in Table 26. Twiddle factor compute time is not shown since, in practice, the twiddles would reside in long term memory and would be loaded once to do any number of GFTs of a particular size.

Separate times are shown for the preshuffle and the actual GFT computation. The former, the data posturing step, takes longer than the computation of the transform and is a function of the V/P ratio. The GFT computation time, on the other hand, is a function only of the transform length. GFTs up to 64 million complex points fit in the memory of the CM-2. The effective computational power shown is based upon the number of arithmetic operations in the Stockham formulation of the FFT and total time.

For a quarter-machine, execution time is approximately four times as long, so that the million point GFTs speed is within the desired range. However, the computation rates in Table 26 understate the effective MFLOPS rate. It should be noted that no computation takes place during the preshuffle time and that the radix-2 SIMD butterfly actually contains

$$N(n) = 8n \cdot \log_2 n$$

real operations. When these factors are taken into account, the effective computational power is 1.77 GFLOPS for any length GFT.

Nor is this the fastest that the CM-2 can operate. To achieve the full performance of the Weitek FPAs, they must be driven by pipeline microcode. Microcoding would be expected to give at least a factor of three speed increase, pushing the compute power to over 5 GFLOPS. The million point GFT would then execute in 50 to 65 ms. Vendors often quote cost efficiency in dollars per MFLOPS. For the GFT the CM-2 is achieving $3K/MFLOP (Fall 1988), and this could drop to $1K/MFLOP for a microcoded version. Vendors typically cost GFT computational performance between $10K and $25K per MFLOP.

Fluid Dynamics Using Cellular Automata

Sometimes rethinking of the algorithm for SIMD takes a dramatic turn. On occasion the entire approach to the algorithm is thrown out in favor of an altogether different approach that lends itself more directly to massive SIMD parallelization. An example of this is fluid flow simulation as implemented by Thinking Machines Corporation.

Fluid flow simulation is a key problem in many technological applications. From the flow of air over an airplane wing to mixing in a combustion chamber, the problem is to predict the performance of a design without building and testing a physical model.

Until recently, fluid flow models were based almost exclusively on partial differential equations, typically the Navier-Stokes equations or approximations to them. These equations are not generally solvable by normal analytical methods. Numerical approximation techniques, such as finite difference methods and finite element methods, have been developed to solve these partial differential equations. All of these methods involve large numbers of floating-point operations which require great amounts of fast memory. In addition, obstructions to the flow must usually be mathematically simple shapes.

Recent physics research has suggested that it is possible to make intrinsically discrete models of fluids. The fluids are made up of idealized molecules that move according to very simple rules, much simpler than the Navier-Stokes equations. The models are examples of cellular automata and are particularly well-suited to simulation on the SIMD computers. Cellular automata are systems composed of many cells, each cell having a small number of possible states. The states of all cells are simultaneously updated at each "tick" of a clock according to a simple set of rules that are applied to each cell. This approach involves only simple logical operations and does not require floating-point arithmetic. It allows for all obstructions regardless of their shape. In addition, mathematical methods can be used to show that the results of such simulations agree with the results that would be obtained from the Navier-Stokes equations.

Discrete simulation can be used to model fluid flow. The technique involves six key elements: particles, cells, time steps, states, obstacles, and interaction rules. Particles correspond to molecules of a fluid. A particle has a speed and a direction which determine how it moves. A time step is a "tick" of a clock that synchronizes the movement of particles. During each time step, particles move one cell in the direction that they are heading. A cell is a specific place in the overall region that is being observed. The region is completely filled with cells. Particles can move into and out of each cell during each time step. A state is a value assigned to each cell that indicates the number of particles within the cell, and in which directions they are heading. An obstacle is a set of special cells that obstruct the natural movement of particles. The interaction rules determine the movement of each particle when it shares a cell with one or more other particles. This movement is carried out by updating the state of the cells to reflect the new positions of the particles within the region.

A discrete simulation typically uses fixed cells. The cells never move or change during the simulation. Particles are completely in one cell during a time step, and move completely into the next cell (determined by the interaction rules) during the next time step. During each time step, every cell gathers data about particles heading in its direction from each of its neighboring cells. Based on the interaction rules, each cell determines the direction of its newly acquired particles and updates its own state.

A simulation designer can choose the cell topology and the interaction rules. The cell topology determines how many sides a cell has, and therefore, the directions by which particles may enter and exit. The simulation designer also determines the number of cells in the region being observed, and the average number of particles in each cell. Cellular automata theory provides the background for the simulation designer's decisions. It suggests that a simple cell topology, a huge number of cells and particles, and simple, local interaction rules are the most likely to be successful.

Thinking Machine Corporation is currently simulating fluid flow using a two-dimensional region that is divided into 16,000,000 hexagonal cells. Each cell is assigned to its own Connection Machine processor (using the virtual processor mechanism). The hexagonal mesh is a simple topology that gives the randomness that is required on a microscopic level to get correct results on the macroscopic level.

One of the fundamental reasons for computer simulation of fluid flow is to observe the behavior of a fluid as it flows past an obstacle. In the discrete model, obstacles are groups of cells that particles can not travel through. When a particle approaches an obstacle cell, it bounces off during the next time step. In order to observe the behavior of a fluid, tens of millions of microscopic particle interactions are simulated. Each individual particle's path through the cells and off of the obstacle cells appears almost random, just as in real fluids. However, when all of the particles' paths are considered, the overall behavior of the model is consistent with the way that real fluids behave.

Individual particles can enter or exit through any of the six sides of each cell. A cell may contain a maximum of one particle heading in each of the six possible directions during a given time step (and so the total number of particles per cell per time step is anywhere from 0 to 6). A particle that has not collided with another particle during a time step will continue moving in the same direction during the next time step. When particles collide, a simple set of rules determines their new directions, conserving both momentum and the number of particles.

At each time step, every cell updates its state by checking all of its adjoining cells, or neighbors, for particles that are heading in its direction. All cells then update their own states based on the information that they have gathered. In the model currently implemented, there are five situations that cause a particle to change directions: 2-way symmetric collisions, 3-way symmetric collisions, 3-way asymmetric collisions, 4-way symmetric collisions, and collisions with an obstacle cell.

Although the algorithm is implemented by modeling the individual movements and collisions of tens of millions of particles at each time step, the behavior of the fluid is observed by averaging the behavior of all of the particles in the entire region and by analyzing the results over many time steps. In a typical simulation, macroscopic results are gathered by averaging particles together in groups of 20,000. Although each individual particle has only one speed and six possible directions, the average of 20,000 particles provides the full range of possible velocities.

There are two available ways for the Connection Machine system to implement the connections among the hexagonal cells. It can use the full router, setting up six connections for each site, one for each adjacent hexagon. Or it can use its grid, which connects four adjacent processors directly. The grid network was chosen for this implementation. It is very fast for small data transfers to nearby processors.

Of course, the grid cannot implement hexagonal connections directly. It connects to four adjacent processors, not six. Therefore, two of the six connections require two-step communication (i.e., up one and over one for the diagonal). The simulation program implements this two-step process. Each site can quickly learn the status of its six neighbors and can determine which ones contain particles that are moving in its direction.

Each cell has only 13 bits associated with it: six bits for incoming state (numbered 0-5), six bits for outgoing state (numbered 0-5), and 1 bit to indicate whether or not it is an obstacle. Each of the six incoming state and six outgoing state bits is dedicated to a particular direction. If a particle is entering or existing through that direction, then the bit is set to 1, otherwise it is set to 0.

/* A cell state is represented by a six-bit unsigned integer, which can also be regarded as an array of six individual bits. */

```
typedef union STATE {UNSIGNED:6 Val; unsigned:1 Bit[6];}
state;
```

/* Each processor in the domain "grid" will contain a cell state (the outgoing state), another state (the incoming state) used for temporary purposes in the calculation, and a bit saying whether or not it is an obstacle cell. */

```
poly state outgoing_state, incoming_state;
poly unsigned:1 obstacle_cell;
```

/* The following declares the actual grid of processors. */

```
processor fluid_grid[ARRAY_X_SIZE] [ARRAY_Y_SIZE];
```

/* Grid is the C pointer type that corresponds to the above array type. "/

```
typedef processor (*grid) [ARRAY_Y_SIZE];
```

At each time step, instructions are broadcast that tell each cell how to gather data about particles handing in its direction. When the cells poll each of their six neighbors for information, they formulate their own 6-bit incoming state. For example, a cell would ask its East neighbor for its outgoing state bit number 3, and would place the answer in its own incoming state bit number 0. It would then ask its North-East neighbor for its outgoing state bit number 4 and would place the answer in its own incoming bit number 1. All cells, in parallel, check the state of all six of their neighboring cells. This extreme data level parallelism allows for a large amount of data to be collected in a small amount of time.

/* This code is executed within each processor. Outgoing state bits from six neighbors are gathered and placed with the local incoming_state array. Note the use of a C case expression ((grid)this) to create a self-pointer that has a two-dimensional array type suitable for double indexing. (This code actually is oversimplified in that it does not handle the boundary conditions for cells on the edge of the grid. Handling these conditions is a bit tedious but conceptually straightforward.) */

```
poly void get_neighbors() {
 incoming_state.Bit[0] = ((grid)this)[ 1][ 0].outgoing_state.Bit[3];
 incoming_state.Bit[1] = ((grid)this)[ 0][ 1].outgoing_state.Bit[4];
 incoming_state.Bit[2] = ((grid)this)[-1][ 1].outgoing_state.Bit[5];
 incoming_state.Bit[3] = ((grid)this)[-1][ 0].outgoing_state.Bit[0];
 incoming_state.Bit[4] = ((grid)this)[ 0][-1].outgoing_state.Bit[1];
 incoming_state.Bit[5] = ((grid)this)[ 1][-1].outgoing_state.Bit[2];
}
```

Once each cell has determined which particles are entering (by collecting its incoming state), it updates its outgoing state to reflect the particle interactions. First, all cells that have their obstacle-bit turned on are instructed to set their outgoing state to be the same as their incoming state (since particles that hit an obstacle bounce back in the same direction).

Next, patterns are broadcast that correspond to each of the possible 6-bit incoming states, followed by the corresponding 6-bit outgoing state. Each cell compares its incoming state to the pattern being broadcast. When there is a match, the cell updates its outgoing state accordingly. For example, a cell with an incoming state of 011011 would then have an outgoing state of 110110.

/* The rule table is indexed by a six-bit incoming-state value and contains the corresponding outgoing-state values. */

```
state rule_table[64];
```

/* Calculate the new outgoing_state for all cells, based on the incoming_state and the obstacle_cell bit. */

```
poly void update_state {
  if (obstacle_cell)
     outgoing_state.Val = incoming_state.Val;
  else outgoing_state.Val =
    rule_table[incoming_state.Val].Val;
}
```

It is important to note that this trivial, non-computational, table look-up is the driving force of the whole simulation. The Connection Machine system has replaced all of the mathematical complexity of the Navier-Stokes equations with this small set of bit-comparison operations. The simulation is successful because the system can perform this operation on huge numbers of particles in very short amounts of time.

A typical "run" of a fluid flow simulation begins by allowing the user to make several choices. The user typically specifies the average number of particles per cell (density) and the average speed and direction of the particles (velocity). Technically this means that the entire region starts out with particles randomly distributed among the cells (based on the density) and moving in a certain overall direction (based on the average velocity). The user also selects or draws one or more obstacles and places them somewhere in the region being observed. All cells that are part of an obstacle have their obstacle bit set. As the simulation runs, new particles are randomly injected from the edges of the region in order to maintain the selected density and velocity. Once the model is running, each cell's state is continually updated, and average results for regions of cells are displayed.

```
/* This is the main computation loop. At each time step, each cell fetches state
   from neighbors and updates its own state; then the results are displayed. */
poly void fluid_flow() {
   for (;;) {
        get_neighbors();
        update_state();
        display_state();
   }
}

/* Execution begins here. */

void start_fluid_flow() {
   /* Initialization. */
   initialize_rule_table();
   initialize_cell();
   /* Activate all processors in fluid_grid
      and then call the function fluid_flow. */
   [[] []fluid_grid]. { fluid_flow(); }
```

A production level version of the algorithm described in this chapter has been implemented and extensively tested. The simulation operates on a 4000 × 4000 grid of cells, typically containing a total of 32 million particles. The Connection Machine system is able to perform one billion cell updates per second. Each time step includes approximately 70 logical operations per cell; a simulation therefore require a total of 100 trillion (10^{14}) logical operations for 100,000 time steps. The complete simulation takes less than 30 min. Current results are very competitive with state-of-the-art direct numerical simulations of the full Navier-Stokes equations.

In addition to providing very accurate simulation of fluid behavior, the cellular automata method for simulating fluid flow allows scientists to continually interact with the model. Any of the user's original choices may be modified during a run of the simulation, without long delays for new results. Since particles are continually moving through the cells, a new density or average velocity may be established by adjusting the particles being randomly injected from the edges. When a new obstacle is added during a run, the obstacle bits in the appropriate cells are set, and those cells begin to reflect particles. Within less than a minute (a few thousand time steps), results based on the new selections become apparent in the displayed flow.

The algorithm for simulating fluid flow is simple. It overcomes problems formerly associated with computer simulation of fluid flow by using a discrete simulation that takes advantage of data level parallelism. During each time step, every particle can move in the direction it is heading, every cell can evaluate its new particles based on collision rules, and every cell can update its state to reflect the direction of the particles it currently contains. The algorithm involves a small number of instructions executed over a large amount of data. Since the system is able to assign a processor to each data element, and to allow all processors to communicate simultaneously, it has provided the computational power required to provide the ideal solution to this applications need.

CONCLUSION

There are and have been many more SIMD computers than have been discussed in this book. The focus has been on the issues and languages and applications for SIMD rather than on an exhaustive survey.

The material has been based upon and drawn from many sources. It has not been possible to include anything like all of the worthy material. Many outstanding topics such as robotic arm path planning, the traveling salesman problem, line-of-sight algorithm, lens design, movie making through ray tracing, and military deployment scheduling have had to be excluded. But it is felt that the essence of SIMD as an approach to practical computing has been fairly represented.

In the future, bigger and faster can be safely forecast. Market forces and new technologies will dictate the path of future evolution. One suspects that a marriage of SIMD and MIMD may well emerge, say an MIMD architecture with a SIMD array at each MIMD node.

SIMD architectures have been described as being suitable for a limited range of applications. The prior material in this book is a partial refutation of that allegation since the applications described are rather diverse, much more so than critics anticipated. But the criticism is partly true. That the architecture is unsuited generally for applications involving only a handful of data items has already been discussed. Moreover, there are some applications that are sequential by nature. For example, to determine the 500th digit of pi, one must already know the 499th digit. Most recursive formulae fall into this category. A third important category may have to be included among the unsuitable for SIMD groups — artificial intelligence. To date very few AI applications have found their way to SIMD implementation. This may remain the case or change with time. One suspects that simulations involving tens of thousands of independent agents such as crowds at a baseball park could be programmed as tens of thousands of expert systems, each in its own processor. Meanwhile, we await the million processor machine.

APPENDIX
FAULT TOLERANCE IN HIGHLY PARALLEL MESH CONNECTED PROCESSORS

INTRODUCTION

The mesh interconnection scheme has been used on several large-scale SIMD parallel processors. This scheme involves organizing the processing elements (PEs) into a two-dimensional matrix such that each PE has data interconnections with its adjacent neighbors. In a typical organization a PE has connections to four near neighbors in the cardinal directions N, S, E and W. In a single instruction data may be shifted in a single specified direction between all adjacent PEs. That is, a distributed matrix of data may be shifted one mesh position in parallel. The main advantage of the mesh scheme is its simplicity and suitability for a large class of scientific applications. Data interconnections only occur between adjacent PEs; this means that they may be kept very short and laid out on a single interconnection plane. The usual disadvantage with this scheme is that data transfers to distant PEs require a large amount of time since the data can only cross between adjacent mesh nodes with each clock cycle. However, there is a large class of problems including physical system modeling using partial differential equations and image processing in which the data needed by a PE are located in its local mesh area and the mesh interconnection scheme is very efficient.

One potential problem with the mesh scheme is that the failure of any node in the mesh renders the whole parallel processor inoperable. Current LSI processor designs involve a mesh with more than 10,000 nodes; VLSI technology systems have 1,000,000 nodes and more may be anticipated. For some applications, e.g., real-time image processing in a remote inaccessable robot system, some fault tolerance is essential.

MESH CONNECTED PARALLEL PROCESSORS

An important large-scale mesh parallel processor is the Illiac IV[1] developed in the late 1960s. It consists of an 8 × 8 mesh connected set of 64 PEs, each PE having an ALU with a 64-bit-wide datapath and floating point capabilities. This architecture is well suited to applications such as partial differential equations. The implementation of Illiac IV was hampered by the technology of the time. Hardware failures were anticipated to occur every few hours. The PEs were regularly subjected to an extensive library of automatic tests and were replaced manually if any faults were detected.

A more recent design, based on LSI technology, is the Massively Parallel Processor (MPP)[2,3] which was developed for NASA by Goodyear. The MPP consists of a 128 × 128 mesh of 16,384 PEs, each of which has a 1-bit-wide data path and can achieve floating-point operations through bit-serial algorithms. This architecture is designed for image processing applications where a single image may be as large as a 6000 × 6000 matrix. The MPP processes such an image as a sequence of 128 × 128 subimages. The MPP involves parity checks on each 8 bits of the PE local memories and has a redundant column of PEs which may be switched in by the host computer to replace a faulty column. With this fault tolerance the MPP can run for several hundred hours before requiring manual intervention.

The fault tolerance concepts in this paper will be considered with respect to a bit-serial PE array scheme or Binary Array Processor (BAP)[4] such as the MPP. These concepts may be extended to word parallel designs such as the Illiac IV type architecture. However, with the constraints that the matrix to be processed is larger than the PE array and that the algorithms to be implemented are well formed for the mesh organization, the bit-serial approach has significant advantages over the word parallel approach for equivalent amounts of hardware.[5]

A general block diagram for a large-scale BAP is shown in Figure 1A. Data processing is achieved with the array of PEs. Data are input to and output from the PE array via the I/O buffer memory which communicates the data to peripherals and bulk auxiliary storage devices. Instructions to the PE array are issued by a single high-speed microprogrammed control unit. The whole system synchronization is maintained by a conventional host computer which issues macro instructions to the control unit. Some feature information may be extracted from the PE array by the global information extraction mechanism.

A typical organization for an MPP-like PE is shown in Figure 2A. Data from adjacent near neighbors are selected by the NN multiplexor. The control lines and local memory address lines are broadcast to all PEs in the array. The OR bus is a line from all PEs to the control unit which has a 1 value if any PE

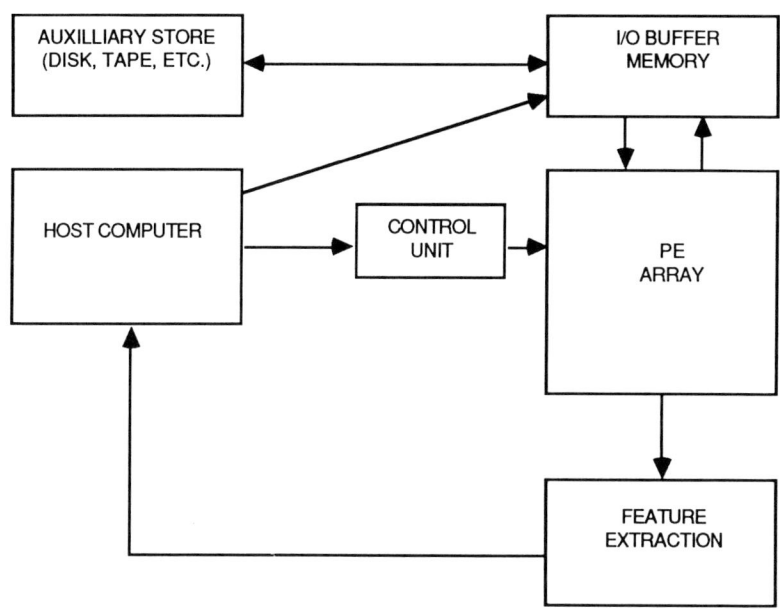

FIGURE 1A. Binary array processor system.

outputs a 1. The I register is used for data I/O; it receives data from the I register of the adjacent PE to the left and transmits data to the PE on the right.

The I/O buffer memory is a vital part of the BAP system; it is responsible for making reformatted data available to the PE array. With the MPP a data matrix is input to the array as a set of bit planes. Each bit plane is input along one edge of the PE array, one column with each clock cycle. Each row of the array acts as a shift register. When the complete bit plane has been input it is stored in the PE local memories in one clock cycle. Fault tolerance in the I/O buffer can be achieved with the single error correction-double error detection (SECDED) schemes common in many recent large memory systems. For the PE array the I/O mechanism is like a one-dimensional mesh connection, and it will not be treated as a separate issue for fault tolerance. For a BAP system the I/O could be achieved by the mesh interconnection hardware; alternatively, a separate but basically similar I/O hardware as shown in Figure 2A could be used, if necessary, to avoid blocking.

The global feature extraction mechanism on the MPP is an OR function over all PE elements, which outputs a 1 if any PE has a 1. If we have an error detection mechanism then a similar global OR function would be needed to report an error to the host processor. Once again these two functions will be considered to be implemented with the same hardware in this paper; such a scheme is used with the MPP.[3] It has been suggested that a more powerful

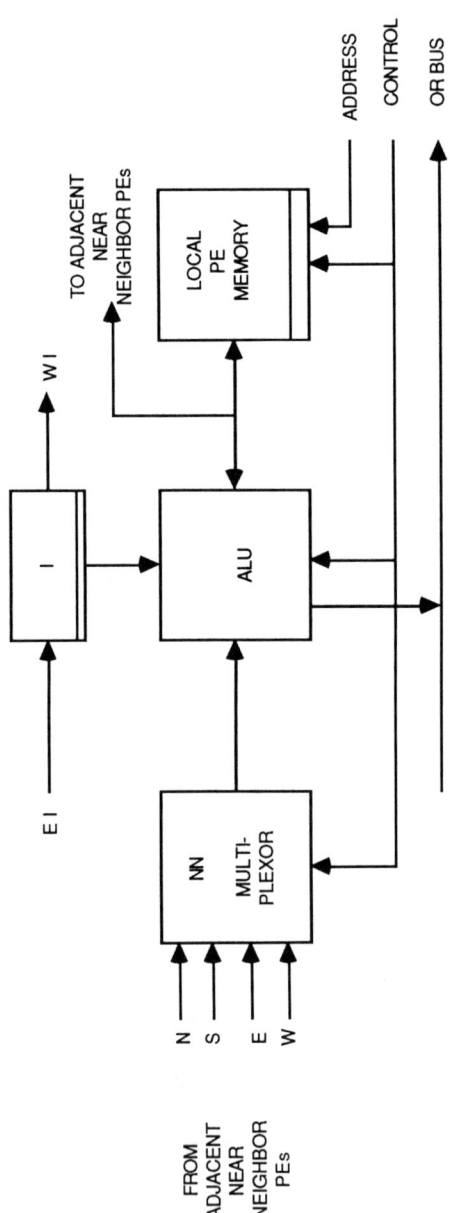

FIGURE 2A. A typical BAP PE organization.

feature extraction mechanism, such as counting the number of bits set in a bit plane may be cost effective for future BAP systems.[6]

The MPP PE array is constructed with two LSI chips: a PE chip and a memory chip. The PE chip contains 8 PEs (without local memories) in a 2 × 4 array. Each PE chip has connections to eight 1-bit memories for the PEs and an additional 1-bit memory for a parity check of the other eight. The total PE array consists of 33 4-PE wide columns, each column consisting of 128 PE chips. A PE chip has a control input which, when activated, disables the chip by connecting corresponding East-West pin data lines together. In this way any one of the 33 columns may be disabled to achieve an operational 128 × 128 PE array. When a fault is detected the faulty column is disabled and the redundant column is used to replace it.

Faults will be considered here to be of two basic types — local and module. A local fault may typically be a broken data line or a faulty memory bit whereas a module fault implies the complete failure of a module, such as a chip, which may result in a set of related PEs being made inoperable.

Since we are dealing with functionally very complex chips, the probability of a local fault may be expected to be significantly higher than a module fault. Therefore the main effort of the work here is concerned with local faults as they are much simpler and cheaper to deal with. However, any practical very large-scale mesh connected processor also needs some fault tolerance at the module level.

For the MPP, the redundant column scheme is effective for any single memory chip failure. It is also effective for most PE local failures, e.g., if a data line breaks in a PE. Therefore the most probable fault causes have been covered. However, if a catastrophic failure occurs to a PE chip (module), then the whole PE array may become inoperable since it is necessary for data to flow through a disabled chip.

A VLSI PE ARRAY ORGANIZATION

For a VLSI system design there are two fundamental chip size limitations: (1) the number of devices which can be put onto a chip and (2) the number of pin connections which may be made to the chip. A usual characteristic of a VLSI design is modularity, i.e., a chip consists of a very large number of identical modules, which is important to minimize the development cost. Finally, with very large functionally complex chips fault tolerance may be effective to significantly increase the production yield and the fault-free lifetime of the chip.

A possible chip organization for a very large VLSI PE array is shown in Figure 3A. Three different VLSI chip types are involved: a PE ALU chip, a local memory chip, and a PE mesh interconnection chip (MIC).

The PE ALU chip consists of a set of PE ALUs, each having a limited amount of local memory. These ALUs share common ALU function and

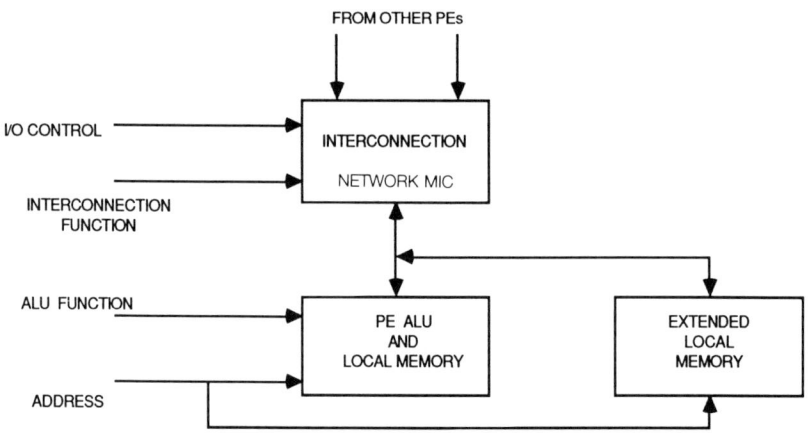

FIGURE 3A. VLSI PE organization.

address lines but do not have any data interconnections between each other. The data access to a PE is achieved by a single pin on the chip which is connected to a bidirectional bus line. The design of an effective PE with this input/output constraint is described in Reference 5. With this design optimal bit-serial processing times for addition, multiplication and logical operations can be achieved. The limited size on-chip local memory may be used for table look-up applications since it may be addressed by an ALU register (unlike the external local memory) or for a cache memory.

The PE ALU chip will be a functionally very dense chip and will contain as much logic as the VLSI technology will allow. There are no pin connection problems since only one pin is required for each additional PE.

The external local memory chip will provide the main local data storage for a PE. With the amount of single chip storage which is becoming available with emerging memory technology, it is possible that one VLSI memory chip could contain adequate storage for one or even several PEs. The 1-bit wide external PE memory is connected to the single-bit PE data bus. Once again the limitation with the memory chip is caused by the functional complexity achievable with the VLSI technology; there is no pin connection problem.

The interconnection chip realizes the mesh interconnections between the PEs and also contains an input/output mechanism for data I/O to the PE array. Unlike the previous chips, this chip is functionally very simple and the size of the mesh which can be contained on a chip is limited by the maximum possible number of pin connections. Each mesh node requires one pin connection to its PE data bus and also, for a $m \times n$ mesh $2(m + n)$ additional data interconnections are needed to adjacent MICs.

All the additional logic to achieve error reconfigurability for the PE array is located in the MICs.

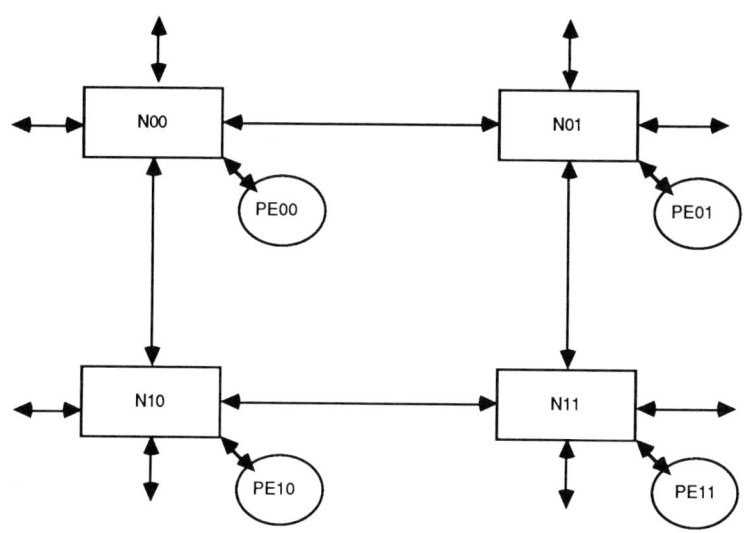

FIGURE 4A. A simple 2 × 2 mesh interconnect chip organization.

PE FAULT TOLERANCE

Both the ALU and external memory chips are functionally very complex and therefore more likely to fail than an MIC. In this section we consider how to reconfigure the array if a single PE-external memory combination fails. This reconfigurability is achieved by modifying the MIC so that is has access to spare PEs which may be switched in to replace the faulty PE.

A basic nonfault-tolerant MIC organization for a 2 × 2 mesh subsection of a PE array is shown in Figure 4A. This chip has a total of 12 data pin connections: 1 to each of the 4 PEs and 8 to adjacent neighbor MICs. In general a m × n mesh MIC would require mn + 2m + 2n data pin connections.

The basic logic device on which the design of the MIC will be based is the selector, illustrated in Figure 5A (upper). A selector has a set of control inputs, C, which specify by a binary code which of the X data items is to be connected to the Y data line. Once connected, data may flow in either direction from X to Y or Y to X. With some logic technologies an additional control input may be needed to specify the data flow direction. However, with designs considered here, the direction information is always locally available; therefore this additional control line is no problem.

Each mesh node of the simple MIC shown in Figure 4A may be implemented with a 5-way selector and a 1-bit register as shown in Figure 5A (lower). Two clock cycles are required, with this design, to transfer data between adjacent PEs. In the first clock cycle the data is output from the PE and loaded into its mesh node P register. Then, in the second clock cycle, the PE reads the value of an adjacent PE's P register. Data may be transferred

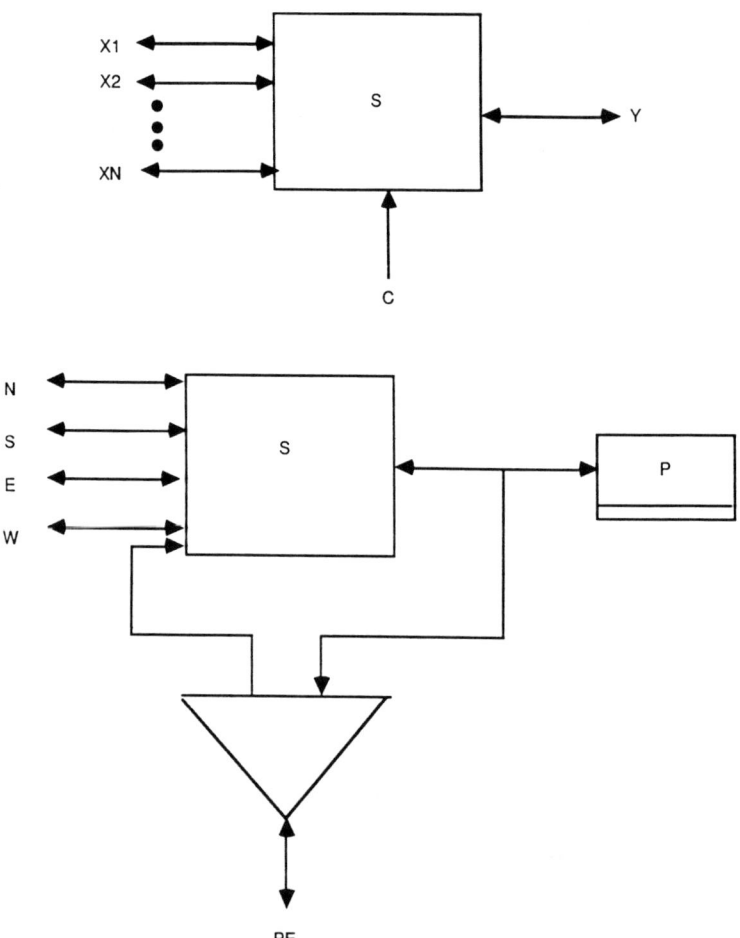

FIGURE 5A. Upper: A selector switch. Lower: A simple MIC mesh node organization.

between more distant PEs by shifting it through a connected sequence of P registers. In general, a data transfer through a path of K stages requires K + 1 clock cycles.

To achieve fault tolerance to a single PE failure we first consider adding a spare PE to each MIC group. A possible organization for such a reconfigurable MIC is shown in Figure 6A. Each mesh node may be connected to a unique, operational PE.

The details of the modified mesh node design are shown in Figure 7A. A Q register and a 2-way selector have been added to each node. The value of the Q register specifies which of the two possible PEs the mesh node is connected to. When a faulty PE is identified the host computer generates a bit mask which

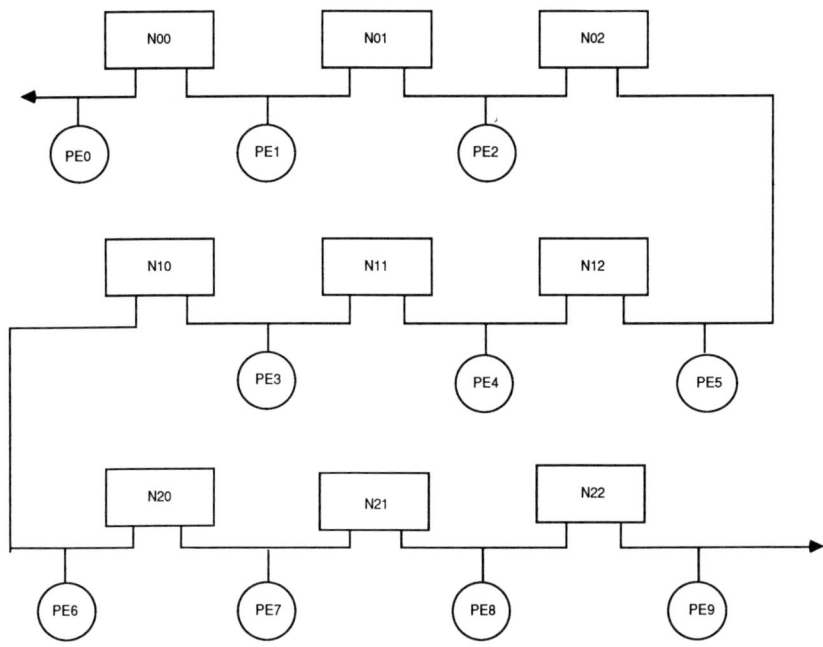

FIGURE 6A. A 3 × 3 matrix of interconnection nodes connected to 10 PEs.

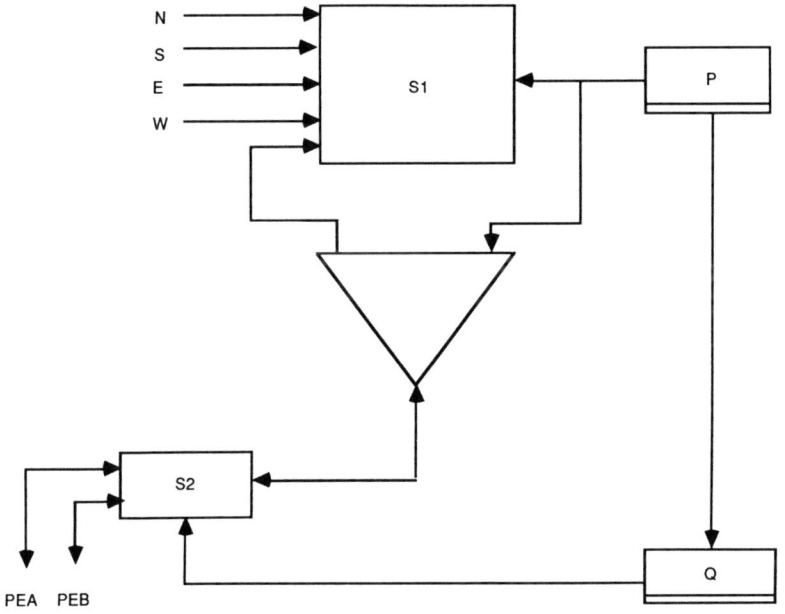

FIGURE 7A. A mech node with PE fault tolerance.

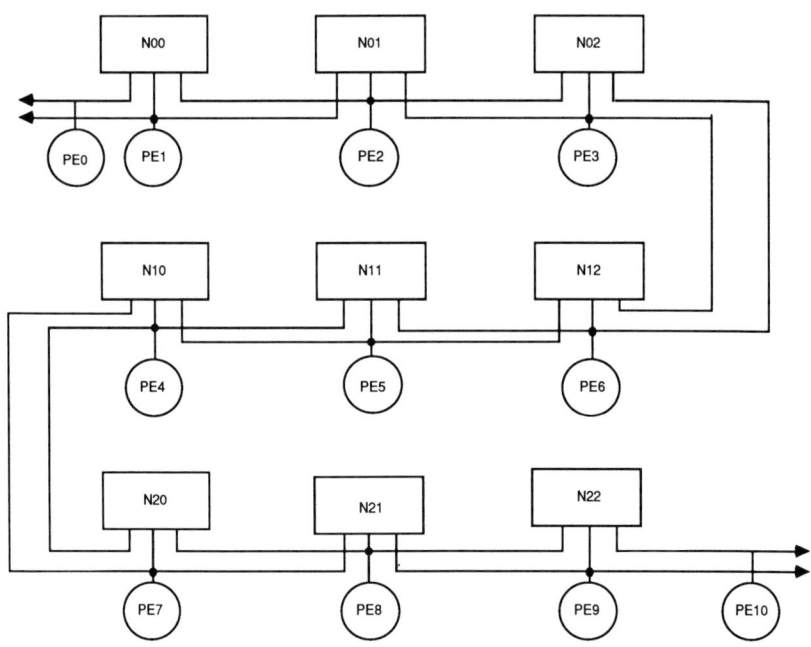

FIGURE 8A. Organization for fault tolerance to any two faulty PEs.

is distributed to the P registers; then the Q registers are loaded from the P registers to isolate the faulty PE. The task which was in progress when the faulty PE was detected must be reloaded or restarted.

The above MIC modifications require only two new pin connections to the MIC. One is the load control and the other is the data connection to the extra PE. One extra PE must be available to each MIC; however, it is possible for MICs to share a PE as indicated by the broken lines in Figure 6A. In this case only one extra PE for a group of MICs is needed.

The above technique is easily extended if protection against more than one faulty PE for each MIC (or group of MICs) is required. For example, protection against any two faulty PEs could be achieved by connecting two extra PEs to the MIC as shown in Figure 8A. The PE selector at each mesh node must select between three PEs, and the Q register must be extended to contain two bits of information. In the general case, fault reconfiguration for the up to K faulty PEs requires K extra PEs; each MIC requires K + 1 more pin connections than for no protection. Each mesh node in the MIC must contain a K + 1 way PE selector and a Q register large enough to address it.

MIC MESH NODE FAULT TOLERANCE

Once the system may be reconfigured for any faulty PE, the next problem area is the very large mesh interconnection network itself. Fault tolerance is considered here for the failure of any mesh node or data interconnection in the interconnection network.

Mesh node fault tolerance on the MIC can be achieved using a similar scheme to the MPP global fault tolerance. That is, have a spare column of mesh nodes which may be utilized when a faulty mesh node is detected. One way in which a spare column of mesh nodes may be incorporated within an MIC is illustrated for a 2×2 mesh MIC in Figure 9A. In the general case with $n + 1$ columns the configuration is specified by a register (not shown in Figure 9A) having two bits for each column. A possible organization of a mesh node for this organization is illustrated in Figure 10A. The two bits from the reconfiguration register are represented by RL and RR. When RL is set it specifies that the left (W) input to the node not be connected to the adjacent column node but to the next node to that, i.e., to skip the node to the left; RR specifies which column node is connected to the (E) input in a similar way. A bit pattern is loaded into the reconfiguration register such that one column is skipped. It does not matter what values a disabled faulty node may have on its interconnection lines since these lines are never used by the other nodes. For the rest of this paper the configuration in Figure 10A will be considered to be implemented by a single 7-way selector.

Any external data interconnections pin may be connected to one of two mesh nodes; therefore it is necessary to have a 2-way selector associated with each node as shown in Figure 9A. The control for these selectors is derived from the reconfiguration register contents.

The simple organization shown in Figure 9A can reconfigure for any faulty mesh node; however, there is no fault tolerance from either a data pin connection failure or a data pin selector failure. Fault reconfigurability for such failures may be achieved by adding spare pin connections and selectors, one for each of the four directions of data connection as shown in Figure 11A. Only the connections to pin selectors are shown; the interconnections between mesh nodes are similar to those shown in Figure 9A. This organization assumes that the MICs are themselves connected in a matrix. Now if any pin selector or connector fails the two remaining pin connections may be used. The MIC connected to this chip must also use the same data connections; therefore, we have fault tolerance to any single selector or data connection failure between the interface of two MIC chips. In the general case, this fault tolerance requires four extra pin connections and pin selectors. Furthermore, except for selectors at the end of the rows or columns, column pin selectors are 3-way and row pin selectors are 4-way.

To complete the reconfigurable mesh node design the PEs must be connected to the enabled mesh nodes. One way of doing this is illustrated in Figure 12A. For a 2×2 active mesh there is one extra column of mesh nodes and one extra PE. Since any mesh node may fail the node-PE selector is associated with the PE rather than the mesh node in contrast to the PE-only fault tolerance

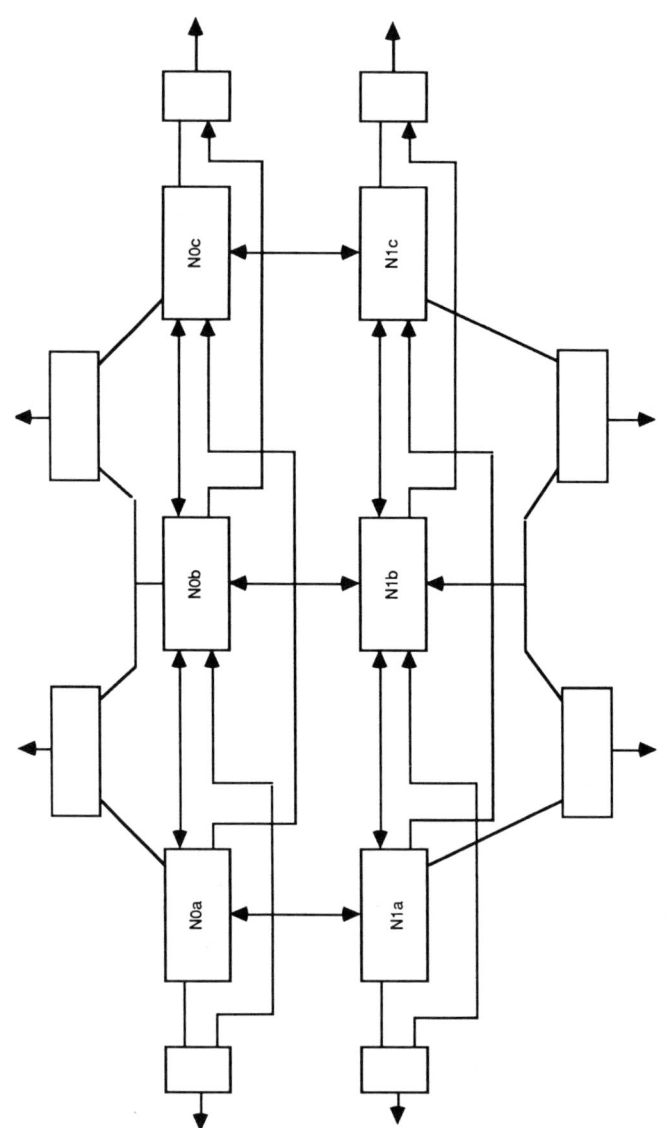

FIGURE 9A. MIC organization for fault tolerance to any single mesh node failure.

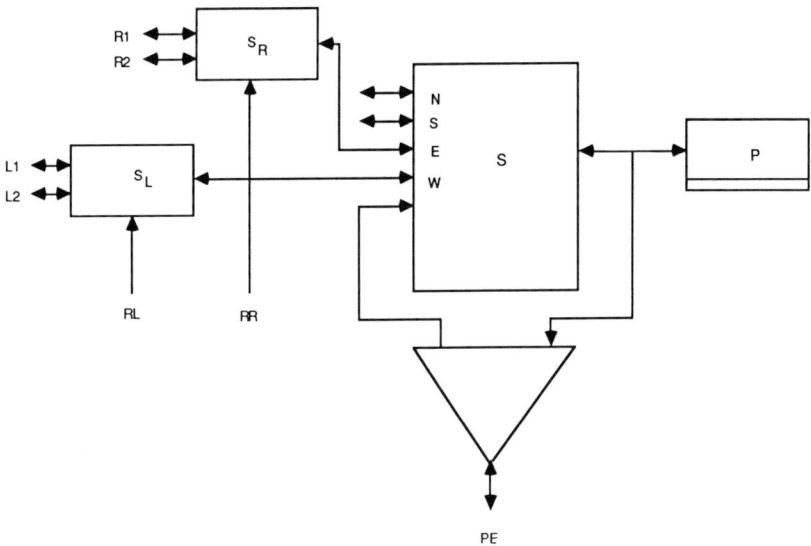

FIGURE 10A. A column reconfigurable mesh node.

shown in Figure 6A. In the general case, each PE must be connected to a mesh node by either a 2-way or a 3-way selector to the mesh nodes.

Finally, we note that there is a simple extension to this scheme to achieve reconfiguration for any two faulty mesh nodes. This may be done by having either two extra columns or one extra row and one extra column. In either case all the mesh nodes require an additional two data connections. In general a second spare column will be cheaper than an additional row. For example, for an $n \times n$ MIC a spare row requires $n + 1$ mesh nodes whereas a spare column only requires n mesh nodes.

MODULE FAULT TOLERANCE

Fault tolerance to catastrophic chip failures such as a broken power line or command line may be achieved by organizing the total array into a set of modules. Each module contains a set of related PEs and fault tolerance is achieved by having a spare module available when one fails.

For the discussions in this section an example array design will be considered; however, the techniques discussed here are general in nature and may be applied to systems with very different design parameters. The example system could be constructed with the present day technology and is for a 1000×1000 PE array. The three PE chip types have the following characteristics: each PE chip contains 16 PEs, each MIC contains a 4×4 matrix of active mesh nodes, and there is a memory chip for every 4 PEs. Fault tolerance will be considered

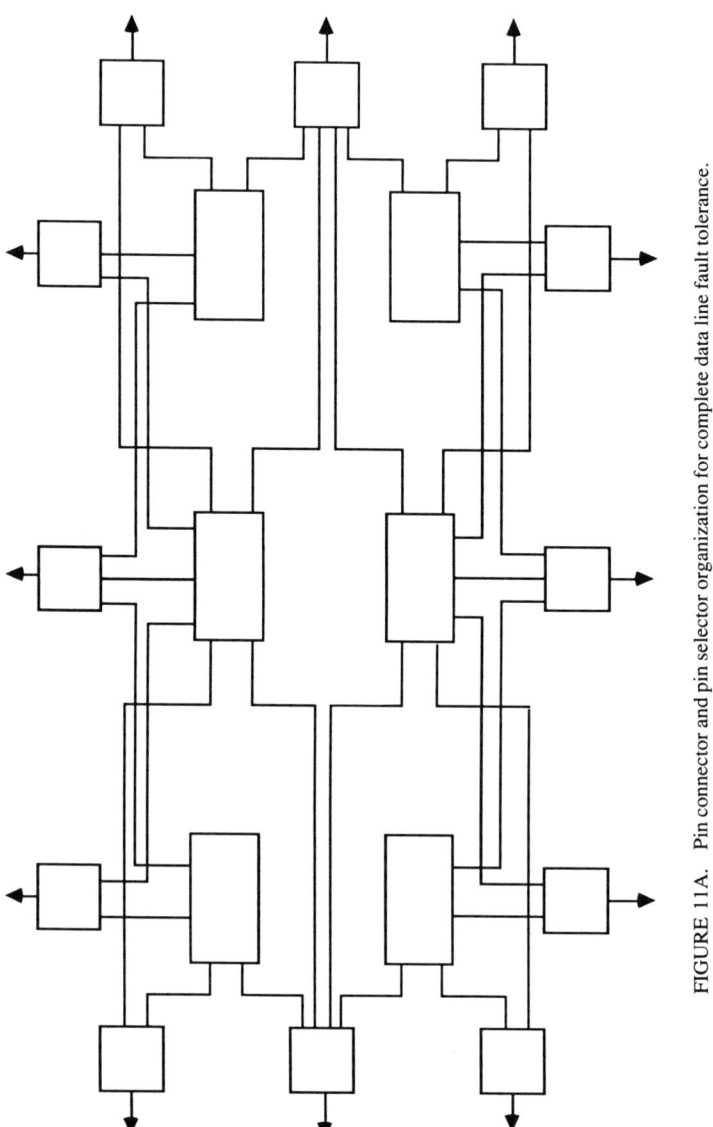

FIGURE 11A. Pin connector and pin selector organization for complete data line fault tolerance.

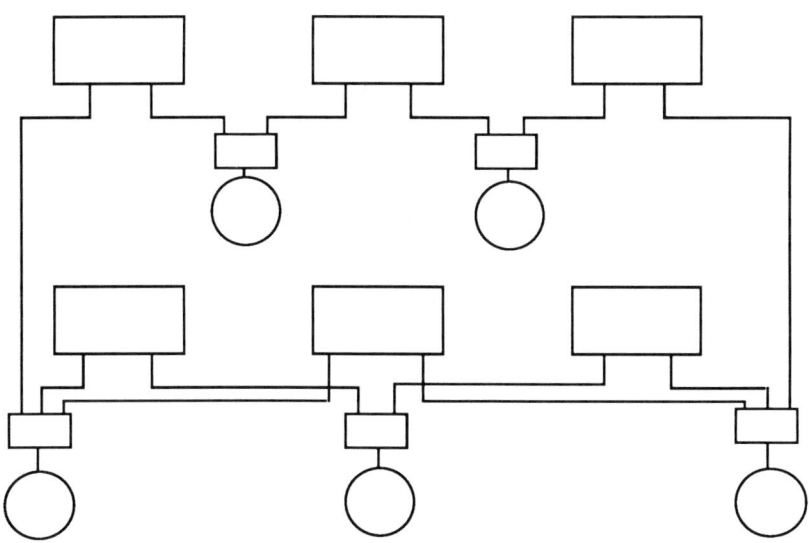

FIGURE 12A. PE — mesh node interconnections for both PE and mesh node fault tolerance.

at two module levels: (a) at the chip level and (b) at the group level where each group consists of a set of chips.

CHIP LEVEL FAULT TOLERANCE

A catastrophic fault could occur in any chip; the PE chip is considered first. The smallest possible group size consists of the following: one MIC, one PE chip (plus an extra PE for fault tolerance), and 4 memory chips (plus 1 bit for fault tolerance). This group, therefore, involves a 4×4 matrix of active PEs. If the PE chip fails, then it is necessary to replace the whole group.

As an alternative, a larger group may be used involving 256 active PEs in which the PE chips are distributed between the MICs. This group consists of 16 MICs, 17 PE chips and 68 memory chips. Each PE chip contributes one PE to every MIC in the group; therefore, since each MIC has fault tolerance to one faulty PE, a single PE chip failure can be handled locally within the group. The cost of PE chip fault tolerance within the group is more complex interchip data routing.

The total failure of a memory chip would render four PEs inoperable in our example. However, the memory chips may be distributed between MICs in a similar way to PE chips in order to achieve group local fault tolerance.

If an MIC fails completely then the whole group is rendered inoperable. Therefore we need a mechanism to selectively enable a group in the total array. One approach is to make a group in the form of a column of PEs and have a spare column of PEs in a similar scheme to the MPP. In our example design, a group could be organized as a 64 × 64 matrix of PEs and 16 groups would constitute one 4 PE-wide column of the PE array; 256 such columns would be required for the complete PE array.

To allow for the disabling of a column each MIC needs to have a spare set of data selectors and data pin connections for all data lines in the E/W directions. This spare set would bypass the adjacent column and link with the spare data connections on the following column. In this way any single column may be completely isolated. Since there are 256 active columns it might be advantageous to have more than one spare column. Then multiple MIC chip failures could be dealt with as long as they do not occur in adjacent columns.

COST OF MIC FAULT TOLERANCE

The cost of implementing the fault tolerance schemes with the MIC has been estimated using three measures: (1) the number of selectors, (2) the number of internal data lines, and (3) the number of data pin connections. The number of selectors may be considered as a measure of the functional complexity of the chip. No weight is attached to the complexity of each selector and the small amount of control logic is not considered. Although the mesh node selectors are more complex for the fault tolerant design this is balanced by the many additional simpler selectors which are used for PEs and data pins. The number of internal data lines is also a measure of chip complexity since they may consume a large proportion of the chip area. The data pin count gives a good indication of the pin requirements of the MIC since less than 10 additional pins will be required for control and power.

The costs for various fault tolerant MIC configurations are expressed for a m x n mesh design in Table 1A. The first row in Table 1A is the cost for the simplest MIC without any falt tolerance. The second row is the single PE fault tolerance as shown in Figure 7A. The third row is for an MIC with complete single mesh node and data interconnection fault tolerance as illustrated in Figures 9A through 12A. The last two rows include the cost of an extra set of left and right data connections and selectors for group fault tolerance. The first set of figures is for a spare column within the MIC and the second set of figures is for a spare row within the MIC. The spare row concept is slightly cheaper than a spare column for a square mesh, i.e., when n = m.

CONCLUSION

The problem of fault tolerance in highly parallel mesh connected processors has been considered and methods of protecting against the most probable faults in the PE array have been proposed.

Fault tolerance at different levels has been considered. It has been shown that fault tolerance to the most error sensitive components, i.e., the functionally complex PE ALUs and local memory chips, may be achieved at a low cost at

TABLE 1A
MIC Cost for an n × m Active Mesh

Fault tolerance	Selectors	Internal data lines	Data pin connections
None	mn	3mn + m + n	mn + 2 + 2n
Single PE	2mn	4mn + m + n + 1	mn + 2m + 2n + 1
Single mesh Node	2mn + 2m + 3n + 5	6mn + 7m + 7n − 1	mn + 2m + 2n + 5
Group (a)	2mn + 4m + 3n + 7	6mn + 17n + 7n − 9	mn + 4m + 2n + 7
Group (b)	2mn + 5m + 2n + 7	6mn + 15m + 7n − 5	mn + 4m + 2n + 7

the local level. More extensive but less common errors such as catastrophic chip failures and broken command lines including a faulty OR bus line usually need to be dealt with at the more expensive module or column level.

The cost of a high degree of fault tolerance can be achieved with a moderate amount of additional hardware. Such hardware may become a very important part of VLSI PE arrays having 1,000,000 or more nodes or in applications for smaller arrays in situations where high reliability and fault tolerance is necessary.

REFERENCES

1. **Bouknight, W. J., Denenberg, S. A., McIntyre, D. I., Randall, J. M., Samen, A. H., and Slotnick, D. L.**, The Illiac IV System, *Proc. IEEE,* 60(4), 369, 1972.
2. **Batcher, K. E.**, Design of a Massively Parallel Processor, *IEEE Trans. Computers,* C-29(9), 836, 1980.
3. **Gilmore, P. A., Batcher, K. E., David, M. H., Lott, R. W., and Burkley, J. T.**, Massively Parallel Processor: Phase 1 Final Report, Goodyear Aerospace Tech. Rep., 1979.
4. **Reeves, A. P.**, Parallel Computer Architectures for Image Processing, 1981 International Conference on Parallel Processing, Bellaire, Michigan, August 25—28, 1981.
5. **Reeves, A. P.**, The anatomy of VLSI binary array processors, in *Languages and Architectures for Image Processing,* Duff, M. J. B. and Levialdi, S., Eds., Academic Press, New York, 1981, 267.
6. **Reeves, A. P.**, On efficient global information extraction methods for parallel processors, *Computer Graphics and Image Processing,* 14, 159, 1980.

ACKNOWLEDGMENTS

SECTION 3:   Illiac IV - The First Supercomputer
   R. Michael Hord, Computer Science Press, Rockville, MD 1982
SECTIONS 3.7 to 3.10: *Special Computer Architectures for Pattern Processing*
   K.S. Fu and Tadao Ichikawa, Eds., CRC Press, Boca Raton, FL, 1982, chap. 6
SECTION 4: Frontiers of Massively Parallel Scientific Computation
   NASA Conf. Publ. 2478, NASA/Goddard Spaceflight Center, Greenbelt, MD 1986
SECTION 4.1: Technical Summary — The Massively Parallel Processor
   Frontiers 86, p. 293
SECTION 4.2: MPP Parallel Fourth
   John Dorband, GSFC, Frontiers 86, p. 275
SECTION 4.3: The Language Parallel Pascal and Other Aspects of the Massively Parallel Processor
   Anthony P. Reeves and John Bruner, School of Electrical Engineering, Cornell University, Ithaca, NY 1982
SECTION 4.4: A Fast MPP Algorithm for Ising Spin Exchange Simulation
   Francis Sullivan and Raymond Mountain, NBS, Frontiers 86, p. 53
SECTION 4.5: Development of a Stereo Analysis Algorithm for Generating Topographic Maps Using Interactive Techniques on the MPP
   James Strong, GSFC, Frontiers 86, p. 141
SECTION 4.6: Sort Computation and Conservative Image Registration
   John Dorband, NASA GSFC, December 1985
SECTION 5.1: DAP Series Technical Overview
   Active Memory Technology, Inc., 16802 Aston St., Irvine, CA
SECTION 5.2: An Overview of Parallel Data Transforms
   Peter M. Flanders, Active Memory Technology, Inc., Irvine, CA
SECTION 5.3: Solution of a Large System of Equations on DAP using a Hybrid Gauss/Gauss-Jordan Method
   D. J. Hunt, International Computers Ltd., Research and Advanced Development Center, Document CM75, July 1981
SECTION 5.4: An Image Understanding Performance Study on the ICL Distributed Array Processor
   D. E. Oldfield and S. F. Reddaway, ICL, Bracknell, Berkshire, U.K.
SECTION 6.1: The Geometric Arithmetic Parallel Processor
   Eugene Cloud, Martin Marietta Corp., Orlando, FL
SECTION 6.2: Systolic Array Chip Matches the Pace of High Speed Processing
   Ronald Davis and Dave Thomas, NCR Corp., Microelectronics Div., Ft. Collins, CO [Reprinted with permission from *Electronic Design,* Vol. 32, No. 22, October 31, 1984. Copyright 1984, Penton Publishing]
   Systolic Array Chip Recognizes Visual Patterns Quicker Than a Wink
   W. W. Smith, Jr., Whitman Engineering, Inc., Maitland, FL
   Paul Sullivan, NCR Microelectronics [Reprinted with permission from *Electronic Design,* Vol. 32, No. 24, November 29, 1984. Copyright 1984, Penton Publishing]
SECTION 6.3: Parallel Array Processor Results in Near Real Time
   S. P. Buchanan, R. E. Wood, and D. A. Simmons, Image and Signal Processing, Martin Marietta Corp., Orlando, FL
SECTION 7.1: CM-2 Technical Summary
   Thinking Machines Corp, Cambridge, MA

SECTION 7.3.1: Numerical Computation of Electromagnetic Scattering on the Connection Machine Using the Method of Moments
Eric Opp, Steven Geyer, Robert Thomas, and Michael Willett, MRJ, Inc.,
Korada Umashankar, Dept. Electrical Engineering, University of Illinois, Chicago
IEEE J. Magnetics, August 1989
    Parallel Implicit Methods for Numerical Physics Using the Connection Machine
Eric Opp, 12th Conf. Numerical Simulation of Plasmas, San Francisco, CA, Sept. 21-24, 1987
SECTION 7.3.2: Massively Parallel Implementations of Some Common 0/1 Knapsack Approximation Algorithms
Scott Weidman and Thomas Gerasch, MRJ, Inc., April 1987
    Nonlinear Network Optimization on a Massively Parallel Connection Machine
S. A. Zenios, Dept. of Decision Sciences, The Wharton School, University of Pennsylvania, Philadelphia
Robert Lasken, MRJ, Inc. [See also *Ann. Oper. Res.*]
SECTION 7.3.3: Automatic Target Detection on the Connection Machine
R. Michael Hord, SPIE Conf. Intelligent Robots and Computer Vision, Cambridge, MA, November 2-6, 1987
SECTION 7.3.4: Neural Network Implementation Approaches for the Connection Machine
Nathan H. Brown, Jr., IEEE Conf. Neural Information Processing Systems, Denver, CO, November 8-12, 1987
    Parallel Free-Text Search on the Connection Machine
Craig Stanfill and Brewster Kahle, Thinking Machines Corp., *Commun. ACM,* 29(12), 1229, 1986
    Orbit Collision Problem Benchmark Study
Kirk Berge, Steven Geyer, and Robert Wall, MRJ, Inc., December 1988
SECTION 8.1: The GAM II Pyramid
Zahi Abuhamdeh, George Mason University, Fairfax, VA
SECTION 8.2: CLIP4
Michael J. B. Duff, University College, London *Special Computer Architectures for Pattern Processing,* K. S. Fu and Tadao Ichikawa, Eds., CRC Press, Boca Raton, FL, 1982
SECTION 9: The Giant Fourier Transform
Steve Bershader, Thomas Kraay, and John Holland, MRJ, Inc., November 1988. Presented at "Scientific Uses of the Connection Machine," NASA Ames Research Center, Fall 1988; and Hypercube Computer Architectures Conf., Spring 1989
    Introduction to Data Level Parallelism
Thinking Machines Tech. Rep. 86.14, April 1986
APPENDIX: The Language Parallel Pascal and Other Aspects of the Massively Parallel Processor
Anthony P. Reeves and John Bruner, Cornell University, Ithaca, December 1982

INDEX

A

Activation update, 287
Activation update cycle times, 289—290
Activation update kernel, 285
Adder Network, GAM II Pyramid, 303
Advanced Data Buffer (ADB), 40
ADVAST (Advanced Station), 34—36
Algorithms, see specific types
AMT DAP, 156—157, 159
APPLE, 48—50
Approximation algorithms, 264—265
Architecture, see specific systems
Arithmetic control unit (ACU), 7—9
Arithmetic logic unit (ALU), 212—213
Arithmetic operations
  CLIP4, 318—323
  GAPP, 188
  parallel FORTH language, 93—94
Arithmetic unit (AU), 7—8
Arithmetic words, 95—97
Array, 41, 108, 116, 185—186
Array addressing, 51—52
Array assignment statement, 109
Array control unit, MPP, 85, 88—89
Array indexing, 108—110, 112
Array instruction generation, 306
Array manipulation, 115
Array memory addresses, 151—152
Array processing, CLIP4, 309—327
Array processors, 5—6, 81
Array reduction, 116
Array stack, 91
Array stack operations, parallel FORTH language, 92—93
Array tests, CLIP4, 325
Array unit, MPP, 85, 87—88, 90
Artificial vision, 1
As-if-serial rule, 230
Assignment statements, 41, 44—47
Associative output control unit (AOCU), 7, 9
Associative processors, 5—6
Automatic target detection, 268—273, see also Image processing
Azimuth correlation, 69
Azimuth vectors, 66—67

B

Back plane, GAM II Pyramid, 302—303

Bad match pixels, 130
Binary Array Processor (BAP), 348—351
Binary processes, 318
Binary-tree summing, 194—195
Blobs, 164, 170, 272—273
Boltzmann's constant, 121
Boolean functions of two binary images, 313—315
Boolean operations, 123—124, 144
Boolean processor, 311—312
Boundary conditions, 59
Branching program control, 43
Burroughs B-6700 computer, 21
Busy bits, Illiac IV, 37—38

C

C* language, 208—209, 227—233
  abstract machine model, 227
  as-if-serial rule, 230
  code types, 228
  data types, 228—229
  domains vs. classes, 229
  memory layout, 227—228
  minimum and maximum operators, 229
  parallel statements, 231—233
  replication rule, 230
  scalar data, 229
Case statement, 117, 119
Case studies, Illiac IV, 75—79
Category assignment, 63
Category migration, 63—64
CDC-7600, 2
Cell description, GAPP, 182—183
Cellular Logic Image Processor, see CLIP4
CFD language, 39—47
  assignment statements, 41, 44—47
  bit setting operations, 45
  bit shifting operations, 45
  Boolean operations, 45
  common blocks, 41
  control statements, 41, 43—44
  disk areas, 41
  FORTRAN contrasted, 41
  input/output statements, 41—43
  integer vector index, 47
  named quantities, classes of, 41
  review, 48—56, see also Illiac IV
  scalar arithmetic statements, 45—46
  specification statements, 41—42
  statements, categories of, 41—47

subprogram statements, 41—42
subprograms, 41
translators, 47
variables, 41
Chip descriptions, GAPP, 183—185
Chip feature sizes, smallness, 1
Chip level fault tolerance, 361—362
Chip performance/mechanics, GAPP, 184—185
Chirped radars, 65
Classification
  DAP, 176—177
  Illiac IV 64—65
CLIP4, 308—327
  arithmetic operations, 318—323
  array processing, 309—327, see also other subtopics hereunder
  array tests, 325
  Boolean functions of two binary images, 313—315
  complete logic circuit, 323
  counting, 323—324
  empty array test, 325
  functions other than processor, 323—326
  input/output of images, 324—325
  INSERT instruction, 324
  integrated circuit, 325—326
  labeled propagation operations, 318—319
  local neighborhood operations, 316—318
  operating speeds, 326
  processor design, general considerations, 311—312
  processor functions, 312—323
  propagation test, 325
  register operations, 325
  serial computer functions, 325
  shifting binary images, 315—316
Clustering technique, 63—64
CM-Lisp language, 208, 222
Coarse grain, 6—7
Common blocks, 41
Comparison operations, parallel FORTH language, 93—94
Comparison words, 98—99
Compatibility, 112, 114
Compiler words, 104
Component and manufacturing technology, Illiac IV, effects of, 80—81
Composite approach, 282—284, 287
Computational algorithm, 59—60
Computational fluid dynamics, 1
Computing technology, Illiac IV, effects of, 79—84

Concurrent computing, 2, see also Parallel processing
Condition code register, 305
Conditional assignment, 118
Conditional evaluation, 119
Configuration, STARAN, 10, 13
Connection machine (CM), 205—300
  applications, 236—298, see also other subtopics hereunder
  arithmetic logic unit, 212—213
  automatic target detection, 268—273
  background, 205—206
  C* language, 208—209, 227—233
  CM-Lisp language, 208, 222
  CM-2 system organization, 209—217
  conditionals, 223
  data processors, 212—213
  data set, 206—207
  data vault, 219—221
  floating point accelerator, 215, 218—219
  front-end system, 208
  hardware, 208
  image processing, 236, 268—273
  input/output channel, 212, 217
  input/output structure, 216
  input/output system, 208
  interprocessor communication, 214, 218
  0/1 Knapsack approximation algorithms, 263—268
  languages, 208—209, 222—236
  linkages among data elements, 222
  *Lisp language, 208, 222—227
  mixed data, 223
  neural nets, 236, 279—291
  nonlinear network optimization, 250—263, see also Operations research
  operations research, 236, 250—268
  orbit analysis, 236, 291—298
  overview, 208—209
  parallel data structures, 208, 222—223
  parallel processing unit, 206—207, 209—212, 219
  Paris language, 233—236
  performance specifications, 217—219
  physics, 236—250
  primary use, 205
  program control on front end, 207—208
  programming, 221—236
  router, 210—211, 217
  router nodes, 214—215
  scalar data, 222
  sites, 207
  sizes, 207

software, 208
system, 206—221
text search, 236, 273—279
Connectivity, 7—9
Constants, 91—92
Context changing words, 95
Context switching, 90—91
Control, GAPP, 183—184
Control/clock, GAPP, 183
Control-flow facilities, parallel Pascal design, 113, 116—120
Control operations, parallel FORTH language, 94—95
Control statements, 41, 43—44
Control unit (CU)
  Advanced Data Buffer, 40
  Illiac IV, 30, 33—35, 39—40
  PEPE, 7—9
  SIMD architectures, 5—6
Control unit overlap, Illiac IV, 34—35
Control words, 102—103
Controllers, GAM II Pyramid, 307—308
Conventional computers, 18, 19
Convolutions, 166—169, 191—199
Corner turning functions, 89
Correlation, 67—69, 191—195
Correlation control unit (CCU), 7
Correlation unit, 7—9
Cost-performance characteristics, 2
Counting, CLIP4, 323—324
CRAY-1, 2
CRAY series of vector supercomputers, 251
CRYSTAL multicomputer, 251
CYBER 205, 124

D

D-memories, 312—315, 318, 321
DAP, 143—179
  array memory addresses, 151—152
  background, 143
  classification, 176—177
  components, 145
  costs in cycles of operation, 171, 173—176
  debugging facility, 154—155
  design, 143—156
  DOG (difference of Gaussians) channel application, 164, 166—172
  fast data channel, 145—146
  feature generation, 173—176
  FORTRAN-Plus language, 144, 154—156
  Gauss-Jordan elimination, 159—164
  hardware, 149—152
  host, 144—145
  host connection unit, 145—146
  image capture, 164—166
  image understanding performance study on, 164—178
  intermediate matrix, 161
  language systems, 154
  languages, 144
  libraries of subroutines, 155
  master control unit, 144—146, 149—152
  matrix after forward elimination, 162
  parallel data transforms, 156—159
  processor array, 144
  processor elements, 143—152
  program parts, 154
  programming software, 154
  programs, 144—145
  registers, 150
  segment labeling, 172—173
  segmentation, 169—172
  simulation system, 154—155
  software, 144—156
  upper triangular matrix, 161
  VME bus, 153
Data bases, 1
Data executive sub-unit, 304—305
Data level parallelism, 329
Data memory, 306
Data processors, connection machine, 212—213
Data set, connection machine, 206—207
Data vault, connection machine, 219—221
Daughter cards, GAM II Pyramid, 301—302
Debugging aids, 86
Defining words, 104
Design, see specific system
Design automation, 80
Design concept, Illiac IV, 18—21
Diagnostic tool simulation, 81—82
Digital Equipment Corporation PDP-10 conventional computer, 21
Digital radar map, 67
Digitizer, GP^2, 200—201
Dilation and erosion, 191
Disabled processors, Illiac IV, 21
Disk areas, 41
Disparity function, 128—130

Display board, GP^2, 200, 203
Distributed Array of Processors, see DAP
Distributed memory architectures, 7, 9
Distributed relaxation algorithm, 252—253
DOG (difference of Gaussians) channel application, DAP, 164, 166—172

E

Early machines, 7—15
ECL development, 80
Economies of scale, 3
Edge enhancement, 191, 197
Edge list approach, 280—282, 285
Edge list interconnect update speeds, 289, 291
Elastic zone, 59
Electromagnetic scattering using method of moments, 237—241
Element memory (EM), 7
Elmental functions, 114
Empty array test, 325
Enable/disable function, Illiac IV, 21, 43
Enabled processors, Illiac IV, 21
Evolution of computer technology, 330
Expand function, 116

F

Fast data channel, 145—146
Fast Fourier transform (FFT), 69—70
Fault model, 58
Fault tolerance concepts, 347—363
Feature generation, DAP, 173—176
Fine grain, 6—7, 241
Finite element analyses, 1
Finite-impulse-response filter, 196—197
FINST (Final Station) overlap, Illiac IV, 34—37
Floating-point accelerators (FPAs), 336
Floating point operations per second (FLOPS), 1
Fluid dynamics using cellular automata, 338—344
Fluid flow simulation, 338—344
For construct, 117, 119
FORTRAN, 49
  CFD contrasted to, 41
  control statements, 43
  specification statements, 42
FORTRAN-Plus language, 144, 154—156
Forward looking infrared (FLIR), 200

Forward looking infrared (FLIR) sensor, 182
Fourier transform, 66—67, 72
Full adder/subtractor, GAPP, 183
Full pivoting, 160
Function call, 117
Future machines, 82

G

GAM II Pyramid, 301—308
  Adder Network, 303
  array instruction generation, 306
  back plane, 302—303
  condition code register, 305
  control buses, 303
  controllers, 307—308
  data executive sub-unit, 304—305
  data memory, 306
  daughter cards, 301—302
  future control system expansion, 307—308
  general purpose registers, 305
  host system, 306—307
  input/output unit, 306—307
  micro memory, 305
  next address generator, 304
  pipeline register, 305
  processing element, 301—302
  program flow sub-unit, 304
  sequencer unit, 303—304
  special purpose registers, 305—306
  structure, 301—302
GAPP, 181—203
  algorithm, 190
  arithmetic operations, 188
  array, 185—186
  array assembly, 185
  array size, 185—186
  background, 181—182
  binary-tree summing, 194—195
  cell description, 182—183
  chip descriptions, 183—185
  chip performance/mechanics, 184—185
  control, 183—184
  control/clock, 183
  correlation, 191—195
  design, 181—186
  edge enhancement, 191, 197
  finite-impulse-response filter, 196—197
  full adder/subtractor, 183
  highest intensity pixels, 194

image enhancement, 191, 196—197
image processing, 192—194
input/output, 185
instruction set for systolic array
 processor, 189
logic operations, 188
processor element diagram, 187
processor element links, 186
programming, 186—199
RAM, 183
register shifting, 192—193
registers, 183
shift register groups, 184
Sobel filter, 197—199
sorting pixels into bins, 195—196
system, 199—203
thresholding, 193
translation, 193
truth table, 188
GAPP array, GP^2, 200, 202
GAPP controller, 200—201
GAPP peripheral processor, see GP^2
Gauss-Jordan method, 159—164
General purpose registers, 305
Geometric arithmetic parallel processor, see GAPP
Giant Fourier transform, 334—338
Glauber dynamics, 123
Global flow adjustments in relaxation steps, 260—261
GLYPNIR, 29, 48—56, see also Illiac IV
Goodyear MPP, 156
Goto statement, 117—118
GP^2, 200—203
 architecture, 200—203
 digitizer, 200—201
 display board, 200, 203
 functional blocks, 200
 GAPP array, 200, 202
 GAPP controller, 200—201
 input buffer, 200—202
 output buffer, 200, 202—203
 system controller, 200—201
Gray-level processes, 318—323

H

Hardware, see specific system
Heat, need to dissipate, 1
Hebbian weight modification kernel, 285
Hebbian weight update kernel, 287
Hierarchical Warp Stereo (HWS) technique, 127

Hopfield spin-glass model, 289
Hopfield weight prescription, 286
Host connection unit (HCU), 145—146
Host processor, MPP, 85, 90
Host system, GAM II Pyramid, 306—307

I

If statement, 117—119
Illiac IV, 2, 17—84, 348
 advantage of, 48
 ADVAST, 34—36
 application, 48, 83—84
 applications programs, 29—30
 architectural features, 18
 architecture of system, 30—39
 array addressing, 51—52
 background, 17
 billion-bit storage subsystem, 21
 branching, 43
 busy bits, 37—38
 case studies, 75—79
 CFD language, 39—47, see also CFD language
 circuitry, 24—26
 classification, 64—65
 clustering, 63—64
 component and manufacturing technology, 80—81
 computational methodology, 60
 computing technology, effects on, 79—84
 conceptual architecture, 31, 40
 control unit, 30—35, 39—40
 control unit overlap, 34—35
 data base design, 60—61
 data management scheme, 61—62
 design concept, 18—21
 difficulties with, 48
 digital processing of synthetic aperture radar data on, 65—69
 emitter-coupled logic circuits, 24—26
 empirical evidence, 56—58
 enable/disable function, 21, 43
 escalating costs, 28—29
 execution rates, 33
 execution time, 32
 fast Fourier transform, 69—70
 FINST overlap, 34—37
 first SIMD supercomputer, 17—84
 forerunner of, 11
 funding of project, 22
 hardware, 24
 history, 17—30

implementation difficulties, 21—30
implementation of, 60—62
input/output peripherals, 21
Instruction Look Ahead, 34—35
interconnection structure, 21
interconnections, 27
interfaces between units, 35
Landsat, 62—65
languages, 29, see also specific languages
   review, 48—56, see also other subtopics hereunder
linear programming image enhancement, 70—74
Logic Unit, 38
machine architecture, 81—82
main memory, 21, 41
management of memory hierarchy, 55—56
megabyte of processor element memory, 21
Memory Service Unit, 34—35
memory subsystem, 21
mode values, 21
model for disaster, 76—78
Monte Carlo methods, 78—79
overlap, types of, 34—37
packaging, 27
performance, 56—58
predecessors, 18
primary memory, 31—32
processing element, 30—39, 41, 70
   variables, 50—51
   variables memory allocation, 51
processing element memory, 19—20, 30, 33—34, 41, 70
processing element number, 31
program development and maintenance tools unavailable to user of, 55
Read-Only-Memory, 36—37
reduction in size, 28—29
research-operational ambivalence, 23—24
ROUTE instruction, 31
routing of operands, 52—56
routing paths, 20—21
routing program, 27
Rowsum operation, 32—33
semiconductor memories, 26
software, 24, 29, 55
sparse matrix multiply, 75—76
subsystems for, 19
system architecture, 82—83
Test and Maintenance Unit, 34
thin-film memory, 25—26
transistor-transistor logic, 27—28
transposition, 68
TRES computer program, 58
14TRES program, 61—62
University of Illinois campus turmoil, 21—23
user input and output, 60
Illiac IV Disk Memory, 21
Illiac IV SAR data processing program, 67
Image capture, DAP, 164—166
Image enhancement, 191, 196—197
Image processing, 199—200
   CFAR annulus sum, 268, 272—273
   connection machine application, 236, 268—273
   first stage, 268—272
   GAPP, 192—194
   gray level, 269, 272
   local area sum, 269, 272
   MAX-MIN texture, 268—271
   second stage, 272—273
   Sobel, 268, 270
   third stage, 272—273
Image understanding performance study, distributed array processor, 164—178, see also DAP
Implementation cost, 2—3
Implementation difficulties, Illiac IV, 21—30, see also Illiac IV
Independent disk drives, synchronization of, 81
Initialization, 63, 67
Input, 68—69
Input blocks of vectors, 67
Input buffer, GP^2, 200—202
Input/output, 120
   CLIP4, 324—325
   connection machine, 208, 216
   GAM II Pyramid, 306—307
   GAPP, 185
   parallel FORTH language, 92
Input/output channel
   connection machine, 212, 217
Input/output control unit, MPP, 89
Input/output peripherals, Illiac IV, 21
Input/output statements, 41—43
Input/output words, 103
INSERT instruction, CLIP4, 324
Instruction Look Ahead (ILA), Illiac IV, 34—35
Integer vector index, 47
Integrated circuit CLIP4, 325—326
Interconnection system, Illiac IV, 21

Interconnection technology, 80
Interconnections, Illiac IV, 27
Interconnects, 7
Interprocessor communication, 2
Ising spin exchange simulations, 121—125
  algorithms, 122—124
  basic model, 121
  $M(RT)^2$ algorithm, 121—122
  vector machines, 124—125
IVTRAN, 48—56, see also Illiac IV

K

Kawasaki dynamics, 123
0/1 Knapsack approximation algorithms, 263—268

L

Labeled propagation operations, 318—319
Laboratory for Applications of Remote Sensing, see LARS
Landsat Multi-Spectral Scanner satellite, 62—65
  data analysis, 63
  Illiac IV implementation, 63—65
Languages, see also specific types
  connection machine, 208—209, 222—236
  goals, 106
LARS, 63
LARSYS, 63—64
Libraries of computational subroutines, 87
Libraries of subroutines, 155
Light, speed, 1, 18
Linear programming image enhancement, Illiac IV, 70—74
*Lisp language, 208, 222—227
Local neighborhood operations, 316—318
Logic circuitry, 80
Logic operations
  GAPP, 188
  parallel FORTH language, 93—94
Logic Unit (LOG), Illiac IV, 38
Logical words, 98
Loop, 68—69, 94, 119—120
Lyapunov energy equation, 286

M

Machine architecture Illiac IV, effects of, 81—82
Main control unit, MPP, 89—90

Main memory, Illiac IV, 21, 41
Mapping, 106, 149
Mapping vectors, 158
Markov process, 122
Marr-Poggio algorithm, 127
Mask bit, 94—95
Mask stack, 91
Mask stack operations words, 105
Mask stack primitives, processing element control unit and, 95
Masked assignment, 118
Masked classification, 65
Massively parallel network optimization, 251—252
Massively Parallel Processor, see MPP
Master control unit (MCU), DAP, 144—146, 149—152
Master/slave control relationship, 86
Matrix inversion, 1, 159—160
MAX function, 330—331
Maze solving, 330—332
Megabyte of processor element memory, Illiac IV, 21
Memory operation words, 101—102
Memory operations, parallel FORTH language, 92
Memory Service Unit (MSU), Illiac IV, 34—35
Merge aggregation, 139—140
Merge distribution, 139—140
Mesh connected parallel processors, 348—351
Mesh interconnection chip, see MIC mesh node fault tolerance
Mesh interconnection scheme, 347—348
Method of moments (MOM), 237—241
MIC fault tolerance cost, 362—363
MIC mesh node fault tolerance, 354—359
Micro memory, 305
Microprogramming, 81
MIMD, 329, see also Parallel processor architectures, 6
  single instruction multiple data distinguished, 5
Mode values, Illiac IV, 21
Module fault tolerance, 359—361
Monte Carlo methods, 121
  Illiac IV, 78—79
MPP, 85—142, 181, 348—349, 351, 357
  array control unit, 85, 88—89
  array unit, 85, 87—88, 90
  corner turning functions, 89

design, 85—90
hardware, 87—90
host processor, 85, 90
input/output control unit, 89
Ising spin exchange simulation, 121—125
libraries of computational subroutines, 87
main control unit, 89—90, 108
master/slave control relationship, 86
multi-dimensional access function, 90
parallel FORTH language, 90—105, see also Parallel FORTH language
parallel Pascal design, 105—120, see also Parallel Pascal design
processing element control unit, 89
processing elements, 88
simulation environments, 87
software, 85—87
sort, 132—141
speed of typical operations of, 88
staging memory, 85, 89—90
stereo analysis, 125—132, see also Stereo analysis
MPP Pascal, 86—87
MPP Simulator, 87
$M(RT)^2$ algorithm, 121—122
Multi-dimensional access function, 90
Multilayer PC cards, 81
Multiple instruction multiple data processors, see MIMD

N

Navier-Stokes equations, 338, 343
Neural nets
  complexity analysis, 287—288
  composite approach, 282—284, 287
  connection machine applications, 236, 279—291
  control structures, 284—287
  data structures, 280—284
  edge list approach, 280—282, 285
  performance comparison, 289
New algorithms, 83
Newton's law, 59
Next address generator, 304
Nonlinear network optimization, connection machine, 250—263

O

Operating speeds, CLIP4, 326

Operations research
  approximation algorithms, 264—265
  connection machine application, 236, 250—268
  CRAY series of vector supercomputers, 251
  CRYSTAL multicomputer, 251
  distributed relaxation algorithm, 252—253
  global flow adjustments in relaxation steps, 260—261
  0/1 Knapsack approximation algorithms, 263—268
  massively parallel network optimization, 251—252
  nonlinear network optimization, 250—263
  parallel relaxation algorithm, 256—260
  parallel relaxation compared with sequential algorithms, 261—263
  parallelization, 265—268
  relaxation algorithm, 253—256
Optical ray trace, 1
Optimization, 1
Orbit analysis, connection machine application, 236, 291—298
Orbit collision
  CM-2 specifications, 293
  conventional supercomputer comparison parameters, 292
  summary results of orbital benchmark, 294—298
Order of algorithm, 330
Otherwise clause, 118
Output, 68—69, see also Input/output
Output blocks of vectors, 68
Output buffer, GP^2, 200, 202—203
Overlap, Illiac IV, 34—37
Overlapping architecture, 18

P

Parallel architecture, 18
Parallel array, 108—109
Parallel data transforms (PDT), 156—159
Parallel element processing ensemble, see PEPE
Parallel FORTH language, 90—105
  arthmetic operations, 93—94
  arithmetic words, 95—97
  array stack, 91
  array stack operations, 92—93

array unit, 90
comparison operations, 93—94
comparison words, 98—99
compiler words, 104
constants, 91—92
context changing words, 95
context switching, 90—91
control operations, 94—95
control words, 102—103
data definition, 91—92
defining words, 104
input/output, 92
input/output words, 103
logic operations, 93—94
logical words, 98
loop, 94
main control unit, 90
mask bit, 94—95
mask stack, 91
mask stack operations words, 105
memory operation words, 101—102
memory operations, 92
PECU and mask stack primitives, 95
PECU primitive words, 104—105
stack operation words, 99—101
staging memory, 90
variables, 91
vocabulary, 91—92
word reference, 95—105
Parallel instruction set, see Paris language
Parallel Pascal, 85
Parallel Pascal design, 105—120
  array, 108, 116
  array indexing, 108—110, 112
  array manipulation, 115
  array reduction, 116
  communicability, 107
  compatibility, 112, 114
  conditional, 117
  control-flow facilities, 113, 116—120
  data structuring facilities, 108
  data types, 107
  elemental functions, 114
  expand function, 116
  extensions to standard Pascal, 107
  goals, 106—107
  goto, 117—118
  implementability, 106—107
  input/output procedures, 120
  loop, 119—120
  motivation, 105—106
  parallel array, 108—109
  parallelism, 106
  pointer types, 108
  power set, 108
  procedures, 117
  record, 108
  repetition, 117
  scalar types, 107
  selection, 117
  specification, 106—120
  specification criteria, 107
  standard functions, 114
  standard procedures, 114
  subrange constant, 111—112
  subrange type, 107—108
  target architecture, 108
  transformational functions, 115
  variant record, 108
Parallel Pascal Translator, 87
Parallel processing, 1
Parallel processing supercomputers, 2
Parallel processing unit, conection machine, 206—207, 209—212, 219
Parallel processor, 18
  classes, 5—6
  coarse grain, 6—7
  connectivity, 7—9
  early machines, 7—15
  fine grain, 6—7
  mesh connected, 348—351
Parallel relaxation algorithm, 256—260
Parallel statements, C* language, 231—233
Parallel variable (pvar), 224—226
Parallelism requirement, 18
Paris language, 233—236
  abstract machine architecture, 235
  addressing modes, 236
  conditionalization, 236
  immediate operands, 236
  purpose, 233
  user interface, 234
Pattern processing, 308—310, see also CLIP4
PECU primitive words, 104—105
PEPE, 7—12
  block diagram, 7, 10—12
  characteristics, 12
  control system, 7
  control units, 7—9
  main application, 7
  processing elements, 7—9
  radar processing, 7—9

Physics
  connection machine application, 236—250
  electromagnetic scattering using method of moments, 237—241
  parallel tri-diagonal matrix inversion, 246—247
  thermal diffusion, 237, 241—250
  wave equation, 237, 241—250
Pipeline architecture, 18
Pipeline processors, 5
Pipeline register, 305
Pixels, 64—65, 129, 191, 194—196
Plastic zone, 58—59
Pointer types, 108
Poisson equation, 122
Power set, 108
Procedure call, 117
Processing element control unit
  mask stack primitives and, 95
  MPP, 89
Processing element memory (PEM), 70
  Illiac IV, 30, 33—34, 41
Processing element number (PEN), Illiac IV, 31
Processing elements (PE), 70
  arithmetic unit, 7—8
  correlation unit, 7—9
  element memory, 7
  GAM II Pyramid, 301—302
  Illiac IV, 30—39, 41
  MPP, 88
  PEPE, 7—9
  programmable registers, 41
  SIMD architectures, 5—6
  variables, 50—51
  variables memory allocation, 51
Processing ensembles, 5—6
Processor arrays
  computation on, 156—157
  DAP, 144
Processor elements
  DAP, 143—152
  fault tolerance, 353—356
  GAPP, 186—187
Processor elements array, 145, 147
Processor elements array memory, 145, 147, 149
Program flow sub-unit, 304
Programming defined, 106
Programming languages, see Illiac IV, Languages, or specific language
Propagation test, 325
Pseudo-stereo-pairs, 125

Q

Quam's algorithm, 127—128

R

Radar processing, 7—9
RAM, 183
Range correlation, 67—68
Read-Only-Memory (ROM), Illiac IV, 36—37
Record, 108
Rectilinear grid, 60
Register operations, CLIP4, 325
Register shifting, 192—193
Registers, see specific types
Relaxation algorithm, 253—256
Repeat-until construct, 117, 119
Replication rule, 230
Research SIMD computers, 301—327
  CLIP4, 308—327
  GAM II Pyramid, 301—308
ROUTE instruction, Illiac IV, 31
Routing of operands, 52—56
Routing paths, Illiac IV, 20—21
Rowsum operation, Illiac IV, 32—33

S

SAR, see Synthetic aperture radar
Scalar arithmetic statements, 45—46
Scalar data, 222, 229
Scalar expressions, 53
Scalar processing, 53
Scalar types, 107
Scalars, 41
Scaling factor, 69
Segment labeling, 172—173
Segmentation, 169—172
Semiconductor memory, 80
Sequencer unit, GAM II Pyramid, 303—304
Sequencing, 117
Sequential operations, 18
Serial computer functions, CLIP4, 325
Serial processing, abandonment, 1
Shared memory architectures, 7—8
Shift register groups, GAPP, 184
Shifting binary images, 315—316

Shuttle Imaging Radar instrument (SIR-B), 125
Signal processing, 1
Signal-to-noise ratio, 72
SIMD, 1, 329, see also Parallel processor
  multiple instruction multiple data distinguished, 5
  parallel organization, 20
SIMD architectures
  control unit, 5—6
  processing elements, 5—6
SIMD supercomputer, see Illiac IV
Simple minimum distance Euclidean classification, 64
Simulation, 1
Single error correction-double error detection (SECDED) schemes, 349
Single instruction multiple data processors, see SIMD
Sobel filter, 197—199
Software, see specific system
SOLOMON I, 11, 14—15
Solvable problems, 83
Sort, 132—141, 332—334
  aggregation, 136—138
  algorithms implemented on MPP, 133—136
  communication primitive, 133
  distribution, 137—139
  Massively Parallel communications, 132—133
  merge aggregation, 139—140
  merge distribution, 139—140
  merge steps, 139—140
  summing, 136—137
  unmerge operation, 140—141
Sorting, see Sort
Sparse matrix multiply, 75—76
Special purpose registers, 305—306
Specification statements, 41—42
Speed
  computer processing, 1
  light, 1, 18
  limits on increases in, 1
  need for higher, 17—18
Stack operation words, 99—101
Staging memory, MPP, 85, 89—90
Standard functions, 114
Standard Pascal, 107—110, 116—117, 120, see also Parallel Pascal design
Standard procedures, 114
STARAN, 10—11, 13—14

applications, 11
array element, 10, 14
configuration, 10, 13
flip network, 10—11
salient features, 10
Statistical maximum likelihood classification, 64
Stereo analysis
  background, 126
  brightness level differences, 126
  detect bad match, 132
  determination of matches, 128—129
  difficulties in stereo matching, 126—127
  initial warp, 132
  interactive operations on MPP, 131—132
  interactive-turnaround time, 132
  interpolate, 130—132
  local distortions, 126
  low contrast areas, 126—127
  Massively Parallel Processor, use of, 125
  match, 132
  matching algorithm on MPP, 127—128
  matching technique, 127
  noise, 126—127
  preprocessing of test image, 128
  removal of "bad match" areas in disparity function, 128—131
  smooth, 132
  smoothing resulting disparity function, 128, 131
  warp, 132
  warping test image, 128, 131
Structured programming, 106
Subprogram statements, 41—42
Subprograms, 41
Subrange constant, 111—112
Subrange type, 107—108
Subroutine call, 117
Supercomputer defined, 2
Synchronous control, 81
Synthetic aperture radar (SAR)
  azimuth correlation, 69
  concepts, 65—69
  digital processing methods for, 65
  Fourier transform, 66—67
  range correlation, 67—68
  transposition of 64×64 subarrays, 68
System architecture, Illiac IV, effects of, 82—83
System controller, GP^2, 200—201
Systems, see specific types
Systolic array processing, 200

T

Tailored algorithms, 159
Team projects, 2
Tensor code, 76
Test and Maintenance Unit (TMU), Illiac
 IV, 34
Text search
 application, 277—278
 benchmark, 278—279
 connection machine application, 236,
  273—279
 data structure, 275—277
 free text, 274—279
 keyword, 274
Thermal diffusion, 237, 241—250
Thresholding, 193
Tightly coupled MIMD multiprocessors,
 6
TRANQUIL, 29
Transformational functions, 115
Translation, 193
Translators, 47
Transpose, 68
TRES computer program, 58—60
 boundary conditions, 59
 computational algorithm, 59—60
 elastic zone, 59
 fault model, 58
 plastic zone, 58—59
14TRES program, 61—62

Truth table, GAPP, 188
Two-dimensional fast Fourier transform
 (TWDFFT), 69—70

U

UNIVAC 1108, 58—60, 62
University of Illinois, Illiac IV development
 and use, 21—23
Unmerge operation, 140—141

V

Variables, 41, 91
 memory allocation, 51
 processing element, 50—51
Variance-covariance calculations, 64
Variant record, 108
VAX, 90, 133, 143—144
Vector aligned arrays, 41
Vector processors, 5
VLSI chip technology, 2
VLSI system design, 351—352

W

Wave equation, 237, 241—250
Where statement, 118—119
While construct, 117, 119
Word reference, parallel FORTH language,
 95—105